PLANT STRESS BIOLOGY

Progress and Prospects of Genetic Engineering

PLANT STRESS BIOLOGY

Progress and Prospects of Genetic Engineering

Edited by

Arindam Kuila, PhD

First edition published 2021

Apple Academic Press Inc.
1265 Goldenrod Circle, NE,
Palm Bay, FL 32905 USA
4164 Lakeshore Road, Burlington,
ON, L7L 1A4 Canada

CRC Press
6000 Broken Sound Parkway NW,
Suite 300, Boca Raton, FL 33487-2742 USA
2 Park Square, Milton Park,
Abingdon, Oxon, OX14 4RN UK

First issued in paperback 2021

Library and Archives Canada Cataloguing in Publication

Title: Plant stress biology : progress and prospects of genetic engineering / edited by Arindam Kuila, PhD.

Names: Kuila, Arindam, editor.

Description: Includes bibliographical references and index.

Identifiers: Canadiana (print) 20200312391 | Canadiana (ebook) 20200312626 | ISBN 9781771889254 (hardcover) | ISBN 9781003055358 (ebook)

Subjects: LCSH: Plants—Effect of stress on. | LCSH: Plants—Molecular aspects.

Classification: LCC QK754 .P63 2021 | DDC 581.7—dc23

Library of Congress Cataloging-in-Publication Data

Names: Kuila, Arindam, editor.

Title: Plant stress biology : progress and prospects of genetic engineering / edited by Arindam Kuila, PhD.

Description: First edition. | Palm Bay, FL : apple Academic Press, [2021] | Includes bibliographical references and index. | Summary: "This unique book covers the molecular aspects of plant stress and the various industrial applications. Chapters cover many important topics in the biology of plant stress, including morphological and physiological changes of plants due to accumulation of pollutants; the types of stress for enhanced biofuel production from plant biomass; plant adaptation due to different types of environmental stresses; potential applications of microRNAs to improve abiotic stress tolerance in plants; plant resistance to viruses and the molecular aspects; photosynthesis under stress conditions; plant responses to weeds, pests, pathogens, and agrichemical stress conditions; and plant responses under the stress of drought. Plant Stress Biology: Progress and Prospects of Genetic Engineering will be useful for researchers in diverse fields as well as for plant biologists, environmental biologists, faculty, and students. The book will be helpful for further advancement of research in the area of plant stress biology. Key features: Details the current and possible applications of plant stress biology Describes the different types of plant stress Presents several case studies that include applications of plant stress Explores plant stress biology for applications in biofuel science"-- Provided by publisher.

Identifiers: LCCN 2020037115 (print) | LCCN 2020037116 (ebook) | ISBN 9781771889254 (hardcover) | ISBN 9781003055358 (ebook)

Subjects: LCSH: Plants--Effect of stress on--Molecular aspects. | Plant molecular biology.

Classification: LCC QK754 .P5847 2021 (print) | LCC QK754 (ebook) | DDC 572.8/2--dc23

LC record available at https://lccn.loc.gov/2020037115

LC ebook record available at https://lccn.loc.gov/2020037116

ISBN: 978-1-77188-925-4 (hbk)
ISBN: 978-1-77463-930-6 (pbk)
ISBN: 978-1-00305-535-8 (ebk)

About the Editor

Arindam Kuila, PhD, is currently working as Assistant Professor in the Department of Bioscience and Biotechnology, Banasthali Vidyapith, Rajasthan, India. Previously, he worked as a research associate at Hindustan Petroleum Green R & D Centre, Bangalore, India. He earned his PhD at the Agricultural & Food Engineering Department, Indian Institute of Technology Kharagpur, India, in 2013, in the area of lignocellulosic biofuel production. He was awarded a Petrotech Research Fellowship. He has an Indo-Brazil collaborative project funded by DBT, India, and has co-authored 18 peer-reviewed research papers, seven review papers, four edited books, and eight book chapters and filed five patents.

Contents

Contributors

Nikita Bhati
Department of Bioscience and Biotechnology, Banasthali Vidyapith, Rajasthan, India

Debalina Bhattacharya
Department of Microbiology, Maulana Azad College, Kolkata – 700013, West Bengal, India

Rupam Kumar Bhunia
National Agri-Food Biotechnology Institute (NABI), Plant Tissue Culture and Genetic Engineering, Sector-81 (Knowledge City), Mohali – 140306, Punjab, India, E-mail: rupamb2005@gmail.com

M. Choudhary
Department of Bioscience and Biotechnology, Banasthali University, Rajasthan, India

Kamalendu De
Microbial Engineering Group, Department of Biotechnology, JIS University Agarpara, 81 Nilgunj Road, Agarpara, Kolkata, West Bengal – 700109, India

Shrestha Debnath
Microbial Engineering Group, Department of Biotechnology, JIS University Agarpara, 81 Nilgunj Road, Agarpara, Kolkata, West Bengal – 700109, India

Nishu Gandass
National Agri-Food Biotechnology Institute (NABI), Sector-81, SAS Nagar, Mohali, Punjab – 140306, India

Dipankar Ghosh
Microbial Engineering Group, Department of Biotechnology, JIS University Agarpara, 81 Nilgunj Road, Agarpara, Kolkata, West Bengal – 700109, India, Phone: +91-7872882337, E-mails: dghosh.jisuniversity@gmail.com, d.ghosh@jisuniversity.ac.in

S. Joshi
Department of Bioscience and Biotechnology, Banasthali University, Rajasthan, India

Megha Katoch
National Agri-Food Biotechnology Institute (NABI), Sector-81, SAS Nagar, Mohali, Punjab – 140306, India

Amandeep Kaur
National Agri-Food Biotechnology Institute (NABI), Sector-81, SAS Nagar, Mohali, Punjab – 140306, India

Ranjeet Kaur
Department of Biotechnology, Mangalmay Institute of Management and Technology (MIMT), Greater Noida – 201306, Uttar Pradesh, India

Nirbhay Kumar Kushwaha
Department of Plant Biology, Swedish University of Agricultural Sciences, Uppsala-75007, Sweden

Mansi
ICAR-National Institute for Plant Biotechnology, Indian Agricultural Research Institute, New Delhi-110012, India

Mainak Mukhopadhyay
Department of Biotechnology, JIS University, Kolkata – 700109, West Bengal, India,
E-mail: m.mukhopadhyay85@gmail.com

Girish Chandra Pandey
Departments of Bioscience and Biotechnology, Banasthali Vidyapith, Rajasthan, India,
E-mail: girish.dwr@gmail.com

Hasthi Ram
National Agri-Food Biotechnology Institute, Mohali-140306, India,
E-mails: hrmundel84@gmail.com, hasthi@nabi.res.in

Ritika
Department of Bioscience and Biotechnology, Banasthali Vidyapith, Rajasthan, India

Shweta Roy
Department of Cell and Molecular Biology, Uppsala University, Uppsala-75236, Sweden

Rituparna Saha
Department of Biochemistry, University of Kolkata, Kolkata – 700019, West Bengal, India

Shritoma Sengupta
Department of Biotechnology, JIS University, Kolkata – 700109, West Bengal, India

Arun Kumar Sharma
Department of Bioscience and Biotechnology, Banasthali Vidyapith, Rajasthan, India,
E-mail: arun.k.sharma84@gmail.com

Shreya
Department of Bioscience and Biotechnology, Banasthali Vidyapith, Rajasthan, India

Chandra Pal Singh
Department of Botany, University of Rajasthan, Jawahar Lal Nehru Marg, Jaipur – 302004,
Rajasthan, India, E-mail: chandrapal203@gmail.com

P. Singh
Department of Bioscience and Biotechnology, Banasthali Vidyapith, Rajasthan – 304022, India

Pankaj Kumar Singh
ICAR-Central Rainfed Upland Rice Research Station, Hazaribagh, Jharkhand, India;
ICAR-Indian Institute of Wheat and Barley Research, Karnal, Haryana, India,
E-mail: singhpankajkumar8@gmail.com

Sunita Singh
ICFRE-Institutes of Forest Productivity, Ranchi, Jharkhand, India

G. Singhal
Department of Bioscience and Biotechnology, Banasthali Vidyapith, Rajasthan – 304022, India

Praveen Soni
Department of Botany, University of Rajasthan, Jaipur-302004, India
E-mail: praveen.soni15@gmail.com

N. Srivastava
Department of Bioscience and Biotechnology, Banasthali University, Banasthali – 304022,
Rajasthan, India, Phone: +91-8741914247, E-mail: nidhiscientist@gmail.com

Vidisha Thakur
Department of Bioscience and Biotechnology, Banasthali Vidyapith, Rajasthan, India

V. Verma
Department of Bioscience and Biotechnology, Banasthali Vidyapith, Rajasthan – 304022, India

Abbreviations

$^1O_2$	singlet oxygen
ABA	abscisic acid
ABRC1	ABA-responsive complex
ACC	aminocyclopropane-1-carboxylic acid
ACCD	aminocyclopropane-1-carboxylic acid deaminases
AcMNPV	*Autographa californica* nuclepolyhedron-virus
ACMV	African cassava mosaic virus
AGO	argonaut
Ago1	argonaute1
AJH	anti-juvenile hormone
ALDH	aldehyde degydrogenase
ALO	arabinono-1,4-lactone oxidase
AlP	aluminum phosphide
Alt	aluminum tolerance
amiRNA	artificial miRNA
AMOPs	marine microbial oxygenic photoautotrophs
AOS	active oxygen species
AP2	apetala2
APS	adenosine 5'-phosphosulfate
APX	ascorbate peroxidase
AQPs	aquaporins
ARFs	auxin response factors
As	arsenic
AsAAO	*Agrostis stolonifera* ascorbic acid oxidase
AsCBP1	AsAAO and copper ion binding protein 1
BCTV	beat curly top virus
BeYDV	bean yellow dwarf virus
BPH	brown planthopper
BRs	brassinosteroids
BSCTV	beet severe curly top virus
Bt	*Bacillus thuringiensis*
bZIP	basic leucine zipper
CaLCV	cabbage leaf curl virus
CaM	calmodulin

CAM	crassulacean acid metabolism
CaMV35S	cauliflower mosaic virus 35S
CAT	catalase
CAX	cation exchange
CBL	calcineurin B-like
CCS	copper chaperon of CSD
Cd	cadmium
CDPKs	Ca^{2+}-dependent protein kinases
CKs	cytokinins
CKX2	cytokinin oxidase
CLCuV	cotton leaf curl virus
CML44	CaM-like 44
CMT3	chromomethylase 3
CMV	cucumber mosaic virus
CNGC	cyclic nucleotide gated calcium
COX	cytochrome C oxidase
CP	coat protein
CPV	cytoplasmic polyhedrosis virus
CRISPR	clustered regularly interspaced short palindrome repeat
CSD	Cu/Zn-superoxide dismutase
CSIs	chitin synthesis inhibitors
Cu	copper
CVYV	cucumber vein yellowing virus
DBH	dibenzoylhydrazine
DC1/PHD	domains-containing protein
DCL	dicer like
DDT	dichlorodiphenyltrichloroethane
DHAR	dehydroascorbate reductase
DML3	demeter-like3
DREB/CRT	dehydration-responsive element binding/C-repeat
DREB1A	dehydration response element B1A
DRI	drought resistance index
DRM1	domain rearranged methyltransferase 1
DRM2	domain rearranged methyltransferase 2
dsRNA	double-stranded RNA
EcR	ecdysteroid receptor
EPA	Environmental Protection Agency
ERF/AP2	ethylene response element binding factor/Apetala2
ET	evapotranspiration

ETI	effector-triggered immunity
FAME	fatty acid methyl esters
Fe	iron
GA	gibberellic acid
GABA	g-aminobutyric acid
GAI	gibberellic-acid insensitive
GAs	gibberellins
GhCHR	*Gossypium hirsutum*cys/His-rich
GHGs	greenhouse gas
GK74	gamma-glutamyl kinase 74
GM	genetically modified
GMC	guard mother cell
GPX	glutathione peroxidase
GR	glutathione reductase
GRF	growth-regulating factor
GSK1	glycogen synthase kinases
GST	glutathione S-transferase
GV	granulosis virus
GWAS	genome-wide association study
H_2O_2	hydrogen peroxide
HATs	histone acetyltransferases
HDACs	histone deacetylases
HDC1	histone deactylase
HDMs	histone demethylases
HD-Zip	homeodomain-leucine zipper
HEN 1	HUA enhancer 1
Hg	mercury
HhH-GPD	helix-hairpin-helix and glycine/proline/aspartate D
HMTs	histone methyltransferases
HPR	host plant resistance
HR	hypersensitive reaction
HSEs	heat shock elements
HSFs	heat shock factors
HSPs	heat shock proteins
HST	HASTY
HYL1	hyponastic leaves1
IAA	indole acetic acid
IDPs	intrinsically disordered proteins
IGR	insect growth regulators

IPT	isopentenyl transferase
JA	jasmonic acid
JH	juvenile hormone
jmjC	jumanji C
K	potassium
KB	karnal bunt
LCR	leaf curling responsiveness
LEA	late embryogenesis abundant
LRR	leucine-rich repeat
LSD1	lysine-specific histone demethylase1
MDH	malate dehydrogenase
MeJA	methyl jasmonate
MET1	methyltransferase 1
Mg	magnesium
miRNA	microRNA
MMC	meristemoid mother cell
Mn	manganese
MNSV	melon necrotic spot virus
MP	movement protein
N	nitrogen
NAC	NAM, ATAF1/2, and CUC2 proteins
NAM	no apical meristem
NbHUB1	*N. benthamiana*HUB1
NBS	nucleotide-binding site
NbUBC2	*N. benthamiana* ubiquitin-conjugating enzyme 2
NCED	9-*cis*-epoxycarotenoid dioxygenase
NCP	nucleocapsid protein
ncRNA	noncoding RNA
NFY	nuclear factor Y
NIb	nuclear inclusion b
NIP	nodulin 26-like intrinsic protein
NLA	nitrogen limitation adaptation
NO	nitric oxide
NPSH	non-protein thiols
NPV	nuclear polyhedrosis virus
O_2^-	superoxide anion
OAA	oxaloacetate
OH^-	hydroxyl radical
OMTN	oryza miR164-targeted NAC genes

OPs	organophosphate
P	phosphorus
PAM	protospacer-adjacent motif
PAMPs/MAMPs	patterns or microbe-associated molecular patterns
Pb	lead
PCL	plastocyanin-like
PDH	proline dehydrogenase
PDR	pathogen-derived resistance
PEPC	phosphoenol pyruvate carboxylase
PepGMV	pepper golden mosaic virus
PGPB	plant growth promoting bacteria
PHO2	pi-responsive genes
PHR1	phosphate starvation response 1
PIF	phytochrome-interacting factor-like 1
PIP	plasma membrane intrinsic protein
piRNA	piwi-interacting RNAs
PLD	phospholipase D
POD	peroxidase
PP2A	protein phosphatase 2A
PPCK	PEP carboxylase kinase
PPO	pure plant oil
PRSV-W	papaya ringspot mosaic virus-W
PTGS	post-transcriptional gene silencing
PTI	PAMP-triggered immunity
PTMs	post-translational modifications
PTTH	prothracicotropic hormone
PVX	potato virus X
PVYO	potato virus Y, O strain
QTL	quantitative trait locus
RDR2	RNA-dependent RNA polymerase 2
RGA	repressor of GAI
RISC	RNA-induced silencing complex
RME	rapeseed methyl ester
RNS	reactive nitrogen species
ROS	reactive oxygen species
RSV	rice stripe virus
S	sulfur
SA	salicylic acid
SBP	squamosa promoter binding protein

SCL	scarecrow-like proteins
SCR	scarecrow
SE	serrate
sgRNA	single guide RNA
SL	strigolactones
SODs	superoxide dismutases
SPCH	speechless
SPL	squamosa promoter binding protein-like
ssRNA	single-stranded RNA
SUTs	sucrose transporters
SWD	soil water deficit
TALEN	transcriptional activator-like effector nuclease
TCP	teosinte branched cycloidea and PCF
TCV	turnip crinkle virus
TFs	transcription factors
TGH	tough
TGS	transcriptional gene silencing
TIP	tonoplast intrinsic protein
TIR1	transport inhibitor response 1
TMM	too many mouths
TMV	tobacco mosaic virus
ToMoV	tomato mottle virus
tracrRNA	trans-activating CRISPR RNA
TuMV	turnip mosaic virus
TYLCV	tomato yellow leaf curl virus
TYMV	turnip yellow mosaic virus
UDPase	UDP glucose pyrophosphorylase
UGC	University Grant Commission
Wsi18	water stress inducible 18
ZFN	zinc finger nuclease
Zn	zinc
ZYMV	zucchini yellow mosaic virus

Preface

This book describes plant stress and its different applications. The uniqueness of this book is that the book describes all possible applications of plant stress and provides different case studies of applications of plant stress. In the book, there are a total of 11 chapters.

The book will be useful for students, and researchers in the diverse field, including plant biologists and environmental biologists.

The first chapter describes morphological and physiological changes of plants due to the accumulation of heavy metals. The second and third chapters deal with different types of stress for enhanced biofuel production from plant biomass. The fourth chapter describes plant adaptation due to different types of environmental stress. The fifth chapter gives information about the potential application of microRNAs to improve abiotic stress tolerance in plants. The sixth chapter describes plant resistance against viruses and their molecular aspects. The seventh chapter describes photosynthesis under stress conditions. The eighth chapter deals with plant responses to weeds, pests, pathogens, and agrichemical stress conditions. The ninth chapter describes plant response under drought stress. The tenth chapter gives us information about the epigenetic regulation of plants under stress and potential application in agriculture. The last chapter deals with the mechanism of drought tolerance in pearl millet.

Introduction

Plant stress biology refers that any hostile condition or toxic substance that affects plant growth through interruption of its metabolic pathway, reproduction, etc. Plant stress can come in different forms and durations. Plant growth is hampered by different types of stresses. Nowadays, researchers are focusing on the development of stress-resistant plants due to climate change-induced increases in drought and other abiotic stresses. The book covers the developments of stress-tolerant plants and their applications in different areas. This book also covers different approaches of genomics, proteomics, etc., of plant stress. The book will be helpful for the further advancement of research in the area of plant stress biology.

CHAPTER 1

Bioaccumulation of Heavy Metal in Plants: Morphological and Physiological Changes

V. VERMA, G. SINGHAL, P. SINGH, and N. SRIVASTAVA

Department of Bioscience and Biotechnology, Banasthali Vidyapith, Rajasthan – 304022, India

ABSTRACT

The bioaccumulation and biomagnification of heavy metals in the environment have become a serious concern for plants and human health. These heavy metals have been a serious threat for the plant production as they lead to the alteration in various physiological, morphological, and biochemical changes in plants after accumulation. When present in higher concentrations, heavy metals interfere with the essential biomolecules of cells as nuclear proteins and DNA that ultimately produce excessive numbers of reactive oxygen species (ROS). These ROS impose severe morphological, metabolic, and physiological abnormalities in plants. The elevated content of such heavy metal in soils is an alarming issue in agricultural production due to the undesirable effects on food safety and marketability.

1.1 INTRODUCTION

Heavy metals have high atomic density and in spite of their concentration, cause toxicity for the plants, humans as well as for animals. There have been two main types of metals, i.e., essential, and nonessential metals. The earth's crush has composed of nonessential metals, which entered into the upper soil sphere during biogeochemical cycles (Tinsley, 1979).

Plants represented the first compartment of the terrestrial food chain and have been the most essential part of the natural ecosystem. The plants bioaccumulating capability of heavy metal has represented the threat for human beings and animals who consumed them. Due to the high atomic density, some metals, i.e., cadmium (Cd), mercury (Hg), lead (Pb) and zinc (Zn) have been known as heavy metals (Oves et al., 2012). In spite of the metalloid nature of arsenic (As), it has been considered as heavy metals because of its high atomic density (Chen et al., 1998). When these elements have exceeded the threshold limit, can caused various anomalies including morphological, physiological as well as genetical anomalies and affected the growth of the plants with high mortality depending on the sensitivity of plants (Khan et al., 2008; Li et al., 2010; Luo et al., 2011). Some plant species might disappear from such lands while others might stimulate by these elements on the contrary.

1.2 SOURCES OF HEAVY METALS AND THEIR EFFECTS

Heavy metals have been typically defined as elements with metallic properties. Cr, Cd, Hg, Cu, Pb, and Zn have been the most frequent heavy metal pollutants present in the environment (Martin and Kaplan, 1998). Some metals including micronutrients have been very essential for plant development such as Mn, Cu, Zn, Ni, and Co. Whereas, other metals like Cd, Pb, and Hg have an unidentified biological function (Khan et al., 2008). Anthropogenic and natural sources have been the major sources of these metals as demonstrated in Figure 1.1.

Metal pollution has a detrimental effect on biological systems and has not undergone bio-decomposition. Metals with a high atomic density such as Pb, Cd, and Co have known as lethal heavy metals can be separated from other pollutants. When these metals get deposited in the human body can lead to various diseases and disorders even when present in lesser concentrations (Williams et al., 1980). They have been acknowledged for the consequences on plant development and have a pessimistic impact on soil micro-flora (Verloo and Eeckhout, 1990). It has been well acknowledged that heavy metals have not environment friendly and could not be degraded easily hence they have required to be physically detached or be altered into eco-friendly compounds (Khan et al., 2008). Figure 1.2 demonstrated the effect of heavy metal stress on different plant parts including morphological and physical changes.

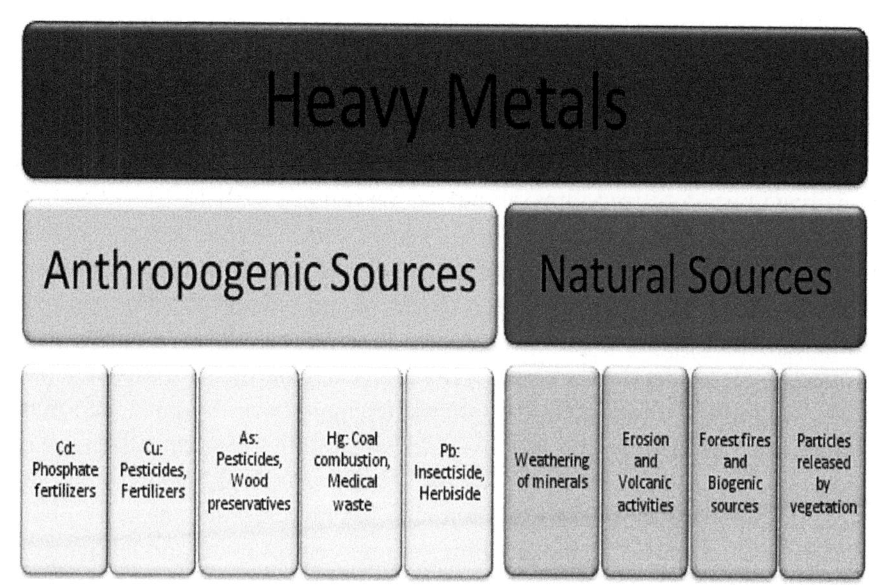

FIGURE 1.1 Various sources of heavy metals.

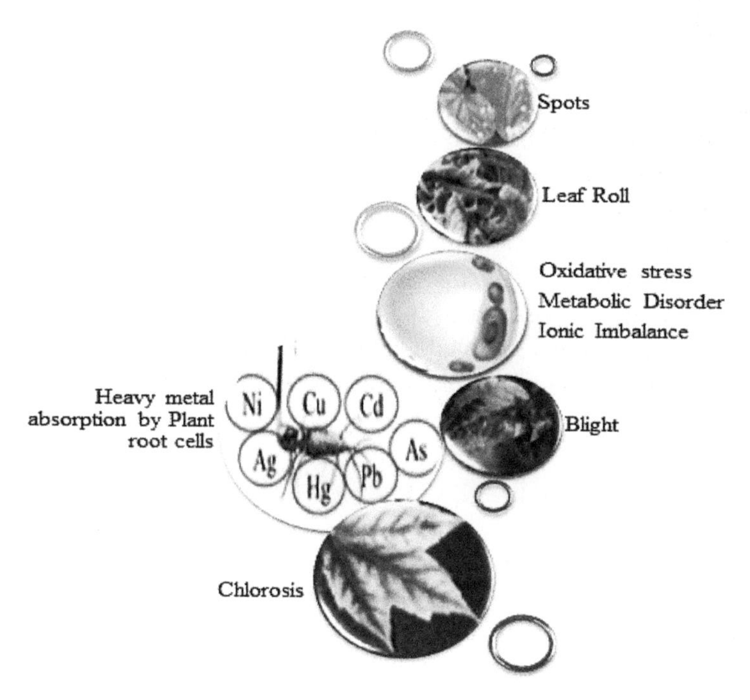

FIGURE 1.2 Effect of heavy metal stress.

1.3 MOVEMENT, UPTAKE, AND ACCUMULATION OF HEAVY METALS

The heavy metal contaminations present in the environment have been transported to the soil with aid of water and air; get deposited and sediment in the soil crust in immobilized form. However, the process of metal to soil binding has considerably be taken a very long period of time. At the beginning of the binding process, a high amount of metal ions have been present in the soil which has been decreased gradually with the time due to the absorption of those ions by plant root system (Martin and Kaplan, 1998). The chemical properties of soil, i.e., pH, soil texture, cation exchange (CAX) capacity, clay content, redox potential, and organic matter have been the most important factor which influenced the solubility and availability of metals to the plant roots (Williams et al., 1980; Verloo and Eeckhout, 1990). Sometimes, soil temperature also played an important role in metal accumulation by plants (Chang et al., 1987). The bioaccumulation process of metal ions movement from soil to plant parts has been explained in Figure 1.3.

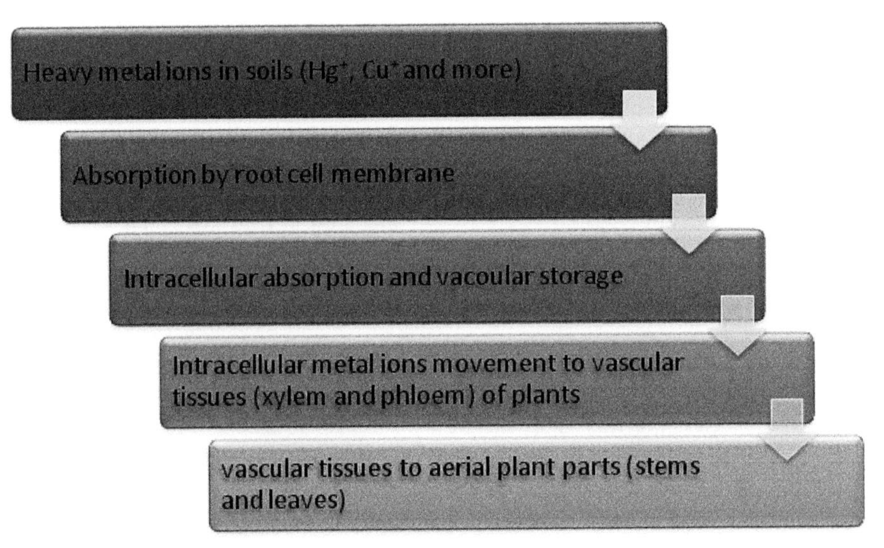

FIGURE 1.3 Movement of metals ions from soil to plant parts.

1.4 METAL TOLERANCE MACHINERY

Due to a rapidly changing environment, plants have adapted a complex process for their metabolism including perception, transduction, and transmission of stress stimulus (Kopyra and Gwozdz, 2004). For the adaptation of stress conditions, plants get resistant and tolerant of the metal ions by immobilized them in the cell walls of roots (Garbisu and Alkorta, 2001). To avoid the toxicity of heavy metals, plants have developed a series of mechanisms including production of reactive oxygen species (ROS), blocking of main functional group, and metal ions displacement from biomolecules (Clemens, 2006). Figure 1.4 showed the adaptation mechanism of plant parts for metal ions' stress.

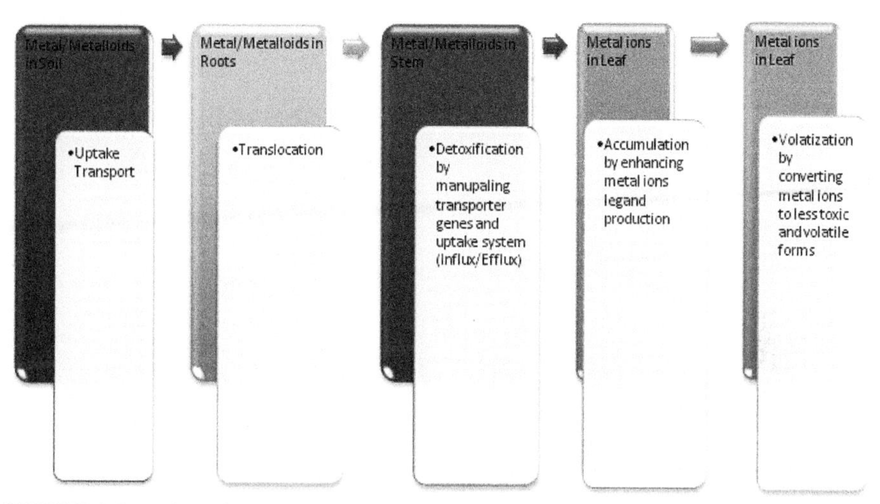

FIGURE 1.4 Adaptation mechanism of plant parts for metal ions stress.

1.5 MORPHOLOGICAL CHANGES IN PLANTS

The normal functioning of the soil ecosystem and plants have been interrupted by the heavy metal accumulation (Khan et al., 2008b). Some metals as copper (Cu), manganese (Mn) iron (Fe) and magnesium (Mg) have been known as essential metals for plant growth and production. These essential metals in specific concentrations have been necessary for the plants while their deficiency or excess concentration can cause toxicity for plant production (Fusconi et al., 2006). Many plant researchers have paid attention to consider the plants as a tool to remove the metals from the soil by a cost-effective and eco-friendly process known as phytoremediation. The morphology and biochemical responses of the plants have been in direct relationship with the chemical properties of soil and concentration of heavy metals (Preeti and Tripathi, 2011). The metal stress of lead has affected the photosynthesis rate by reducing the chlorophyll content in leaves. Lead has reduced the uptake of essential elements for chlorophyll, i.e., Mg, and Fe, therefore, affected essential enzymatic processes for photosynthesis and resulting in the closing of stomata (Sharma and Dubey, 2005). Lead has also showed a significant effect on root-shoot length, weight, seedling dry mass; respiration rate, and plant metabolism (Paolacci et al., 1997). Figure 1.5 showed the relationship of heavy metals with soil properties and plant response.

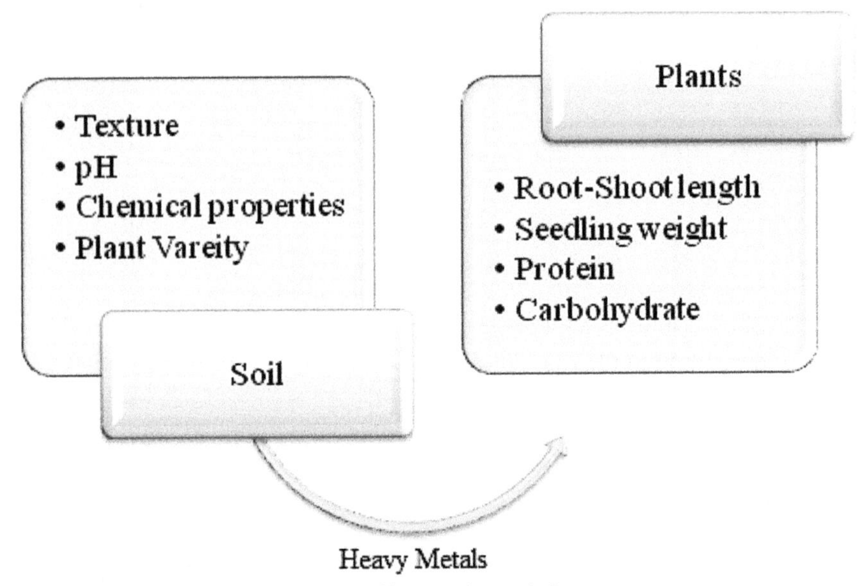

FIGURE 1.5 Relationship between soil properties and plant response.

Generally, all heavy metals as Pb, Cd, Hg, and As have been first localized in the root than leaf and stem while Cr has been first localized in the root and then stem and leaf respectively. The heavy metal concentration in the plant tissues has been indirectly proportional to the suspended solid matter of soil (Ndimele and Jimoh, 2011).

1.5.1 EFFECT ON GROWTH AND DEVELOPMENT

Heavy metal stress has retarded the growth and production of either the whole plant or a plant part (Shafiq and Iqbal, 2005). Generally, the plant part that has direct contact with contaminated soil showed the most sensitive changes in the growth pattern. However, a variety of heavy metals, i.e., Ni,

Hg, Cu, Cd, Cr, Pb, and Fe have been significantly affect the growth of plant parts that have not been in contact with contaminated soil. Heavy metal ions have generated free radical ions and ROS which damaged the important cellular components and in turn affect the growth of plants. In cucumber plant, heavy metal, i.e., Cu has limited the uptake of potassium by leaves and accumulated the sugar by inhibiting photosynthesis. These processes of events have been lead to the blockage of cell expansion resulting in a shortening of leaves (Alaoui-Sosse et al., 2004). Similarly, Cd and Ni metal ions exposed rice seedlings as well as Cd and Cu treated runner bean plants have shown a hike or quantitative increment in carbohydrate content and decline/decreased photosynthesis rate which caused the inhibition of growth in these plants. The Cd metal stress showed some typical symptoms in the rice plants including browned roots and root tips, progressive chlorosis in some leaves, wilted leaves, and growth inhibition of whole plant (Chugh and Sawhney, 1999). In addition, Cd also inhibited plant growth in maize. Tomato plants irrigated with heavy metal-rich water showed some phenotype abnormalities including less branching, undeveloped growth, and less fruiting. However, it has been shown that heavy metals accumulation has been significantly low in fruits in comparison to lower portion like roots, leaves, and shoots (Gupta et al., 2008).

1.5.2 *GERMINATION*

The changing environmental conditions have greatly affected the germination rate of seeds. Therefore, the germination rate and growth of the seedlings have used to affect by the plant tolerance to metal stress (Peralta et al., 2001). In barley, rice, and wheat seedlings, higher concentrations of heavy metals, i.e., Zn, Cu, and Mg have significantly inhibited seed germination as well as early growth of these plants.

1.5.3 *ROOT*

Roots have been the first part of the plant that has come in direct contact with toxic heavy metals and usually accumulated more metals than shoot. The first visible effect of metal toxicity has been observed as the inhibition of root length that can be reduced either by inhibition of root cell division or by reduction of cell expansion (Han et al., 2007). The heavy metal response in roots has been broadly studied in herbaceous plants as well as in the trees. It has been reported that the translocation in the plant has been avoided in the *Brassica juncea* plant due to the precipitation of Cr metal ion. The major morphological changes caused by metal stress in roots has been summarized as damage of root tip, collapsing of root hair, decreased number of roots, biomass, and root vessel diameter reduction, blockage in root elongation, reduction in lateral root formation and structural alterations of hypodermis and endodermis. The plant response to metal stress has been varied according to the metal ions as Cr ion affected the root length severely than any other metal ion (Prasad et al., 2001).

1.5.4 STEM

Some metal ions show varied affect on plant growth like the height of plants and the growth of the shoot portion adversely. Reduction and regulation of root growth, less water transport to aerial parts, and availability of lesser nutrients to the shoot have been the main reasons for the plant height reduction. The transport of Cr ion to the aerial plant part has directly affect the cellular metabolism of shoots and caused the plant height reduction (Shanker et al., 2005).

1.5.5 LEAF

A proper crop yield contributed to total leaf numbers, the leaf is developed and its healthy growth. However, the metal stress of Cd has induced some morphological alterations as necrosis of younger leaves, drying of older leaves, and chlorosis (Shanker et al., 2005). Lettuce, a leafy vegetable has been considered as a potential hyper-accumulator of heavy metals. Whereas the heavy metals accumulation in green, leafy vegetables did not show any visible toxicity symptoms (Intawongse and Dean, 2006). Some plants like *Datura innoxia*, grown in Cr rich environment exhibited toxic symptoms in the form of wilting and leaf fall. In seedlings of *Albizzia lebbek*, the high concentration of Cr showed toxic effects on leaf area and seedling's biomass. In bush bean plants, the additional concentration of Cr showed a 45% reduction in dry leaf yield (Wallace et al., 1976).

1.5.6 EFFECT ON ENZYMES AND OTHER COMPOUNDS

The enzyme activity has played a crucial or vital role for the boosted or enhanced stress response in plants through signaling molecule synthesis.

According to the literature cited and recent reports available, it has been mentioned that metal stress hindered/delayed the enzymes functionality which are linked with photosynthetic cycle and Calvin cycle (Krupa and Baszynski, 1995).

1.6 PHYSIOLOGICAL CHANGES IN PLANTS

Heavy metals have been the main part of all the major pollutants in the surroundings. Adjacent to the normal activities, approximately all human deeds also have prospective inputs to generate heavy metals. Relocation of these pollutants to non-pollutants zones as leachates or dust through the soil and smattering of heavy metals including sewage sludge have been considered as some examples for the environmental contamination (Gaur and Adholeya, 2004).

Now, several methods have been introduced for the decontamination of environment from these contaminants (heavy metal) however, most of these methods lack behind due to their relatively high cost. The traditional decontamination method for contaminated soils has included either on site execution or digging and consequent dumping to landfill site. The method of dumping exclusively moved the infectivity problem away along with the hazards connected with the transportation of polluted soil and the relocation of contaminants from landfills into an adjoining environment (Gaur and Adholeya, 2004).

Many plant species have been effectively engrossed various noxious contaminant metals such as chromium, lead, arsenic, cadmium, and assorted radioactive compounds from soils. One of the well-established methods to remove these heavy metal contaminations involved the phytoremediation technique (Cho-Ruk et al., 2009).

1.6.1 ARSENIC (AS)

Arsenic has been a semi-metallic (metalloid) element with an atomic number of 33. It has a brittle crystalline solid with silver-grey color and symbolized as 'As.' Its physical properties made it tasteless, odorless and transformed into organic and inorganic arsenicals when merged with the other elements (National Ground Water Association, 2001). In environments, As has formed the inorganic arsenic compounds in combination with chlorine, oxygen, and sulfur. These inorganic arsenicals have been mainly used for wood preservation while organic arsenicals have been used in the form of pesticide for the prevention of cotton plants (U.S. Department of Health and Human Services, 2005). Arsenic has acted like a hard acid and preferred to make complexes with nitrogen and oxygen. As has been a major, heavy metal found in the environment and caused toxicity in all living organisms (Chutia et al., 2009). The main sources of As metals have included contaminated food and water that has been consumed by human beings. The uptake of As rich food can cause toxicity at the cellular and genetic level that has been demonstrated in Figure 1.6.

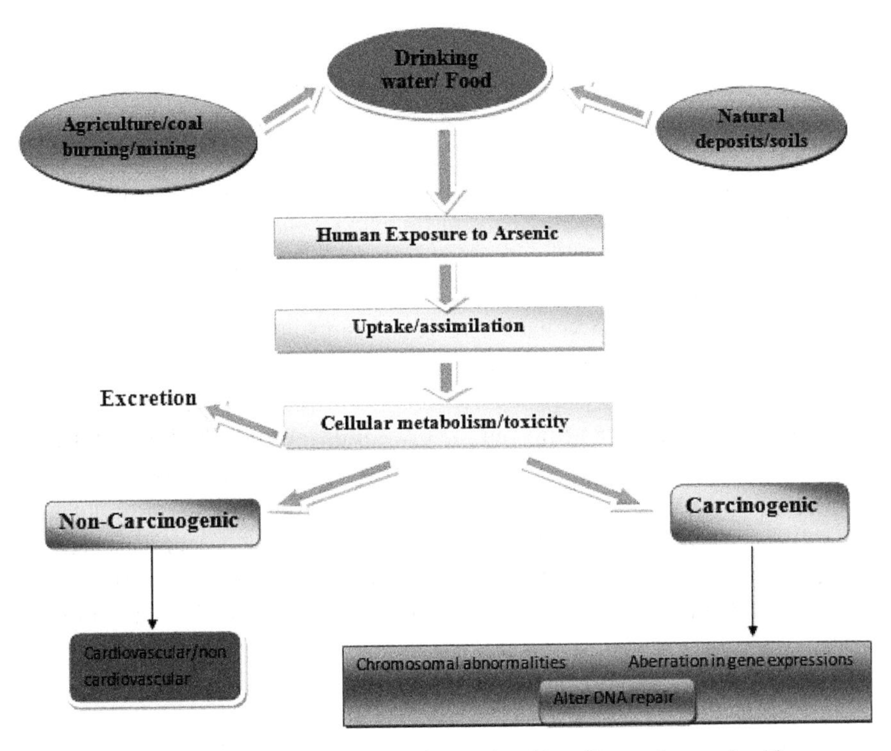

FIGURE 1.6 Sources and uptake of arsenic metal and its effect on human health.

Various physiological changes have been observed with the exposure of this heavy metal. The plant exposed to this metal has led to the reduction of photosynthetic rate (Hartley-Whitaker et al., 2001). Arsenic has also induced oxidative stress in plants which caused the inhibition of re-adaptation, transportation, various metabolic processes, and developments of plants. The reduction in these processes has also developed a toxic system during the electron transport processes. Abundant toxic ROS have been produced in the cell wall during this process, which affected the various factors of plants including biomass, enzyme activity, membrane permeability, leaf chlorosis, metabolic pool, and necrosis (Nguyen et al., 2003). During transportation across the plasma membrane via phosphate transport systems arsenate acted as a phosphate analog and within the cytoplasm, it reacted with phosphate and replaced the ATP from ADP-As specifically due to disturbance of energy flow in cells (Meharg, 1994).

1.6.2 LEAD (PB)

Lead (Pb) has been bluish or silvery-grey colored metal found majorly in four naturally occurring isotopes and has very poor solubility in water (WHO Regional Office for Europe, 2001). In natural sources, it existed as several forms and now presented as one of the widely and uniformly scattered trace metal. The soil has been contaminated by lead due to the dust, car exhaust, and gas emission from different industrial sources. Lead has also been a well-known contaminated or toxic element for microorganisms, plants, and animals. The toxic effects of lead have been usually restricted to particularly polluted areas (European Commission DG ENV, 2002). Figure 1.7 showed the exposure of lead stress and physiological alterations inside the plant cells that caused the disruption of protein, DNA, lipids, and resulted in cell death.

In the environment, the lead infection has existed in the insoluble form and its toxicity caused severe human diseases including brain retardation and damage (Cho-Ruk et al., 2009). In plants, the lead toxicity has significantly affect the root growth by inhibited the cell division in root tips (Eun et al., 2000). Lead exposure has also caused various degrees of changes in terms of root diameter and volume as well as lateral roots development that have been inhibited. As the concentration of lead ions has increased in the soil, the root cell viability has been affected which leads to cell death (Huang and Huang, 2008). The cell toxicity mechanism of lead in plants has not been entirely understood however, it has been reported that lead has overproduced the ROS including hydrogen peroxide (H_2O_2) and superoxide radicals followed by the death of the plant root cells (Eun et al., 2009; Reddy et al., 2005). These ROS has caused lipid peroxidation, oxidative stress, and membrane damage of the plant cells where the photosynthesis process has been unfavorably affected by lead toxicity (Liu et al., 2010). Plants exposed with lead ions have shown reduced chlorophyll synthesis, partial changes

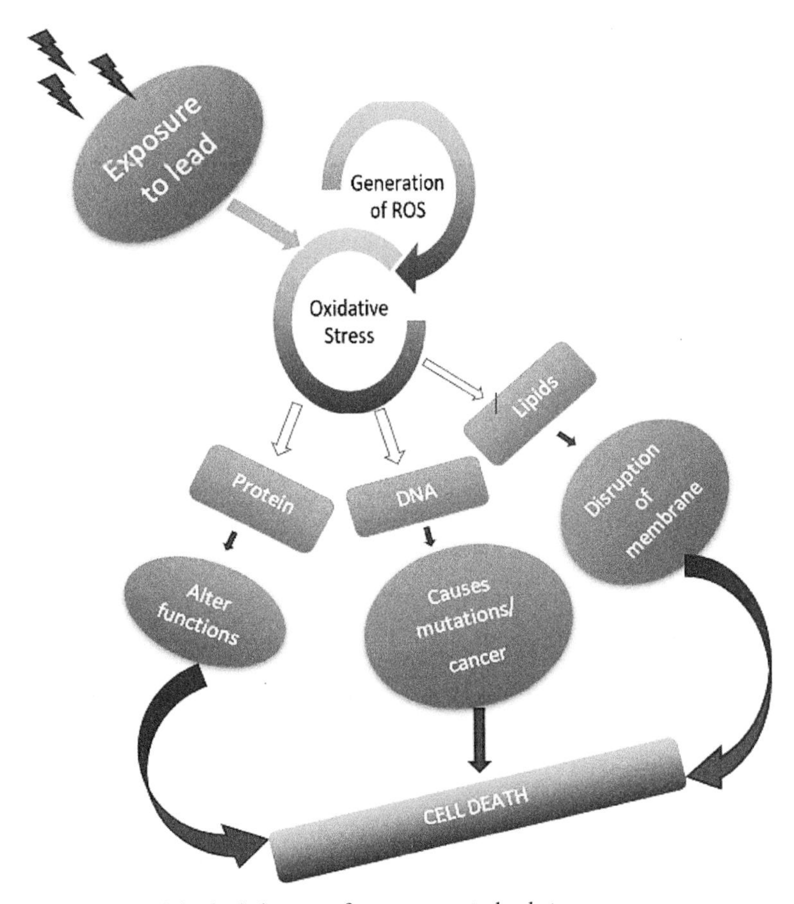

FIGURE 1.7 Physiological changes after exposure to lead stress.

in the ultrastructure of the chloroplast, blocked electron transport system, inhibition of Calvin cycle enzymes, carbon dioxide deficiency resulting in the stomata closure and declined rate of photosynthesis (Sharma and Dietz, 2009). The lead ions affected plant leaves have shown the reduction in grana stacks with stroma as well as the absence of starch grains and lipid composition of thylakoid membranes.

1.6.3 MERCURY (HG)

Mercury has been present in several forms in the environment as it has been a naturally occurring metal. The mercury as a metal present in the odorless, shiny, and silver-white liquid form however when it combined with the other elements such as oxygen, chlorine, and sulfur, converted into the white powdery inorganic compounds and salts. Mercury has also occurred in the soil in various forms like any other metals. It has been dissolved as a free ion or soluble complex and nonspecifically adsorbed by binding to the soil due to chelation, electrostatic forces, and precipitated in the form of carbonate, phosphate, sulfide, and hydroxide. Mercury has been a persistent environmental pollutant and accumulated in plants, fish, animals, and the humans. Generally, the exposure of mercury metal has altered the cell membrane, generated the ROS, as well as damaged the photosynthetic complexes inside the plant cells (Figure 1.8).

Generally, terrestrial plants have been not affected by the harmful effects of this metal however, in some plants; mercury affected the oxidative

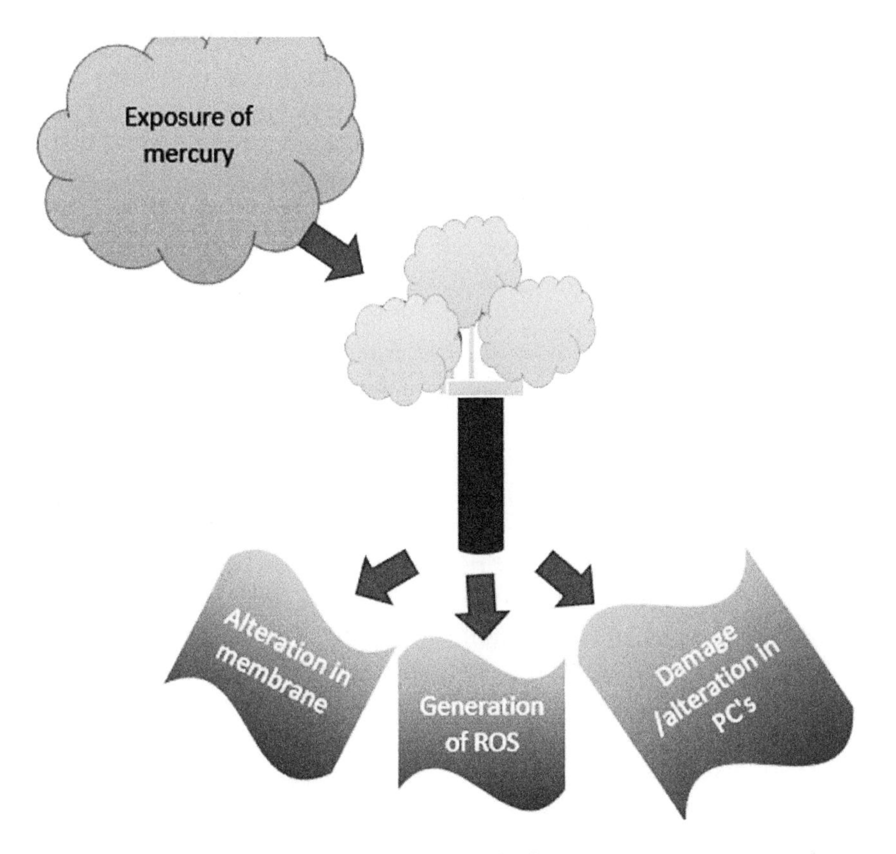

FIGURE 1.8 Physiological alterations in plant cells after exposure to mercury metal.

metabolism and photosynthesis rate by blocking the electron transport chain in mitochondria and chloroplast (Chang et al., 2009). This metal has damaged the possible mechanisms at the cellular level by blocking the important molecule such as enzymes, transport of essential ions, denaturing of protein, displacement of metal ions from molecules such as Mg from chlorophyll and cell membrane disruption (Sas-Nowosielska et al., 2008). Mercury has also affected the antioxidant defense system by interfering with the non-enzymatic antioxidant modulation like glutathione (GSH), enzymatic antioxidant superoxide dismutase, non-protein thiols (NPSH), ascorbate peroxidase (APX) and glutathione reductase (GR) (Patra et al., 2000; Ortega-Villasante et al., 2005).

1.6.4 COPPER

Copper is one of the micronutrients that occur in plants and it always functions in concurrence with various other enzymes that are somehow related to processes like respiration and photosynthesis in plants (Israr et al., 2006; Marschner, 1974). The foremost visual symptoms of the copper toxicity have been seen as the impaired root and shoot growth, chlorosis, nutrient deficiency, tissue necrosis and plant death (Israr et al., 2006; Yruela et al., 2000). Copper stress mainly reduced the grain yield, the biomass of the plant, root, and shoot length while altered the ultrastructure and anatomy of plant cells (Figure 1.9).

These symptoms have been shown on the plant after direct or indirect uptake of excess copper ion. The damage in the cellular membranes of the roots resulted in a substantial decline in the minerals, nutrients, and water uptake (Kopsell and Kopsell, 2007). Copper toxicity has affected the root cells directly and indirectly. The indirect or hidden symptoms like decreased branch growth and chlorosis have been caused by the deficiency in nutrients and water (Yruela et al., 2000). The direct symptoms at the cellular level have been produced by the increased concentration of ROS, such as superoxide anion (O^{2-}), singlet oxygen ($^1O_2$), hydrogen peroxide (H_2O_2) and hydroxyl radical (OH^-). These ROS could have damaged the biomolecules. However, its most important effect has been observed as lipid peroxidation of cell membranes. The damage of the cell membrane has lowered the selectivity of root cells towards other metal ions and caused membrane breakage with leakage of the cell contents (De Vos et al., 1989).

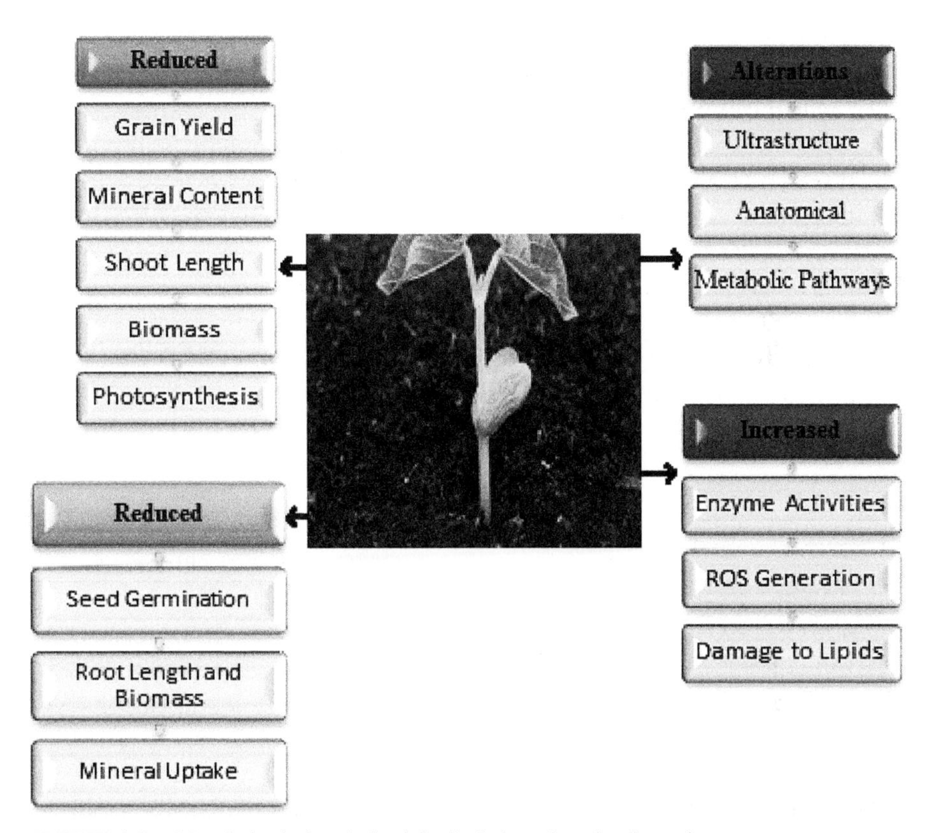

FIGURE 1.9 Morphological and physiological alterations in plants due to copper stress.

1.6.5 CADMIUM

Cadmium (Cd) has also known an environmental toxic metal deposited excessively in soil mainly from anthropogenic activities like mining, metallurgy, and agriculture (Davidson, 2013). This heavy metal has been extremely mobile and has got collected in the lower organisms and transported to higher levels through the food chain. Therefore, Cadmium has been considered as a major contaminant among three main pollutants as they have been the greatest threat including mercury and lead for the environment by the US Environmental Protection Agency (EPA) (Jamers et al., 2013). Figure 1.10 showed the physiological effects after exposure to cadmium stress.

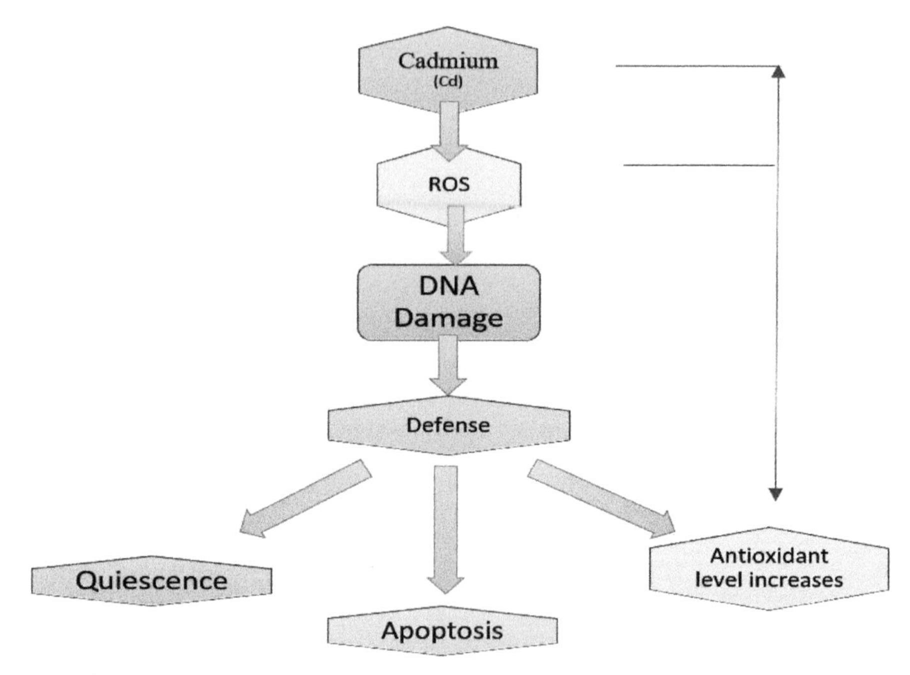

FIGURE 1.10 Physiological effects after cadmium exposure.

1.7 CONCLUSION

Green leafy vegetables and other plants have been the major dietary part of being consumed all over the world. These plants have been the rich sources of mineral and nutrition hence played an important role in the balanced diet. However, they have been the main accumulators of heavy metals when exposed to a contaminated environment. The heavy metals accumulation, directly and indirectly, affected the morphological and physiological activities of plants. These heavy metals have changed the activities mostly in a negative manner as they did not play any valuable role in the development of plants. However, when provided in less concentration, heavy metals can also enhance the physicochemical and morphological activities. The removal of excess heavy metals from the soil using bioremediation has been the most effective solution for this problem. The heavy metal accumulation in plants also showed the potential risk to human health. These health effects can be direct or indirect including ingestion and reduction in nutritional components.

KEYWORDS

- **bioaccumulation**
- **heavy metals**
- **morphological alterations**
- **non-protein thiols**
- **physiological alteration**
- **reactive oxygen species (ROS)**

REFERENCES

Adriano, D. C., et al., (1995). Soil contamination and remediation philosophy, science and technology. In: Prost, R., (ed.), *Contaminated Soils* (pp. 466–504). INRA, Paris.

Alaoui-Sosse, B., et al., (2004). Effect of copper on growth in cucumber plants (*Cucumis sativus*) and its relationships with carbohydrate accumulation and changes in ion contents. *Plant Sci., 166*, 1213–1218.

Chang, A. C., Page, A. L., & Warneke, J. E., (1987). Long-term sludge application on cadmium and zinc accumulation in Swiss chard and radish. *J. Environ. Qual., 16*, 217–221.

Chang, T. C., et al., (2009). Treating high-mercury-containing lamps using full-scale thermal desorption technology. *Journal of Hazardous Materials, 162*(2/3), 967–972.

Chen, H. M., et al., (1998). Heavy metal pollution in soils in China: Status and countermeasures. *Ambio, 28*, 130–134.

Cho-Ruk, K., et al., (2009). Perennial plants in the phytoremediation of lead contaminated soils. *Biotechnology, 5*(1), 1–4.

Chugh, L. K., & Sawhney, S. K., (1999). Photosynthetic activities of *Pisum sativum* seedlings grown in the presence of cadmium. *Plant Physiol. Biochem., 37*(4), 297–303.

Chutia, P. S., et al., (2009). Arsenic adsorption from aqueous solution on synthetic zeolites. *Journal of Hazardous Materials, 162*(1), 440–447.

Clemens, S., (2006). Toxic metal accumulation, responses to exposure and mechanisms of tolerance in plants. *Biochimie, 88*, 1707–1719.

Davidson, C. M., (2013). Methods for the determination of heavy metals and metalloids in soils. *Heavy Metals in Soils* (pp. 97–140). Springer.

De Vos, C. R., et al., (1989). Copper-induced damage to the permeability barrier in roots of *Silene cucubalus*. *Journal of Plant Physiology, 135*(2), 164–169.

Eun, S., et al., (2000). Lead disturbs microtubule organization in the root meristem of *Zea mays*. *Physiol. Plant, 110*, 357–365.

European Commission DG ENV, (2002). *E3, Heavy Metals in Waste*, Final Report Project ENV.E.3/ETU/2000/0058.

Fusconi, A., et al., (2006). Effects of cadmium on meristem activity and nucleus ploidy in roots of *Pisum sativum* L. cv. frisson seedlings. *Environ. Exp. Bot., 58*, 253–260.

Garbisu, C., & Alkorta, I., (2001). Phytoextraction: A cost-effective plant-based technology for the removal of metals from the environment. *Biores. Technol., 77*, 229–236.

Gaur, A., & Adholeya, A., (2004). Prospects of arbuscular mycorrhizal fungi in phytoremediation of heavy metal contaminated soils. *Current Science, 25*, 528–534.

Gupta, S., et al., (2008). Assessment of heavy metal accumulation in macrophyte, agricultural soil and crop plants adjacent to discharge zone of sponge iron factory. *Environ. Geol., 55*, 731–739.

Han, Y. L., et al., (2007). Cadmium tolerance and accumulation by two species of iris. *Ecotoxicology, 16*, 557–563.

Hartley-Whitaker, J., Ainsworth, G., & Meharg, A., (2001). Copper-and arsenic-induced oxidative stress in *Holcus lanatus* L. Clones with differential sensitivity. *Plant, Cell and Environment, 24*, 713–722.

Huang, T. L., & Huang, H. J., (2008). ROS and CDPK-like kinase-mediated activation of MAP kinase in rice roots exposed to lead. *Chemosphere, 71*, 1377–1385.

Intawongse, M., & Dean, J. R., (2006). Uptake of heavy metals by vegetable plants grown on contaminated soil and their bioavailability in the human gastrointestinal tract. *Food Addit. Contam., 23*(1), 36–48.

Israr, M., et al., (2006). Bioaccumulation and physiological effects of mercury in *Sesbaniadrum mondii. Chemosphere, 65*(4), 591–598.

Jamers, A., et al., (2013). An omics based assessment of cadmium toxicity in the green alga *Chlamydomonas reinhardtii. Aquatic Toxicology, 126*, 355–364.

Khan, S., et al., (2008). Accumulation of polycyclic aromatic hydrocarbons and heavy metals in lettuce grown in the soils contaminated with long-term wastewater irrigation. *J. Hazard. Mater., 152*, 506–515.

Khan, S., et al., (2008b). Health risks of heavy metals in contaminated soils and food crops irrigated with wastewater in Beijing, China. *Environ. Pollut., 152*, 686–692.

Kopsell, D. E., & Kopsell, D. A., (2007). Copper. In: Barker, A. V., & Pilbeam, D. J., (eds.), *Handbook of Plant Nutrition* (pp. 293–328). Boca Raton, Taylor, and Francis Group.

Kopyra, M., & Gwozdz, E. A., (2004). The role of nitric oxide in plant growth regulation and responses to abiotic stresses. *Acta Physiol. Plant, 26*, 459–472.

Krupa, Z., & Baszynski, T., (1995). Some aspects of heavy metals toxicity towards photosynthetic apparatus—Direct and indirect effects on light and dark reactions. *Acta Physiol. Plant, 17*, 177–190.

Li, Q., et al., (2010). Toxic effects of heavy metals and their accumulation in vegetables grown in a saline soil. *Ecotoxicol. Environ. Saf., 73*, 84–88.

Liu, C. N., et al., (2010). *Ecotoxicol. Environ. Saf., 73*(6), 1238–1245.

Luo, C., et al., (2011). Heavy metal contamination in soils and vegetables near an e-waste processing site, south China. *J. Hazard, 186*, 481–490.

Marschner, H., (1974). Mechanisms of regulation of mineral nutrition in higher plants. *Bull. R. Soc. NZ.*

Martin, H. W., & Kaplan, D. I., (1998). Temporal changes in cadmium, thallium, and vanadium mobility in soil and phytoavailability under field conditions. *Water Air Soil Pollution, 101*, 399–410.

Meharg, A., (1994). Integrated tolerance mechanisms: Constitutive and adaptive plant responses to elevated metal concentrations in the environment. *Plant, Cell Environment, 17*, 989–993.

National Ground Water Association, (2001). Copyright. Arsenic.

Ndimele, P. E., & Jimoh, A. A., (2011). Water hyacinth (*Eichhornia crassipes* (Mart.) Solms.) in phytoremediation of heavy metal polluted water of Ologe Lagoon, Lagos,Nigeria. *Res. J. Environ. Sci., 5*, 424–433.

Nguyen, T., et al., (2003). Increased protein stability as a mechanism that enhances Nrf2-mediated transcriptional activation of the antioxidant response element. Degradation of Nrf2 by the 26 S proteasome. *The Journal of Biological Chemistry, 278*, 4536–4541.

Ortega-Villasante, C., (2005). Cellular damage induced by cadmium and mercury in *Medicago sativa, Journal of Experimental Botany, 56*(418), 239–2251.

Oves, M., (2012). Soil contamination, nutritive value, and human health risk assessment of heavy metals: An overview. *Toxicol. Heavy Metals Leg. Biorem.*, 1–27.

Pandey, V., Dixit, V., & Shyam, R., (2005). Antioxidative responses in relation to growth of mustard (*Brassica junceacv.* Pusa Jai Kisan) plants exposed to hexavalent chromium. *Chemosphere, 61*, 40–47.

Paolacci, A. R., Badiani, M., Damnibale, A., Fusari, A., & Matteucci, G., (1997). Antioxidants and photosynthesis in the leaves of *Triticum durum* desf seedlings acclimated to non-stressing high temperature. *J. Plant. Physiol., 150*, 381–387.

Patra, M., & Sharma, A., (2000). Mercury toxicity in plants. *Botanical Review, 66*(3), 379–422.

Peralta, J. R., Torresdey, J. L. G., Tiemann, K. J., Gomez, E., Arteaga, S., & Rascon, E., (2001). Uptake and effects of five heavy metals on seed germination and plant growth in alfalfa (*Medicago sativa*) L. *B Environ. Contam. Toxicol., 66*, 727–734.

Prasad, M. N. V., Greger, M., & Landberg, T., (2001). *Acacia niloticaL.* bark removes toxic elements from solution: Corroboration from toxicity bioassay using *Salix viminalis* L. in hydroponic system. *Int. J. Phytoremed, 3,* 289–300.

Preeti, P., & Tripathi, A. K., (2011). Effect of heavy metals on morphological and biochemical characteristics of albiziaprocera (Roxb.) Benth. seedlings. *International Journal of Environmental Science, 1*, 5.

Rebechini, F., and, H. M., & Hanzely, L., (1974). Lead-induced ultrastructural changes in chloroplasts of the hydrophyte *Ceratophyllum demersum. Z. Pf Lanzen Physiology, 73*, 377–386.

Reddy, B., et al., (2005). Lead induced changes in antioxidant metabolism of horsegram (*Macrotyloma uniflorum* (Lam.) Verdc.) and bengalgram (*Cicer arietinum* L.). *Chemosphere, 60*(1), 97–104.

Rout, G. R., Samantaray, S., & Das, P., (2001). Differential lead tolerance of rice and black gram genotypesin hydroponic culture. *Rost. Výroba(Praha), 47*, 541–548.

Sas-Nowosielska, A., et al., (2008). Remediation aspect of microbial changes of plant rhizosphere in mercury contaminated soil. *Environmental Monitoring and Assessment, 137*(1–3), 101–109.

Sersen, L., et al., (1998). Action of mercury on the photosynthetic apparatus of spinach chloroplasts. *Photosynthetica, 35*, 551–559.

Shafiq, M., & Iqbal, M. Z., (2005). Tolerance of *Peltophorump terocarpum* D. C. Baker Ex K. Heyne seedlings to lead and cadmium treatment. *J. New Seeds, 7*, 83–94.

Shanker, A. K., et al., (2005). Chromium toxicity in plants. *Environ. Int., 31*, 739–751.

Sharma, P., & Dubey, R. S., (2005). Lead toxicity in plants. *Braz. J. Plant. Physiol., 17*, 35–52.

Sharma, S. S., & Dietz, K. J., (2009). The relationship between metal toxicity and cellular redox imbalance. *Trends in Plant Science, 14*(1), 43–50.

Tinsley, I. J., (1979). *Chemical Concepts in Pollutants Behavior*. John Wiley and Sons Inc, NY.

U.S. Department of Health and Human Services, (2005). *Public Health Service Agency for Toxic Substances and Disease Registry.* Division of Toxicology and Environmental Medicine. Arsenic.

Verloo, M., & Eeckhout, M., (1990). Metal species transformations in soil: An analytical approach. *Int. J. Environ. Anal. Chem., 39*, 170–186.

WHO Regional Office for Europe, (2001). *Air Quality Guidelines: Chapter 6.7* (2nd edn.). Lead, Copenhagen, Denmark.

Williams, D. E., et al., (1980). Trace element accumulation movement and distribution in the soil profile from massive applications of sewage sludge. *Soil Sci., 1292*, 119–132.

Yruela, I., et al., (2000). Copper effect on the protein composition of photosystem II. *Physiologia plantarum.*, *110*(4), 551–557.

CHAPTER 2

Different Environmental Stress for Enhanced Biofuel Production from Plant Biomass

S. JOSHI, M. CHOUDHARY, and N. SRIVASTAVA

Department of Bioscience and Biotechnology, Banasthali University, Rajasthan, India

ABSTRACT

In today's world, the major challenges for agriculture is to accomplish improved plant biomass and biofuel production yet environmental conditions are unfavorable. For the fostering of biofuel production appropriate temperature, salinity, pH, and nutrients these are the desired environmental conditions. Beneath these environmental-stress conditions of luminosity, heat, nutrient, and salt, these circumstances are not constantly well-suited through the situations important to biofuel production. As maintaining elevated yields of biomass production latent avenues of reaping the remuneration of improved biofuel production. A complicated compound material poised of a variety of polymer networks as well as frequent polysaccharides, cell walls, and the polyphenol lignin all are consist of plant biomass.

2.1 INTRODUCTION

Efforts for the development of renewable energy sources are needed to complete the demand of global energy as the rapid increase of its demand. As biofuels are sustainable, environment-friendly, and renewable are considered to be one of the majorly feasible substitutes of energy sources. Technological and scientific interest has been well-known in the earlier (Svanberg, 1995). Nowadays, the mainly employed are individuals, which develop the fluorescence emission from the plant leaves generated in the photosynthesis process.

There are so many challenges for agricultural productivity and plant growth and one of the key is stress in arable land. There is a loss of billions of dollars estimated annually. While for the production of agricultural biomass, 86% of the freshwater is utilized. Through the long-term transition from the era of 'fossil energy' to the moment of 'energy from biomass,' further pressure on freshwater resources is likely to arise (Boyer, 1982). There is a prominent strategy, which is used for the improvement of the structural design of the plants and their biomass (Bray et al., 2000; Cushman and Bohnert, 2000). The strategy is to cut apart the regulatory network involved in cell wall biosynthesis (Moffat, 2002; Heyne, 1940). During current days, significant consideration has been given to polymers from the plant cell wall and the most important constituent of the biomass of plant is the form through it.

On the other hand, get through the plant biotechnology is needed and basically depends on it for the use of lignocellulosic ethanol the same as a feasible option for petroleum-based transportation fuels. There are so numerous challenges which are faced in the field of bioenergy production. Biotechnology has a potential impact on renewable energy. Biotechnologies are uses in many ways for the production of bioenergy such as abiotic stress resistance, lignin modification, feedstock establishment, nutrition usage, basic research, biomass production, biocontainment of transgenes, and metabolic engineering (Craufurd and Peacock, 1993).

As bioenergy has several preferences recently there has been a comeback of interest in it. It refers as renewable energy from biological resources with the purpose of can be used for fuel, heat, and electricity and as well as for co-products of these (Jiang and Huang, 2001; Thomashow, 1999; Lee et al., 2005; Rizhsky et al., 2002).

Cellulose, hemicelluloses, and lignin are the lignocelluloses the major constituents in most plant and it comprises many different polysaccharides, phenolic polymers, and proteins for liquid biofuel production (Figure 2.1). "Environmental stress is used across multiple fields in biology; the inherent ambiguity associated with its definition has caused confusion when attempting to understand organismal responses to environmental change. To be a viable alternative, a biofuel should provide a net energy gain, have environmental benefits, be economically competitive, and be producible in large quantities without reducing food supplies." Oil-product substitution, climate change mitigation, and economic growth for achieving these targets

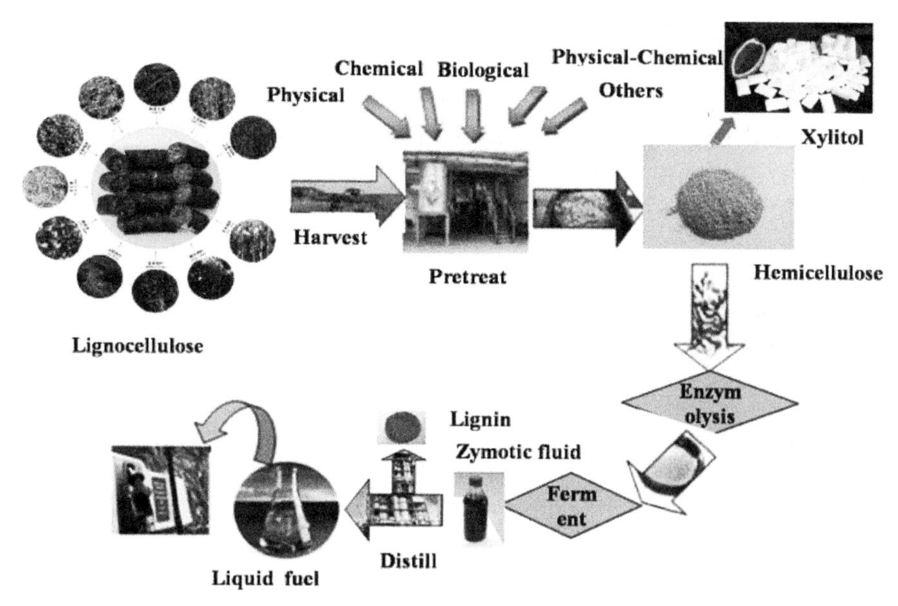

FIGURE 2.1 Liquid fuel production from cellulosic biomass.

biofuels are limited. Their sustainable fabrication is beneath scanner, as is the prospect of creating excessive rivalry for land and swatter used for foodstuff and thread construction. "Biomass crops, such as miscanthus, are being developed for the production of biofuels to replace our fossil fuel-based energy supply chain with a renewable and more sustainable biomass-based alternative. Environmental modulation of homeostasis may be defined as biological stress." Lignocellulosic feedstocks could be a substitute for bioethanol production from this alternative for the disposal of these residues (Figure 2.2).

2.2 RESPONSES OF PLANTS TO DIFFERENT STRESSES

There are many microbes, which are helpful in biomass production with the plant. Plant and microbes partnerships are increasing biomass production. Rhizobacteria are mainly concentrated by researchers for the promotion of plant growth (Whipps, 2001; Barea et al., 2005; Van Loon, 2007). At the present time, on the other hand, consideration has to pay attention to the endophytes for the plant growth-promoting ability of (Taghavi et al., 2009; Mano and Morisaki, 2008).

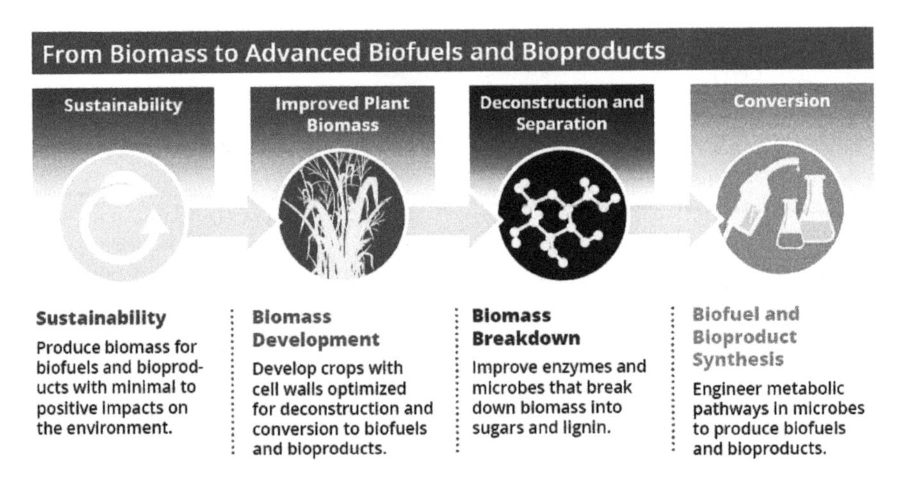

FIGURE 2.2 From biomass to advanced biofuel and byproducts.

Although there are many plant-associated bacteria, which have advantageous effects on their host but still these are still undervalued. An enhanced considerate of these bacteria is accommodating for the increment of biomass for feedstocks and biofuel production as well as for sustainable development of feed and food crops. In this manner, these partnerships of the plant with microbes are encouraging in industrial processes. "In a more sustainable manner, such plant growth-promoting mechanisms may ease the elevated production of energy crops even on marginal land, and thus contribute to avoiding conflicts between food and energy production. Furthermore, because many bacteria show a natural capacity to cope with contaminants, they could be exploited to improve the efficiency of phytoremediation or to protect the food chain by reducing levels of agrochemicals in food crops."

Different food producing plants have been used for biofuel production including sugar crops (sugarcane, sweet sorghum), starch plants (cassava) and vegetables (corn, sorghum, and wheat), oilseed plants (soybean, oil palm) from the past few years (http://www.iddri.org/Activites/Ateliers/081009_Conf-Ethanol_Presentation_Andre_Nassar.pdf)).

Devoid of infringing on food provisions and without fertile soil or food plants supplementary ways of emergent plants as energy sources need to be explored. "The drawbacks of phytoremediation, particularly those associated with low productivity and limitations to the use of contaminant-containing biomass, could be addressed through novel biotechnological approaches that harness recent advances in our understanding of chemical interactions between plants and microorganisms in the rhizosphere and within plant tissues."

2.3 ENVIRONMENTAL STRESS FOR ENHANCED BIOFUEL PRODUCTION FROM PLANT BIOMASS

Several plants are affected by drought and other stresses, such as heat or salinity (Gonzalez et al., 2012; Dkhar and Pareek, 2014). "Abiotic stress conditions cause extensive losses to agricultural production worldwide (Gerbens et al., 2009) Individually, stress conditions such as drought, salinity or heat have been the subject of intense research (Risopatron et al., 2010; Szechyńska et al., 2016) and in the last decades, the idea of using biomass (especially when this is converted to liquid fuels, or biofuels) for sustainable energy production has been attracting more and more interest from policymakers, scientists, and investors, in the hope of providing an answer to the energy crisis and to the need to reduce greenhouse gas (GHG) emission." Through superior acceptance to meadow stress circumstances in which diverse aspects of stress, amalgamation is required to connection for improved biofuel production.

A combination of dissimilar stresses like abiotic in the field and crops are consistently subjected (Gonzalez et al., 2012; Dkhar and Pareek, 2014; Tiller et al., 2008; Stewart, 2007).

2.3.1 DROUGHT AND HEAT STRESS

There are many abiotic stress but **the** drought is one of them, due to weather and temperature modification the frequency of confined and regional drought events is growing globally (Sheffield and Wood, 2008; Dai, 2013).In the field consecutively themost prevalent abiotic stress, symbolize a good example of two diverse stress circumstances (Chaves et al., 2003; Farooq et al., 2009). High-quality lignocellulosic feedstocks are critical to cellulosic biorefinery achievement (Perlack et al., 2005; Van Der Weijde et al., 2013). Transversely assorted environments, crops must be elevated yielding and have constant performance. From the last few years, it have seen that blend of drought and heat stress compared with each of the different stresses applied individually drought and heat stress had a significantly greater for the biofuel production from the agricultural waste Figure 2.3 (Heyne, 1940; Brunson, 1940; Jiang and Huang, 2001; Wang and Huang, 2004; Perdomo et al., 1996).

The stress combination has quite a lot of exclusive aspects. Heat stress or the combination of drought and heat stress influenced the development and productivity of these plants and crops considerably more and it reveals the several aspects, i.e., with short photosynthesis combining high respiration, Physiological categorization of plants subjected to drought, closed stomata and high leaf temperature (Rizhsky et al., 2002).

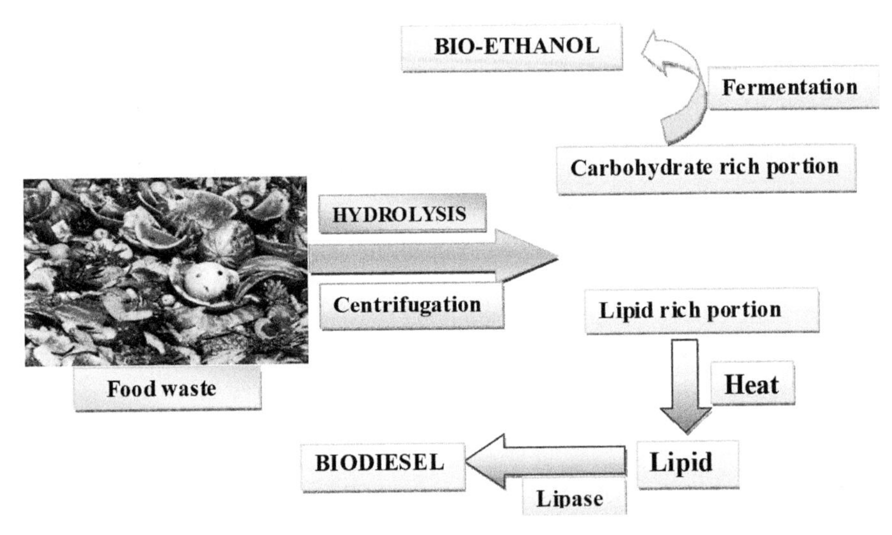

FIGURE 2.3 Biodiesel production from food waste.

2.3.2 *TEMPERATURE STRESS*

Designed for the biodiesel production under the temperature stress fatty acids are favored sources, it might be sufficient to exercise moderately elevated illumination intensities to get better biodiesel to acquiesce. "At the molecular level, responses to heat and cold stresses vary and are mediated by different sets of stress-related genes, many of which are transcription factors (TFs) temperature extremes affect not only the morphology and eco-physiology but, the cellular mechanisms of the plants as well and Hormones and its signaling also play a key role in dealing with stress and accumulation and signaling through abscisic acid (ABA), salicylic acid (SA), and ethylene are important in response to high-temperature stress" (Figure 2.4) (Thomashow, 1999).

Light as stress commonly induces the gathering of antioxidant pigments. In addition, luminosity, other radiation such as UV also induces oxidative stress. In their expression, numerous stress-responsive genes induced by cold, drought, and high salinity have a strong association.

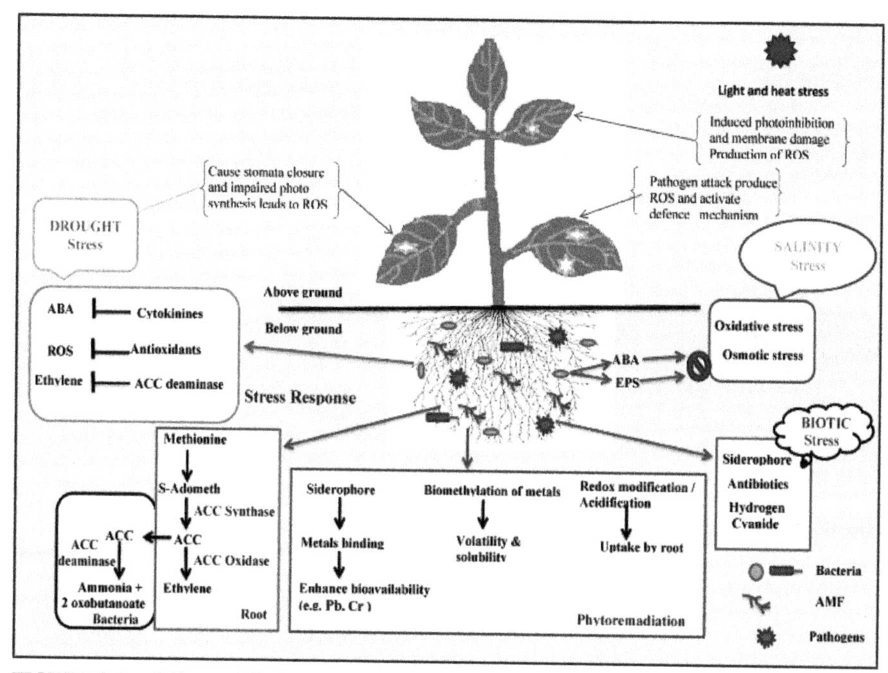

FIGURE 2.4 Effects of light and heat stress, drought stress, salinity stress.

2.3.3 SALT STRESS

In the fabrication of a variety of biofuels, salt stress has been playing a significant character. Amplify the efficiency of biodiesels, elevated salinity tends to persuade the dissemination of fatty acid, (Xu and Beardall, 1997; Chen et al., 2008). In 2010, Carrieri et al. establish that for ethanol production high salt deliberation amplified ethanol production by 121-fold compared to stumpy (low) salt concentration. Several microalgae generate carbohydrates in rejoinder to salt stresses (Figure 2.5) (Warr et al., 1985; Stal and Reed, 1987; Page-Sharp et al., 1998; Rao et al., 2007).

In Northern China, switchgrass and its biomass have been used for biofuel production and possessions of salt stress on the biomass fabrication and fuel distinctiveness of energy crops. Switchgrasses and Leymas individually give acquiesce but switchgrasses yielded a superior aboveground biomass per plant than Leymus under low-salt conditions ($\leq$ 2 g NaCl kg^{-1}). However, at elevated salt concentrations, their biomass fabrication declined considerably.

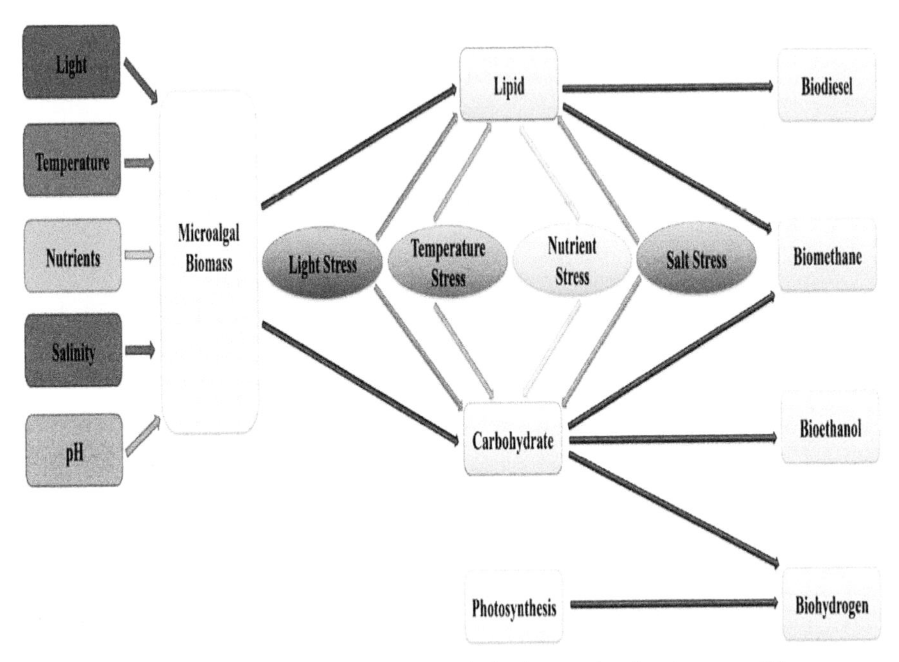

FIGURE 2.5 Several microalgae generate carbohydrate under the stress conditions.

2.4 CHOICES OF CROPS AND FEEDSTOCKS

In the two groups, we can classify the bioenergy crops:

1. conventional sugar-producing crops; and
2. conventional cereal crops.

2.5 SUGAR-PRODUCING PLANTS

During direct fermentation, the two main crops involved are Sugarcane (*Saccharum officinarum*) and sugar beet (*Beta vulgaris*), and both the crops require a favorable condition (grown in warm temperate to tropical and temperate areas) for its growth. Therefore, the two sugar crops are located in diverse geological niches. The one country which has flourished illustration of a nation that has condensed its petrol and diesel convention by producing bioenergy is none other than "Brazil."

In a nation of Brazil ethanol agenda, which is based exclusively on sugar-cane, the net production of ethanol per annum is around 4.2 billion gallons. "a year, although the resultant ecological and environmental effects are still debatable (Searchinger et al., 2008) Europe becomes the main hub for the maximum ethanol production using sugar beet; however, utilization of the sugar beet to produce ethanol could possibly upsurge/increase soil erosion and lower the net energy balance whereas the other sugar-producing crops like energy can help in the enhancement of cultivars of sugarcane and varieties of sweet sorghum."

With the exemption of sugarcane, the rest of the exceeding crops is generally annual crops. For the purpose of bioenergy feedstock, perennials are considered more beneficial/helpful compared to the annuals because they do not require to be reseeded every emergent period and therefore farming expenses are inferior (Fargione et al., 2008).

2.6 TRADITIONAL CEREAL CROPS

A number of cereal crops are also a part of lignocellulosic biomass and the main resource for starch-based ethanol production. In the last few years, *Zea mays* have been used as a bioenergy crop. Maize also has very traditional importance used as processed food, a vegetable, oil, fodder, food, and feed crop and byproducts. Maize widely used in industry as a bioenergy crop in two ways: lignocellulosic ethanol can be produced from starch in seeds, and the crop residuals (termed stover).

Globally *Sorghum bicolor* is the most civilized cereal crop. It is an alternative source of lignocellulosic biomass that is utilized for biofuel and bioenergy fabrication/production on a large/huge scale/extent and is grown and cultivated for grain, forage, sugar, and fiber. There are several advantages associated with this crop, "First no strong competition for the utilization of land for cultivation of food or for energy as the seeds can be considered for food and feed and the stems could be optimized and utilized for different platforms of ethanol production, which is predominantly important for heavily populated developing countries including China and Second, sorghum is drought and heat tolerant, which would enable the usage of marginal land that is not suitable for the cultivation of many other crops"

Using accessible technology platforms, for lignocellulosic ethanol production the grain, sorghum, sugar maize, wheat, and rice (Perennial bioenergy feedstocks)) also could be feedstock and crop residuals could be useful (De Oliveira et al., 2005) "Once the technologies for biorefineries are established and commercialized, a wide range of chemicals (e.g., olefins, plastics, solvents, many chemical intermediates) and biofuels (e.g., biogasoline, alcohols, biodiesel, JP-8, and FT liquids) could be produced from lignocellulosic biomass and biodiesel is a mixture of diesel fuel with oils from plant seeds, algae or other biological sources such as animal renderings that have been transesterified for removal of glycerol."

Switchgrass (*Panicum virgatum*) is a widely adapted, efficient use of nitrogen and water. In Europe hybrid *Miscanthus*, including *Miscanthus giganteus*, is a different extremely preferential biomass feedstock.

2.7 PLANT-MICROBE INTERACTION

Rhizosphere and root-associated microbes to last in deadly low-input biotechnology for the renaissance of the ecosystem countless crops in relationships (Weyens et al., 2009) and nutrient-limited environments. In phytoremediation, systems for this plant-microbe interaction earlier mechanism have tinted the prospective of microbes (Paul, 2007).

In the surroundings, vegetation can be genetically customized to improve chemical interventions in the environment and plant-microbial signaling Figure 2.6 (Badri et al., 2009).

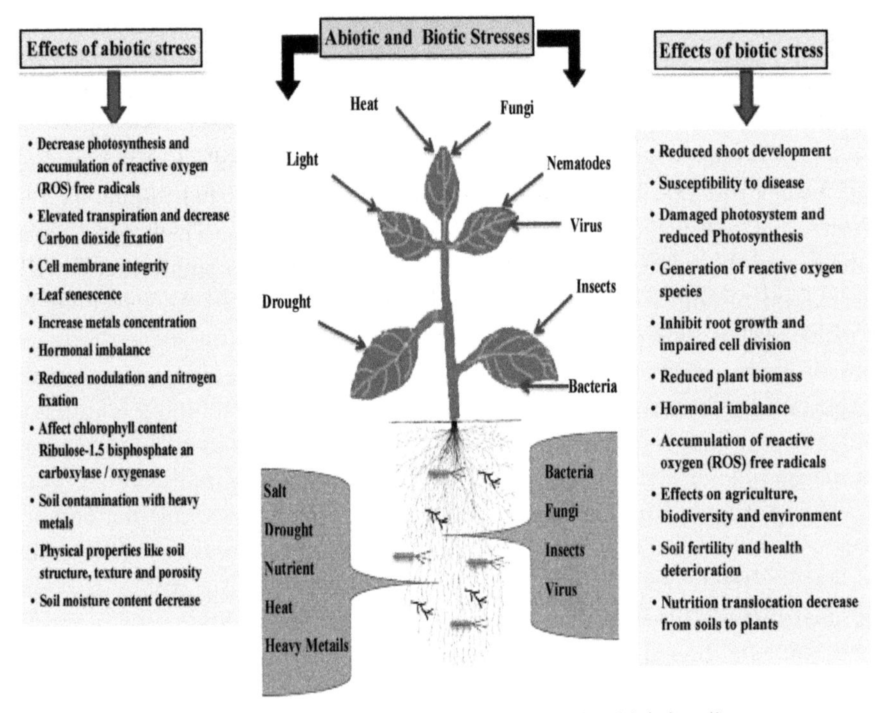

FIGURE 2.6 Effects of abiotic and biotic stress on plant microbial signaling.

We recommend the use of biofuels derivative from marine microbial oxygenic photoautotrophs (AMOPs), to alleviate numerous of the potentially lethal ecological and rural cost allied with existing terrain based-biofuels feedstock, further frequently identified as diatoms, cyanobacteria, and algae. "Herein we review their demonstrated productivity in mass culturing and aspects of their physiology that are particularly attractive for integration into renewable biofuel applications and Compared with terrestrial crops, AMOPs are inherently more efficient solar collectors, use less or no land, can be converted to liquid fuels using simpler technologies than cellulose, and offer secondary uses that fossil fuels do not provide. AMOPs pose a new set of technological challenges if they are to contribute as biofuel feedstocks" (23). In the harvested plant, parts there are extensive concerns on the subject of the employ of harvested biomass and the providence of perilous toxins (Abhilash and Yunus, 2011).

Manufacturing ethanol requires much easier processing technology. Various plant species are presently used for biodiesel manufacturing (Ma and Hanna, 1999).

2.8 CONCLUSION

At the present time for the improvement of sustainable bioenergy, there are several biotechnology-based different raised areas. From these different areas, lower production costs and higher net energy gain are the two platforms, which have advantages for ethanol production using lignocellulosic biomass. For transportation and to reach the target to reinstate petroleum-based shipping fuels bioenergy is a healthier substitute. Temperature, salinity, pH, and nutrients are the desired environmental stress conditions for augmentation in biofuel and bioenergy production.

KEYWORDS

- **abscisic acid**
- **biofuel**
- **biomass**
- **environmental-stress**
- **marine microbial oxygenic photoautotrophs**
- **polymer**

REFERENCES

Abhilash, P. C., & Yunus, M., (2011). Can we use biomass produced from phytoremediation? *Biomass Bioenergy, 35*, 1371–1372.

Abhilash, P. C., Powell, J. R., Singh, H. B., & Singh, B. K., (2012). Plant-microbe interactions: Novel applications for exploitation in multipurpose remediation technologies. *Trends in Biotechnology, 30*(8), 416–420. doi: 10.1016/j.tibtech.2012.04.004.

Badri, D. V., et al., (2009). Rhizosphere chemical dialogues: Plant-microbial interactions. *Curr. Opin. Biotechnol., 20*, 642–650.

Barea, J. M., et al., (2005). Microbial co-operation in the rhizosphere. *J. Exp. Bot., 56,* 1761–1778.

Boyer, J. S., (1982). Plant productivity and environment. *Science, 218*, 443–448.

Bray, E. A., et al., (2000). Responses to abiotic stresses. In: Gruissem, W., (ed.), *Biochemistry and Molecular Biology of Plants* (pp. 1158–1249). American Society of Plant Physiologists.

Carrieri, D., Momot, D., Brasg, I. A., Ananyev, G., Lenz, O. B., Bryant, D. A., et al., (2010). Boosting autofermentation rates and product yields with sodium stress cycling, application to renewable fuel production by cyanobacteria. *Appl. Environ. Microbiol., 76*, 6455–6462. doi: 10.1128/AEM.00975-10.

Chaves, M. M., Maroco, J. P., & Pereira, J. S., (2003). Understanding plant responses to drought-from genes to the whole plant. *Functional Plant Biology, 30*, 239–264.

Craufurd, P. Q., & Peacock, J. M., (1993). Effect of heat and drought stress on sorghum. *Exp. Agric., 29,* 77–86.

Cushman, J. C., & Bohnert, H. J., (2000). Genomic approaches to plant stress tolerance. *Curr. Opin. Plant Biol., 3*, 117–124.

De Oliveira, M. E. D., et al., (2005). Ethanol as fuels, energy, carbon dioxide balances, and ecological footprint. *Bioscience, 55*, 593–602.

Dismukes, G. C., Carrieri, D., Bennette, N., Ananyev, G. M., & Posewitz, M. C., (2008). Aquatic phototrophs: Efficient alternatives to land-based crops for biofuels. *Current Opinion in Biotechnology, 19*(3), 235–240. doi: 10.1016/j.copbio.2008.05.007.

Dkhar, J., & Pareek, A., (2014). What determines a leaf's shape? *Evo. Devo., 5*, 47.

Fargione, J., et al., (2008). Land clearing and the biofuel carbon debt. *Science, 319*, 1235–1238.

Farooq, M., Wahid, A., Kobayashi, N., Fujita, D., & Basra, S., (2009). Plant drought stress: Effects, mechanisms and management. In: *Sustainable Agriculture* (pp. 153–188). Springer, Dordrecht.

Gerbens, L. P. W., Hoekstra, A. Y., & Van, D. M. T. H., (2009). The water footprint of energy from biomass: A quantitative assessment and consequences of an increasing share of bio-energy in energy supply. *Ecol. Econom., 68*, 1052–1060.

Gonzalez, N., Vanhaeren, H., & Inzé, D., (2012). Leaf size control: Complex coordination of cell division and expansion. *Trends Plant Sci., 17*, 332–340.

Heyne, E. G., & Brunson, A. M., (1940). Genetic studies of heat and drought tolerance in maize. *J. Am. Soc. Agro., 32*, 803–814.

Ho, S. H., Chen, C. Y., & Chang, J. S., (2012). Effect of light intensity and nitrogen starvation on CO_2 fixation and lipid/carbohydrate production of an indigenous microalgae *Scenedesmus* obliquus CNW-N. *Bioresour. Technol., 113*, 244–252. doi: 10.1016/j.biortech.2011.11.133.

http://www.iddri.org/Activites/Ateliers/081009_Conf-Ethanol_Presentation_Andre_Nassar.pdf (accessed on 13 June 2020).

Jiang, Y., & Huang, B., (2001). Drought and heat stress injury to two cool season turfgrasses in relation to antioxidant metabolism and lipid peroxidation. *Crop Sci., 41*, 436–442.

Lee, B. H., Henderson, D. A., & Zhu, J. K., (2005). The *Arabidopsis* cold-responsive transcriptome and its regulation by ICE1. *Plant Cell, 17*(11), 3155–3175.

Ma, F. R., & Hanna, M. A., (1999). Biodiesel production: A review. *Biores. Tech., 70*, 1–15.

Mano, H., & Morisaki, H., (2008). Endophytic bacteria in rice plant. *Microbes Environ., 23*, 109–117.

Moffat, A. S., (2002). Finding new ways to protect drought-stricken plants. *Science, 296*, 1226–1229.

Page-Sharp, M., Behm, C. A., & Smith, G. D., (1998). Cyanophycin and glycogen synthesis in a cyanobacterial Scytonema species in response to salt stress. *FEMS Microbiol. Lett., 160*, 11–15. doi: 10.1111/j.1574-6968.1998.tb12883.

Paul, E., (2007). *Soil Microbiology, Ecology, and Biochemistry* (3rdedn.). Academic Press.

Perdomo, P., et al., (1996). Physiological changes associated with performance of Kentucky bluegrass cultivars during summer stress. *Hort. Sci., 31*, 1182–1186.

Perlack, R. D., Wright, L. L., Turhollow, A. F., Graham, R. L., Stokes, B. J., & Erbach, D. C., (2015). *Biomass as Feedstock for a Bioenergy and Bioproducts Industry: The Technical Feasibility of a Billion-Ton Annual Supply*. Oak Ridge National Laboratory, Oak Ridge.

Rao, A. R., Dayananda, C., Sarada, R., Shamala, T. R., & Ravishankar, G. A., (2007). Effect of salinity on growth of green alga *Botryococcus braunii* and its constituents. *Bioresour. Technol., 98*, 560–564. doi: 10.1016/j.biortech.2006.02.007.

Rass-Hansen, J., et al., (2007). Bioethanol: Fuel or feedstock? *J. Chem. Tech. Biotechnol., 82*, 329–333.

Risopatron, J. P. M., Sun, Y., & Jones, B. J., (2010). The vascular cambium: Molecular control of cellular structure. *Protoplasma., 247*, 145–161.

Rizhsky, L., et al., (2002). The combined effect of drought stress and heat shock on gene expression in tobacco. *Plant Physiol., 130*, 1143–1151.

Searchinger, T., et al., (2008). Use of U.S. croplands for biofuels increases greenhouse gases through emissions from land-use change. *Science, 319*, 1238–1240.

Sims, R. E. H., et al., (2006). Energy crops: Current status and future prospects. *Glob. Change Biol., 12*, 2054–2076.

Somerville, C., (2007). Biofuels. *Curr. Biol., 17*, R115–R119.

Stal, L. J., & Reed, R. H., (1987). Low-molecular mass carbohydrate accumulation in cyanobacteria from a marine microbial mat in response to salt. *FEMS Microbiol. Lett., 45*, 305–312. doi: 10.1111/j.1574-6968.1987.tb02381.

Stewart, C. N., (2007). Biofuels and biocontainment. *Nat. Biotechnol., 25*, 283–284.

Sun, X., Gao, Y., Xu, H., Liu, Y., Sun, J., Qiao, D., et al., (2014). Effect of nitrogen starvation, light intensity and iron on triacylglyceride/carbohydrate production and fatty acid profile of *Neochlorisoleoabundans* HK-129 by a two-stage process. *Bioresour. Technol., 155*, 204–212. doi: 10.1016/j.biortech.2013.12.109.

Szechyńska, H. M., Czarnocka, W., Hebda, M., & Karpiński, S., (2016). PAD4, LSD1, and EDS1 regulate drought tolerance, plant biomass production, and cell wall properties. *Plant Cell Rep., 35*, 527–539.

Taghavi, S., et al., (2009). Genome survey and characterization of endophytic bacteria exhibiting a beneficial effect on growth and development of poplar. *Appl. Environ. Microbiol., 75*, 748–757.

Thomashow, M. F., (1999). Plant cold acclimation: Freezing tolerance genes and regulatory mechanisms. *Annu. Rev. Plant Physiol. Plant Mol. Biol., 50*, 571–599.

Tilman, D., et al., (2006). Carbon-negative biofuels from low-input high diversity grassland biomass. *Science, 314*, 1598–1600.

Van, D., Weijde, T., Torres, A. F., Dolstra, O., Dechesne, A., Visser, R. G. F., & Trindade, L. M., (2016). Impact of different lignin fractions on saccharification efficiency in diverse species of the bioenergy crop *Miscanthus*. *Bioenergy Research, 9*, 146–156.

Van, L. L. C., (2007). Plant responses to plant growth-promoting rhizobacteria. *Eur. J. Plant Pathol., 119*, 243–254.

Wang, Z. L., & Huang, B. R., (2004). Physiological recovery of Kentucky bluegrass from simultaneous drought and heat stress. *Crop Sci., 44*, 1729–1736.

Warr, S. R. C., Reed, R. H., & Stewart, W. D. P., (1985). Carbohydrate accumulation in osmotically stressed cyanobacteria (blue-green algae), interactions of temperature and salinity. *New Phytol., 100*, 285–292. doi: 10.1111/j.1469-8137.1985.tb02779.

Weyens, N., et al., (2009). Exploiting plant-microbe partnerships to improve biomass production and remediation. *Trends Biotechnol., 27*, 591–598.

Whipps, J., (2001). Microbial interactions and biocontrol in the rhizosphere. *J. Exp. Bot., 52*, 487–511.

Yaun, J. S., Tiller, K. H., Al-Ahmad, H., Stewart, N. R., & Stewart, C. N., (2008). Plants to power, bioenergy to fuel the future. *Trends in Plant Science, 13*(8), 421–429. doi: 10.1016/j.tplants.2008.06.001.

CHAPTER 3

Application of Plant Stress for Enhanced Biofuel Production

SHRITOMA SENGUPTA[1], DEBALINA BHATTACHARYA[2], and MAINAK MUKHOPADHYAY[3]

[1]*Department of Biotechnology, University of Calcutta, Kolkata- 700019, West Bengal, India.*

[2]*Department of Microbiology, Maulana Azad College, Kolkata- 700013, West Bengal, India*

[1]*Department of Biotechnology, JIS University, Kolkata – 700109, West Bengal, India,*

ABSTRACT

Bioenergy production in the form of high energy biofuels from sustainable resources with the help of green conversion and microbial-based technologies plays an essential role in the replacement of petroleum-based fuels with economic sustainability. Biofuel can be a viable alternative in reducing long-term carbon dioxide emissions. But the major challenge in agriculture is to achieve enhanced plant growth and biomass even under adverse environmental stressed conditions, both biotic and abiotic. Biotic, abiotic stresses, and suboptimal water are important limiting factors for biomass production. So stress-tolerant traits are important to enable feedstock and food crops. Plant-microbe interactions and male sterility preventing transgene escape create biotic stresses by silencing specific gene expressions of both protein-coding, non-protein-coding genes. Various abiotic stresses such as drought, metal, salt, cold, and heat can induce similar responses in plants, but each of these stresses actually induces a different set of genes. In this chapter, the study describes the various environmental stressed conditions and their potential avenues of reaping benefits of biofuel over production by environmental stresses while maintaining high yields of biomass production and transports.

3.1 INTRODUCTION

Conventional fuels are non-sustainable and are currently having a few foremost issues such as the exhaustion of natural reserves in the near future and the significant environmental impacts associated with the natural reserve uses. There has been a recent alteration in the pattern which drives out conventional nonrenewable fuels with renewable, sustainable, and eco-friendly energy sources. Out of all the various sources of energy, biomass-derived energy appears to be the most striking and operative one (Klass, 1998; Lucian et al., 2007; Srirangan et al., 2012). The production of first-generation biofuels, which is primarily produced from food crops such as sugar crops, cereals, and oilseeds have already been extensively studied and are present in the high amount on the market. It mainly includes the biogas and bio-ethanol from starch and sugar, pure plant oil (PPO) and fatty acid methyl esters (FAME) such as rapeseed methyl ester (RME) (Sims et al., 2010). There has always been a long-term economic and environmental concern that has focused on an enormous emphasis in the past couple of decades on the research of renewable sources for liquid fuels to replace fossil fuels. One of the major causes of global warming is the utilization of fossil fuels such as oil and coal which basically releases CO_2 (Yat et al., 2008). Hence, over the past few decades, there has been a growing concern about global warming; so an extensive increase in the research and development in the field of bioenergy and biofuels has encouraged the global market in using biomass for energy. The collective impacts for sustainability have enhanced the interest in developed biofuel production from different sources such as food or non-food plant biomass (Sims et al., 2010). Biofuels are becoming increasingly important in the reduction of CO_2 emissions. Currently, worldwide researches emphasize on various ways to produce these biofuels. If environmental and economic sustainability is considered; along with biofuels, bioenergy goals to be an essential part of the replacement of petroleum-based transportation which may be a feasible alternative in reducing long-term CO_2 emissions (Yuan et al., 2008).

Bioenergy chiefly refers to the renewable energy that can be harnessed from biological sources and may be used for heat, electricity, fuel, and their co-products generation (Somerville, 2007). Bioenergy is a renewable form of energy, is derived from biological sources and may be used to generate heat, electricity, and fuels (Himmel and Bayer, 2009; Yuan et al., 2008). Bioenergy production in the form of high energy biofuels from sustainable resources plays an essential role in replacement of the petroleum-based fuels with help of green conversion, microbial-based technologies, and economic sustainability (Tilman et al., 2006; Rass-Hansen, 2007; Somerville, 2007; Stewart, 2007). Biomass energy is considered to be close to 'carbon neutral,' which implies that it produces energy while the only releasing carbon to the atmosphere has been arrested during the plant growth cycle, rather than emitting carbon to the atmosphere like the fossil fuel reserves that have been produced for millions of years (Sims et al., 2006).

The conventional fossil fuels are on the verge of getting exhausted, which were obtained from plants that grew millions of years ago. So, as a viable alternative, biofuels are being produced from plants and are gaining importance gradually. Biofuels are important for future sustenance in the reduction of carbon dioxide emissions and the production of the next generation of fuels by obtaining security of supply for sustainability. They generally have a short growth cycle of plants, being a cleaner-burning fuel than fossil fuels as they do not add carbon dioxide to the atmosphere. For improving energy security and reducing greenhouse emissions, conversion of the hugely available lignocellulosic biomass to biofuels may be a viable option for transportation fuels. It has been reported that cellulosic ethanol and ethanol produced from various biomass resources have the potential to cut greenhouse gas (GHG) emissions by 86% (Wang et al., 2007). Every year energy crops typically contribute a quite small proportion to the total energy produced from biomass, but the proportion may tend to grow in the next few decades.

Biodiesel is a mixture of diesel fuel obtained from oils or lipids of plant seeds, algae, or other biological sources such as animal renderings that have been trans-esterified for removal of glycerol (Yuan et al., 2008). Biodiesel is produced from biofuel that requires an easier processing technology compared with to the one required for bioethanol production. On the contrary, the increasing use of natural gas for the energy supply mix required for the infrastructure and technological benefits for its transportation, liquefaction, and conversion to alcohols has displaced both oil and coal in developing and developed countries. There should be an awareness of reducing the complete dependence on natural gas by the biofuels and biogas (Sims et al., 2006). The sources and forms of biofuels are diverse, so it is substantial to consider the different environmental and economic factors that apply in the production of different types of biofuel (Ma and Hanna, 1999). Different environmental stress such as abiotic stress (water deficit, salinity, heat, heavy metals soil contamination, intense light, etc.) affects significantly crop growth and yield in agricultural areas all over the world. Thus, it is imperative to study their effect upon the crops and discriminate among abiotic stresses using new noninvasive and nondestructive diagnostic techniques, so higher biomass yield may lead to greater biofuel production.

This study details the various options to produce biofuel under the various environmental stressed conditions. The study also emphasizes their potential avenues of reaping benefits of biofuel over-production by environmental stresses while maintaining high yields of biomass production and transports. It also details about the production and the marketing requirements and potential application of these biofuels.

3.2 PRODUCTION OF BIOFUEL FROM PLANT BIOMASS

For the generation of biofuels and chemicals, plant cell walls represent an enormous biomass resource. It also represents the most abundant renewable biomass resource for biofuels on the earth. The conversion of lignocellulose to ethanol involves the following key stages: physical and chemical pretreatments to enhance cell wall destruction, enzymatic digestion to release soluble sugars, and yeast fermentation to produce ethanol (Carroll and Somerville, 2009; Wang et al., 2016). The genetic modification of plant cell walls has been postured as a powerful solution for such biofuel production, because the lignocellulose property of plant biomass predominantly governs the biomass recalcitrance. The several sources of the renewable form of energy are lignocellulosic materials such as agricultural remains, forest products, and dedicated crops (Kumar et al., 2009). Agricultural remains or residues include wheat straw, sugarcane bagasse, corn stover, etc.; whereas forest products mostly include the hardwood and softwood; and the dedicated crops include switchgrass, salix, etc. As alternative fuel sources, there has been an increasing demand for the development of maize ethanol and soybean biodiesel in the markets of the western countries (Sims et al., 2006).

From biomass, biofuel production commonly includes primary, secondary, and tertiary sources of biomass. The primary resources for biomass are produced directly by photosynthesis and are taken directly from the land, which includes the seeds of oil crops, perennial short-rotation woody crops, and herbaceous crops. It results from the harvesting of agricultural crops such as wheat, corn, and also the tops, limbs, and bark from forest trees. Secondary resources of biomass result from the processing of primary biomass resources either physically, i.e., produced from the sawdust in mills or chemically, i.e., produced from black liquor by pulping processes, or biologically, i.e., by manure produced by animals. Tertiary resources of biomass are obtained from the residue after a human utility which may include used vegetable oils, animal fats, greases, packaging wastes and construction, demolition waste debris. In plant biomass biofuel production, liquid biofuels usually include PPO, biodiesel, and bioethanol. The worldwide production of biofuels comprises mostly of ethanol, followed by the production of biodiesel. Biodiesel is formed from plant oils through an esterification reaction. Ethanol is chiefly derived from crops such as sugar, maize, and other starchy crops. According to the global scenario, Brazil has been considered to have the world's first sustainable biofuels economy and the biofuel industry, which may be a policy model for other countries. The sugarcane ethanol produced in Brazil is considered the most successful alternative fuel to date. About alternative options to renewable energy, biofuels are basically based on the utilization of plant oil as fuel in both forms of stationary and mobile engines, which are the subject gaining much attention recently (Gouveia-Neto et al., 2011).

Interest in using biofuels as an alternative energy source (Cassman and Liska, 2007) is now high on the agenda of policymakers in many countries. In response to the global increase in the use of biofuels, there has been a rapid expansion of biofuel production capacity from food crops in some North and South American, European, and several Southeast Asian countries. Food crops currently used for biofuel production include grains (maize, sorghum, and wheat), sugar crops (sugarcane, sweet sorghum, and sugar beet), starch crops (cassava), and oilseed crops (soybean, oil palm, castor bean, and rapeseed). To meet the rapidly increasing demand without infringing on food supplies, additional ways of growing plants as energy sources without using fertile land or food crops are being explored (Weyens et al., 2009). The increasing amount of cropland is being dedicated to growing crops used in fuel production rather than in food usage. This may be a sustainable method of producing biofuel crops which are urgently needed nowadays. One way of avoiding the conflict between food and energy crops is to produce biofuel feedstocks on marginal land that is not suitable for agriculture. This marginal land comprises soils that lack nutrients, receive little rain, or have been contaminated by previous industrial or agricultural activities. Plant-associated bacteria can be of great value in enabling plants to establish or to grow better on marginal land and could aid in the economic production of biofuels. For the production of biofuels from plant, biomass is generally done by converting biomass to fermentable sugars by using pretreatment processes that disrupt the lignocellulose and remove the lignin. This further allows the access of microbial enzymes for cellulose deconstruction. On the contrary, both the pretreatments and the production of enzymes in microbial tanks are expensive in nature, recent advances in plant genetic engineering could reduce biomass conversion costs by developing process where ligninase enzymes will help in lignin degradation and crop varieties with less lignin that can self-produce cellulase enzymes for degradation of cellulose and plants that have increased cellulose or an overall biomass yield (Sticklen, 2006).

In the past few decades, there has been a widespread scientific and technological interest in laser-induced remote techniques to monitor the status of terrestrial vegetation for having high biomass yield (Svanberg, 1995).

3.3 ROLE OF DIFFERENT ENVIRONMENTAL STRESS ON PLANTS

Due to environmental issues, currently, for energy production, one of the major scientific issues aims to the replacement of fossil-based fuel by biofuel. One of the foremost issues in the present agricultural condition is to attain heightened plant growth and biomass even under hostile environmental conditions. There are a number of biotic and abiotic stresses that impose damage to crop and aggravate a decrease of yield efficiency. The main study is to utilize these stresses for high biomass generation for biofuel production. Several earlier studies have established that plant-microbe interactions influence abiotic stress and tolerance in the system along with growth and biomass. Plants form highly complex and diverse links with co-evolved microbial communities that help in the production of biotic stress for plants (Lau and Lennon, 2011).

3.3.1 ABIOTIC STRESS

To the natural environment and agriculture, abiotic stresses have always been an excessive challenge (Pereira, 2016). Abiotic stresses play one of the most crucial challenges for plant growth and agricultural throughput in arable lands that lead to an assessed annual loss of billions of dollars (Pareek et al., 2009). The challenges from the abiotic stress may grow enormously with the expansion of agricultural activities in less fertile and peripheral areas, which may play a vital role in satisfying the growing food demands (Koh et al., 2013). Amongst several abiotic stresses, soil salinity and water scarcity are the most frequently explored part of the study owing to the extent of cultivated area affected by them. The early evolution of plants in land occurred under dry conditions of harsh sunlight with extremes of temperature. The domestication of crops happened later in more favorable environments (Pereira, 2016). Various abiotic stresses and suboptimal water levels may also be the limiting factors for biomass production. Hence, stress tolerance traits are important factors for food crop production, that is required to enable feedstock for production on marginal or sub-marginal lands, which may not be a favorable one (Yuan et al., 2008). One of the common and important abiotic stresses comprises salt-stress induced accumulation of ROS within plants. They are detrimental to cells at high concentrations because they cause the oxidative damage to the membrane lipids, proteins, and nucleic acids (Smirnoff, 1993; Gomez et al., 1999; Hernandez et al., 2001). Plants employ antioxidants (e.g., ascorbate, glutathione, α-tocopherol, and carotenoids) and detoxifying enzymes, such as superoxide dismutase, catalase (CAT), and enzymes of the ascorbate-glutathione cycle to combat oxidative stress.

Biofuel productivity from the plant is also affected by environmental stresses also. The components of plant biomass which include biofuel-related lipids and carbohydrates can differ in composition with the environmental conditions. Several studies have described the utilization of different environmental stresses such as light stress, temperature stress, nutrient stress, salt stress, etc., for improving the biofuel production:

1. **Light Stress:** Remarkable changes in the chemical composition of plants are observed in those plants growing under different light conditions. Under low light conditions, the amount of polyunsaturated fatty acids in the system increases, whereas the presence of high light promotes the accumulation of saturated and mono-unsaturated fatty acids for storage of lipids (Spoehr and Milner, 1949). The saturated and mono-unsaturated fatty acids are the chosen sources for biodiesel, which may be a viable form to use at comparatively high light intensities to improve biodiesel yield (Cheng and He, 2014).
2. **Temperature Stress:** The lipid content in the plant cell may be affected by temperature changes. Generally, lower temperatures lead to an increase in the content of unsaturation in fatty acid, but higher temperatures increase the saturation content of fatty acids in the plant. To promote the production of high quantity of biodiesel, an appropriate high temperature is important because of the high total lipid yield and high amount of saturated fatty acids is required for biodiesel production (Cheng and He, 2014).

3. **Nutrient Stress:** A plant may change its metabolic strategies and biochemical composition to combat the stress conditions, when they are grown under nutrient-stressed conditions. Therefore, by manipulating nutrient conditions in the plant, an improved production of the chosen biofuels can be attained.
4. **Water Stress:** Water shortage and increased competition for water resources in several sectors of the production segment like agriculture, industry, hydroelectric energy, etc., and also for an increase in human basic necessities, enforces the study of varying new concepts of irrigation. It is done in order to adapt the crops in water shortage conditions and maintain satisfactory levels of its productivity (Gouveia-Neto et al., 2011).
5. **Salt Stress:** Several studies showed that for the production of various biofuels, salt plays a significant role. The high salt content in the system enhances the content of saturated fatty acid, which helps in increasing the biodiesel productivity (Chen et al., 2008). Whereas on the other side, salinity affects 7–9% of the world's land area (Szabolcs, 1996), and now this area is increasing gradually (Ghassemi et al., 1995).
6. **Role of Plant Growth Hormone Regulators:** To overcome the abiotic stressed conditions, the microbial diversity enhances its local or systemic inducible response mechanisms in plants, while, on the contrary, they aid in sustaining the biomass and growth through various processes like-cellular uptake, utilization of nutrients and its synthesis. Out of the various factors, by manipulating the plant hormone content under abiotic stressed condition is one of the utmost resourceful methods for controlling the plant biomass production. Hormones such as auxins, cytokinins (CKs), gibberellins (GAs), brassinosteroids (BRs) and abscisic acid (ABA) along with other organic compounds can stimulate plant cell growth, division, and biomass production under various environmental stresses have also been implicated in such complex interactions (Fahad et al., 2015). Under salt stress and heavy metal stressed condition, indole acetic acid (IAA) helps in the growth of both shoots as well as root in plants (Fahad et al., 2015; Joshi et al., 2018).

The plant-microbe interactions might affect the plant growth, carbon confinement into the system, nutrient cycling, and productivity along with the regaining plant responses against abiotic stresses (Joshi et al., 2018). From various studies, it was observed that varied bacterial species that belong to different genera, cumulatively help the host plants in tolerance against various abiotic stresses, ensuing a higher biomass production (Grover et al., 2011). By regulating the plant growth under several abiotic stressed conditions, the plant growth regulators normally have an undeviating important role. Controlling the various intrinsic plant hormone content, it can improve the abiotic stress tolerance capability of the plant as well as the biomass content of plants. The studies of the various master switches that help the plant survive in a stressed condition, may advance the understanding for developing new strategies instantaneously. These strategies may help in genetically modifying the quantitative components in the stressed system of lignocellulosic biomass, along with growing tolerance in abiotic stressed conditions (Joshi et al., 2018).

Various abiotic stressed conditions such as drought, metal, salt; temperature stress all persuade some comparable similar responses in plants. But in contrast, on varying stressed condition each of these stresses may induce a diverse set of genes for its sustenance (Maggio et al., 2006). Some instances of abiotic stresses on various plants and the means by which they are avoided are as follows:

- The upstream pathways for salt and drought stresses are well-characterized in *Arabidopsis thaliana*, but in recent days, it has led to only limited success in translational research to produce field crop abiotic stress tolerance (Chinnusamy et al., 2005; Song et al., 2005; Hirayama and Shinozaki, 2007).
- Improved cold and drought tolerance have been reported in tobacco and potato by transformation with the gene encoding DREB1A (dehydration response element B1A), which is driven by a promoter of a stress-responsive water channel, RD29A (Kasuga et al., 2004; Behnam et al., 2007).
- Rice plants with induced expression of a NAC (for NAM, ATAF, and CUC)-type transcriptional factor, OsNAC6, have been shown to enhance tolerance to both high salinity and plant pathogens (Nakashima et al., 2007).
- Switchgrass shows large phenotypic variation for water and cold-stress tolerance for biofuel production (Kerckhoffs and Renquist, 2013).
- Bacteria, such as *Pseudomonas* spp., *Burkholderia caryophylli*, *Achromobacter piechaudii*, showed a reduction in endogenous ethylene levels in plants by producing 1-aminocyclopropane-1-carboxylic acid (ACC)-deaminase enzyme, thus resulting in increased root growth, and improved tolerance of salt and water stress (Wu et al., 2009; Joshi et al., 2018).

- *Burkholderia phytofirmans* strain PsJN improves leaf area, chlorophyll content, photosynthetic rate, and water-use efficiency, ultimately resulting in the enhanced shoot and root biomass under various abiotic stresses in a wide spectrum of host plants including potato, tomato, and grapevine (Mitter et al., 2013).

By keeping all these factors in mind, it is necessary to explore the various effects of water deficit, salinity stress, metal, salt, cold, and heat stress conditions occurring in plant species with a high potential in an application for huge amount manufacture of non-fossil-based fuels. Plant-microbe interactions may also be exploited to handle with the abiotic stress response leading for elevated biomass production as several studies have already demonstrated that plant-microbe interactions generally influence abiotic stress tolerance along with growth and biomass yield (Joshi et al., 2018). Certain studies have evidences that support the plant-microbe interaction in mitigating abiotic stress response during varied climatic conditions.

3.3.2 BIOTIC STRESS

Biotic stresses are considered to cause damage to the plant, which may be triggered by other various living organisms, for instance, bacteria, fungi, viruses, algae, nematodes, diatoms, plasmodium, insects, and several others. Several biotic stresses on plants have been thoroughly studied earlier, some of such is the great famine of 1943 in Bengal, India; rust plant disease in the coffee plant was caused by the fungus *Hemileia vastatrix* in Brazil; the potato blight in Ireland and leaf blight in maize plant caused by *Cochliobolus heterostrophus* in the United States (Hussain, 2015; Onaga and Wydra, 2016). Plant-microbe interactions and their relationships can be utilized to evade the separation between food and energy crops. By promoting plant growth, plant-associated bacteria can promote sustainable production of energy crops on both arable and less arable lands. The incidence of new pathogen types and insect biotypes can create an increased hazard in the production of food and energy crops. But plants have specialized molecules and mechanisms that resist the infection by several pathogens, insects, and pests. These infecting organisms are known to comprise a range of modifications in plants such as morphological, genetic, biochemical, and molecular processes. But the resistance of the plant may be expressed constitutively against the biotic stresses, or they may be induced and starts its function only after pathogen attack (Howe and Jander, 2008). Short-term beneficial effects of plant growth-promoting microorganisms will result in improved plant establishment. These effects include accelerated root development, resulting in better access to nutrients and water and thus faster initial growth, which facilitates plants to outcompete weeds for available resources, thereby reducing the need for herbicides. Long-term beneficial effects of plant growth-promoting microorganisms might result in improved plant growth, health, and survival, leading to environmentally and economically sustainable food, feed, and biofuel feedstock production. This could be achieved by counteracting stress responses caused by drought and contamination, by protecting against pathogens via competition for available resources and by assisting the plant's defense response against pathogenic invasions, thus reducing pesticide requirements (Weyens et al., 2009). The importance of altering plant growth and development to increase the biomass production for bioenergy cannot be over-emphasized. The microbes induce local or systemic inducible response mechanisms in plants to overcome both biotic as well as abiotic stress, while, on the other hand, they contribute towards sustaining the biomass and growth through-uptake, mobilization, and synthesis of nutrients (Joshi et al., 2018). Male sterility is another desirable feature for feedstock development to prevent transgene escape from genetically modified (GM) feedstock to have high biomass yield. Engineered plants and their beneficial symbionts will pave the way for future strategies to modify them for higher biomass production to meet the demands of a growing population in a changing climate scenario.

Considering the different approaches for studying microbe mediated stress mitigation strategies may further strengthen the knowledge on the mechanisms of plant-microbe interactions with the naturally associated or artificially inoculated microorganisms that has a direct or indirect role in environmental stress generation and/or mitigation. The prime targets for optimizing beneficial plant-microbe interactions generally include quorum sensing, bacterial motility, biofilm formation, and their signaling pathways. At present, a wide range of bacterial formulations is readily available commercially for their use as 'bioprotection agents' or 'biofertilizers' which improves biomass and yield under biotic stresses (Joshi et al., 2018).

3.4 CROPS AND FEEDSTOCKS THAT ACT UNDER STRESSED CONDITION

Bioenergy crops can be classified into the following four groups: traditional cereal crops, traditional sugar-reducing crops, dedicated lignocellulosic biomass feedstocks (Yuan et al., 2008). The traditional cereal crops, traditional sugar-reducing crops are generally summer annual species while the dedicated lignocellulosic biomass feedstocks are the perennial species.

3.4.1 TRADITIONAL CEREAL CROPS

Cereal crops are a major source for starch-based ethanol production. Maize (*Zea mays*) is high yielding and its agronomy is well defined; therefore, it is a good species for assessment as a gasification feedstock in the planned engineering model in the research project (Kerckhoffs and Renquist, 2013). Maize is an important food and feed crop, used as processed food, oil, fodder, a vegetable, and byproducts. Maize can be used as a bioenergy crop in two ways: the starch in seeds can be used to produce ethanol, and the crop residuals (termed stover) could potentially be used to produce lignocellulosic ethanol. Sorghum *(Sorghum bicolor)* is the fifth most cultivated cereal crop in the world and is grown for grain, forage, sugar, and fiber. Sorghum could also be used for bioenergy in several ways.

3.4.2 TRADITIONAL SUGAR-PRODUCING CROPS

Sugar can be used for direct fermentation of ethanol. Sugarcane (*Saccharum officinarum*) and sugar beet (*Beta vulgaris*) are the major sugar-producing plants; both have an immense role in biofuel production and are discussed later in this chapter. Other sugar-producing crops include energy cane, improved cultivars of sugarcane and varieties of sweet sorghum, etc. Most of these crops are annual crops, with an exception of sugarcane which is a perennial crop. Perennials are more desirable than annuals as bioenergy feedstocks because they do not need to be reseeded each growing season and therefore cultivation costs are lower.

1. **Sugarcane:** The first use of sugarcane ethanol as a fuel was in Brazil that dates back to the late twenties and early thirties of the twentieth century, with the introduction of the automobile in the country. Sugarcane refers to any of 6 to 37 species of tall perennial grasses of the genus *Saccharum officinarum* (family *Poaceae*) commonly known as sugarcane. Native to warm temperate to tropical regions of Asia, they basically have stout, jointed, fibrous stalks that are rich in sugar, and have a height of two to six meters tall. All sugarcane species interbreed, and the major commercial cultivars are complex

hybrids sugarcane products include table sugar, falernum, molasses, rum, bagasse, and mainly ethanol (Gouveia-Neto et al., 2011).

2. **Sugar Beet:** *Beta vulgaris* commonly known as sugar beet along with sugarcane are the major sugar-producing plants, which help in biofuel production from biomass. Sugar beet is generally seen to grow in temperate areas. Therefore, the two sugar crops occupy different geographical niches. Most ethanol production using sugar beet takes place in Europe; however, using sugar beet to produce ethanol could potentially increase soil erosion and lower the net energy balance. Other sugar-producing crops include energy cane, improved cultivars of sugarcane, and varieties of sweet sorghum. All the above crops are annuals, with the exception of sugarcane. Perennials are more desirable than annuals as bioenergy feedstocks because they do not need to be reseeded each growing season and therefore cultivation costs are lower.

3.4.3 DEDICATED LIGNOCELLULOSIC BIOMASS FEEDSTOCKS

Dedicated bioenergy feedstocks are woody plants, including hybrid poplar, willow, and pines. Hybrid poplar is considered a model woody biomass feedstock because of its broad adaptation, available genome sequence and transformation techniques, and fast growth. The biomass accumulation of hybrid poplar is reported to be between 7 to 20 Mg/ha/yr depending on the nutrition and environmental conditions (Christersson, 2006; Yuan et al., 2008; Pinzi and Dorado, 2011). From the perspective of biomass production, switchgrass, and hybrid *Miscanthus* seem to have the potential to produce more biomass compared with that produced by poplar.

Perennial bioenergy feedstocks are important sources of lignocellulosic biomass production. Switchgrass has been proposed as the major perennial feedstock in some countries because it is widely adapted, has high biomass production as well as high C-4 photosynthetic efficiency and efficient consumption of water and nitrogen. Hybrid *Miscanthus*, including Giant miscanthus (*Miscanthus* x *giganteus)*, is another highly favored biomass feedstock, mainly in Europe (Price et al., 2003; Danalatos et al., 2007; Yuan et al., 2008).

1. **Switchgrass:** *Panicum virgatum* has been proposed as the major perennial feedstock in some countries because it is widely adapted has high biomass production as well as high C-4 photosynthetic efficiency and efficient consumption of water and nitrogen. Switchgrass

yield is around 10 to 25 Mg/ha/yr depending on latitude, nutrition, and other factors. The plant has a lesser amount of nitrogen requirement and requires lower water content which can withstand for several years and has easy maintenance capacity (Kerckhoffs and Renquist, 2013).

2. **Giant Miscanthus:** *Miscanthus* x *giganteus* is another highly favored biomass feedstock, mainly seen in Europe. From earlier studies, it has been seen that giant miscanthus to be a consistently high performer with irrigation in some European countries. It has low nitrogen content, which is environmentally advantageous because it requires less nitrogen fertilizer to grow. Hence, combustion of the biomass produces less reactive nitrogen than burning fossil fuels or other crop species that have higher nitrogen contents (Ceotto and Di Candilo, 2010). The biofuel produced from Giant miscanthus may replace fossil fuels which have a positive effect on GHG emissions (Clifton-Brown et al., 2004).
3. **Physic nut:** *Jatropha curcas* (family *Euphorbiaceae*) is one of the most versatile plants with many attributes and notable potential. Jatropha is a small tree, easily settled, grows fast, and is hardy, and in some way drought tolerant. Thus, it remedies soil degradation, desertification, and deforestation. *Jatropha* being a native of tropical America, now flourishes in many parts of the tropical and sub-tropical areas in Asia or Africa. Various parts of the plant are of medicinal value for both human and veterinary purposes, and are under intensive scientific investigation. The oil is a strong purgative, widely employed as an antiseptic for cough, skin diseases, and a pain reliever from rheumatism (Gouveia-Neto et al., 2011).

3.5 IMPORTANT ASPECTS AND APPLICATION OF BIOFUEL

Studies from various literature showed that in 2009, Brazil, and the USA lead in the industrial production of ethanol fuel, which together accounted for 89% of the total world's production. Brazil was considered to have the world's first sustainable biofuels economy and the leader in the biofuel industry, where its sugarcane ethanol proved to be the most effective substitute fuel. In 2009, Brazil was the world's second-largest producer and the world's largest exporter of ethanol fuel where the country produced 37.7% of the world's total ethanol that was used as fuel (Renewable Fuel

Association report, 2010). Concerning alternative proponents to renewable energy, biofuels based on the use of plant oil as a fuel in stationary and mobile engines are the subject of much attention recently (Gouveia-Neto et al., 2011). The production of biofuel and especially ethanol from plant biomass is receiving increased attention and developments in plant genetic engineering are working some way in reducing the costs of biomass conversion. Several avenues are being investigated, including the engineering of plants that self-produce enzymes like cellulase and ligninase, which plays an essential role in biomass degradation and biofuel production. The development of plants with reduced lignin content, and the production of crops with increased cellulase or overall biomass enhance better production of biofuel (Sticklen, 2006). From previous studies, it was observed that there are strong evidences for anthropogenic contributions to climatic changes. The displacement of fossil fuels with biofuels from plant biomass is a technology change that might probably have a beneficial and critical to the current and following generations. This may be the basis for active bioenergy research programs also internationally. The carbon sequestration and the use of the biomass in replacement to fossil fuels may be an advantageous impact on net GHG emissions (Lemus and Lal, 2005; Kerckhoffs and Renquist, 2012). There has been a huge research emphasis on new biomass species that has an importance on the agronomic aspects of their production. The focus on dry mass yield, suggests that there may be a good potential for the production of fuels and other types of energy from biomass crops (Kerckhoffs and Renquist, 2012). Much work has been also done in the past years on the pathways modulating hormone biosynthesis and signaling and dissecting out their role in response to varied climatic conditions for enhanced biomass production.

3.6 CONCLUDING REMARKS

With the apparent changes in the global environment, agricultural production systems are likely to change. Thus, there may be a need to produce cost-effective biomass to replace the existing fossil fuel in the near future. Biomass will be playing an important role in the near future global energy infrastructure not only for its power and heat generation, but also for the production of biofuels and chemicals. For bioenergy production, plant biomass can be used in multiple forms. There is a very large potential supply of biomass for the land area that could be available for biofuel production. The major technological goals of energy production are the replacement of biofuel, mainly due to environmental issues. In abiotic stresses, soil salinity and water scarcity are the frequently explored part of the study owing to the extent of cultivated area affected by them. Various bioenergy crops enhance biofuel production even in various environmental stressed conditions, where plant-microbe interaction may play an important role. So these microorganisms may be utilized which may further enhance their tolerance against various abiotic stresses, thus establishing a novel and promising techniques for sustainable agriculture. A strategy may be designed to target plant-microbe interaction and the plant susceptible factor on the environmental stress for high biomass yield and biofuel production at different locations.

CONFLICTS OF INTEREST

The authors declare that they have no conflicts of interest.

KEYWORDS

- **abiotic and biotic stress**
- **biofuel**
- **cytokinins**
- **plant biomass**
- **plant growth regulators**
- **plant-microbe interactions**

REFERENCES

Behnam, B., Kikuchi, A., Celebi-Toprak, F., Kasuga, M., Yamaguchi-Shinozaki, K., & Watanabe, K. N., (2007). *Arabidopsis* rd29A: DREB1A enhances freezing tolerance in transgenic potato. *Plant Cell, 26*, 1275–1282.

Cassman, K. G., & Liska, A. J., (2007). Food and fuel for all: Realistic or foolish? *Biofuels. Bioprod. Bioref., 1*(1), 18–23.

Ceotto, E., & Di Candilo, M., (2010). Sustainable bioenergy production, land, and nitrogen use. In: Lichtfouse, E., (ed.), *Biodiversity, Biofuels, Agroforestry and Conservation Agriculture* (Vol. 5, pp. 101–122). Springer Science, Dordrecht.

Chen, G. Q., Jiang, Y., & Chen, F., (2008). Salt-induced alterations in lipid composition of diatom *Nitzschia laevis* (*Bacillariophyceae*) under heterotrophic culture condition. *J. Phycol., 44*, 1309–1314.

Chinnusamy, V., Jagendorf, A., & Zhu, J. K., (2005). Understanding and improving salt tolerance in plants. *Crop Sci., 45*, 437–448.

Christersson, L., (2006). Biomass production of intensively grown poplars in the southernmost part of Sweden: observations of characters, traits and growth potential, Biomass. *Bioenerg.*, 30, 497–508.

Clifton-Brown, J. C., Stampfl, P. F., & Jones, M. B., (2007). Miscanthus biomass production for energy in Europe and its potential contribution to decreasing fossil fuel carbon emissions. *Glob Chang Biol., 10*, 509–518.

Dan, C., & Qingfang, H., (2014). Assessment of environmental stresses for enhanced microalgal biofuel production: An overview. *Front. Energy Res., 2*, 26.

Danalatos, N. G., Archontoulis, S. V., & Mitsios, I., (2007). Potential growth and biomass productivity of *Miscanthus* x *giganteus* as affected by plant density and N-fertilization in central Greece. *Biomass. Bioen., 31*, 145–152.

Fahad, S., Hussain, S., Bano, A., Saud, S., Hassan, S., Shan, D., Khan, F. A., et al., (2015). Potential role of phytohormones and plant growth-promoting rhizo bacteria in abiotic stresses: Consequences for changing environment. *Environ. Sci. Pollut. Res., 22*(7), 4907–4921.

Ghassemi, F., Jakeman, A. F., & Nix, M. A., (1995). *Salinization of Land and Water Resources: Human Causes, Extent, Management and Case Studies* (p. 381). Wallingford, Oxon, UK: CAB International, 085198990X.

Gomez, J. M., Hernandez, J. A., Jimenez, A., Del Rio, L. A., & Sevilla, F., (1999). Differential response of antioxidative enzymes of chloroplast and mitochondria to long term NaCl stress of pea plants. *Free Radic. Res., 31*, S11–S18.

Gouveia-Neto, A. S., Da Silva, Jr. E. A., Cunha, P. C., Filho, R. A. O., Silva, L. M. H., Da Costa, E. B., Câmara, T. J. R., et al., (2011). *Abiotic Stress Diagnosis via Laser Induced Chlorophyll Fluorescence Analysis in Plants for Biofuel* (pp. 3–22). Biofuel production-recent developments and prospects. Intech Open.

Grover, M., Ali, S. Z., Sandhya, V., Rasul, A., & Venkateswarlu, B., (2011). Role of microorganisms in adaptation of agriculture crops to abiotic stresses. *World J. Microbiol. Biotechnol., 27*, 1231–1240.

Hernandez, J. A., Ferrer, M. A., Jimenez, A., Barcelo, A. R., & Sevilla, F., (2001). Antioxidant systems and O_2^{-}/H_2O_2 production in the apoplast of pea leaves. Its relation with salt-induced necrotic lesions in minor veins. *Plant Physiol., 127*, 817–831.

Himmel, M. E., & Bayer, E. A., (2009). Lignocellulose conversion to biofuels: Current challenges, global perspectives. *Curr. Opin. Biotechnol., 20*(3), 316–317.

Hirayama, T., & Shinozaki, K., (2007). Perception and transduction of abscisic acid signals: Keys to the function of the versatile plant hormone ABA. *Trends Plant Sci., 12*, 343–351.

Howe, G. A., & Jander, G., (2008). Plant immunity to insect herbivores. *Annu. Rev. Plant Biol., 59*, 41–66.

Hussain, B., (2015). Modernization in plant breeding approaches for improving biotic stress resistance in crop plants. *Turk J. Agric. For., 39*, 515–530.

Joshi, R., Pareek, S. L. S., & Pareek, A., (2018). Engineering abiotic stress response in plants for biomass production. *J. Biol. Chem., 293*(14), 5035–5043.

Kasuga, M., Miura, S., Shinozaki, K., & Shinozaki, K. Y., (2004). A combination of the *Arabidopsis* DREB1A gene and stress-inducible rd29A promoter improved drought-and low temperature stress tolerance in tobacco by gene transfer. *Plant Cell Physiol., 45*, 346–350.

Kerckhoffs, H., & Renquist, R., (2012). Biofuel from plant biomass. *Agronomy for Sustainable Development* (Vol. 33, No. 1, pp. 1–19.). Springer Verlag/EDP Sciences/INRA.

Kerckhoffs, H., & Renquist, R., (2013). Biofuel from plant biomass; *Agron. Sustain. Dev., 33*, 1–19.

Klass, D. L., (1998). *Biomass for Renewable Energy, Fuels, and Chemicals.* San Diego: Academic Press.

Koh, L. P., Koellner, T., & Ghazoul, J., (2013). Transformative optimization of agricultural land use to meet future food demands. *Peer J., 1*, e188.

Kumar, P., Barrett, D. M., Delwiche, M. J., & Stroeve, P., (2009). Methods for pretreatment of lignocellulosic biomass for efficient hydrolysis and biofuel production. *Ind. Eng. Chem. Res., 48*(8), 3713–3729.

Lau, J. A., & Lennon, J. T., (2011). Evolutionary ecology of plant-microbe interactions: Soil microbial structure alters selection on plant traits. *New Phytol., 192*(1), 215–224.

Lemus, R., & Lal, R., (2005). Bioenergy crops and carbon sequestration. *Crit. Rev. Plant Sci., 24*, 1–21.

Lucian, L. A., Argyropoulos, D. S., Adamopoulos, L., & Gaspar, A. R., (2007). Chemicals, materials, and energy from biomass: A review. In: Argyropoulos, D. S., (ed.), *ACS Symposium Series 954. Materials, Chemicals, and Energy for Forest Biomass* (pp. 2–30). Washington: American Chemical Society.

Maggio, A., Zhu, J. K., Hasegawa, P. M., & Bressan, R. A., (2006). Osmogenetics: Aristotle to *Arabidopsis*. *The Plant Cell, 18*(7), 1542–1557.

Mitter, B., Petric, A., Shin, M. W., Chain, P. S., Hauberg-Lotte, L., Reinhold-Hurek, B., Nowak, J., & Sessitsch, A., (2013). Comparative genome analysis of *Burkholderia phytofirmans* PsJN reveals a wide spectrum of endophytic lifestyles based on interaction strategies with host plants. *Front. Plant Sci., 4*(120), 1–15.

Nakashima, K., Tran, L. S., Van, N. D., Fujita, M., Maruyama, K., Todaka, D., Ito, Y., et al., (2007). Functional analysis of a NAC-type transcription factor OsNAC6 involved in abiotic and biotic stress responsive gene expression in rice. *Plant J.*, 51, 617–630.

Onaga, G., & Wydra, K., (2016). Advances in plant tolerance to biotic stresses. *Plant Genomics, 10*, 229–272.

Pareek, A., Sopory, S. K., Bohnert, H. J., & Govindjee, (2009). *Abiotic Stress Adaptation in Plants: Physiological, Molecular, and Genomic Foundation.* Springer Netherlands.

Pereira, A., (2016). Plant abiotic stress challenges from the changing environment. *Front. Plant Sci., 7*, 1123.

Pinzi, S., Dorado, M. P., (2011). Vegetable-based feedstocks for biofuels production, Handbook of Biofuels Production, editor: Rafael luque, Juan Campelo, James Clark, *Woodhead Publishing Series in Energy.* 61–94.

Price, L., Bullard, M., Lyons, H., Anthony, S., & Nixon, P., (2003). Identifying the yield potential of Miscanthus x giganteus: An assessment of the spatial and temporal variability of *M. x giganteus* biomass productivity across England and Wales. *Biomass Bioen., 26*, 3–13.

Rass-Hansen, J., Falsig, H., Jørgensen, B., & Christensen, C. H., (2007). Bioethanol: Fuel or feedstock? *J. Chem. Tech. Biotechnol., 82*(4), 329–333.

Sims, R. E. H., Hastings, A. F. S. J., Schlamadinger, B., Taylor, G., & Smith, P., (2006). Energy crops: Current status and future prospects. *Glob. Change Biol., 12*, 2054–2076.

Sims, R. E. H., Mabee, W., Saddler, J. N., & Taylor, M., (2010). An overview of second generation biofuel technologies. *Bioresour. Technol., 101*, 1570–1580.

Smirnoff, N., (1993). The role of active oxygen in the response of plants to water deficit and desiccation. *New Phytol., 125*, 27–58.

Somerville, C., (2007). Biofuels. *Curr. Biol., 17*, R115–R119.

Song, C. P., Agarwal, M., Ohta, M., Guo, Y., Halfter, U., Wang, P., & Zhu, J. K., (2005). Role of an *Arabidopsis* AP2/EREBP-type transcriptional repressor in abscisic acid and drought stress responses. *Plant Cell, 17*, 2384–2396.

Spoehr, H. A., & Milner, H. W., (1949). The chemical composition of Chlorella; effect of environmental conditions. *Plant Physiol., 24*, 120–149.

Srirangan, K., Akawi, L., Young, M. M., & Chou, C. P., (2012). Towards sustainable production of clean energy carriers from biomass resources. *Appl. Energy, 100*, 172–186.

Stewart, C. N., (2007). Biofuels and biocontainment. *Nat. Biotechnol., 25*, 283–284.

Sticklen, M., (2006). Plant genetic engineering to improve biomass characteristics for biofuels. *Curr. Opin. Biotechnol., 17*,315–319.

Svanberg, S., (1995). Fluorescence lidar monitoring of vegetation status. *PhysicaScripta, 58*, 79–85.

Szabolcs, I., & Greenland, D. J., (1996). *Soil Resilience and Sustainable Land Use* (Vol. 52, No. 1, pp. 137–138). CAB International, Wallingford, UK.

Tilman, D., Hill, J., & Lehman, C., (2006). Carbon-negative biofuels from low-input high diversity grassland biomass. *Science, 314*, 1598–1600.

Wang, M., Wu, M., & Huo, H., (2007). Life-cycle energy and greenhouse gas emission impacts of different corn ethanol plant types. *Environ. Res. Lett., 2*, 1–13.

Wang, Y., Fan, C., Hu, H., Li, Y., Sun, D., Wang, Y., & Liangcai, P., (2016). Genetic modification of plant cell walls to enhance biomass yield and biofuel production in bioenergy crops. *Biotechnol. Adv., 34*(5), 997–1017.

Weyens, N., Van, D. L. D., Taghavi, S., Newman, L., & Vangronsveld, J., (2009). Exploiting plant-microbe partnerships to improve biomass production and remediation; *Trends Biotechnol., 27* (10), 591–598.

Wu, C. H., Bernard, S. M., Andersen, G. L., & Chen, W., (2009). Developing microbe-plant interactions for applications in plant-growth promotion and disease control, production of useful compounds, remediation, and carbon sequestration. *Microb. Biotech., 2*, 428–440.

Yat, S. C., Berger, A., & Shonnard, D. R., (2008). Kinetic characterization of dilute surface acid hydrolysis of timber varieties and switchgrass. *Bioresour. Technol., 99*, 3855–3863.

Yuan, J. S., Tiller, K. H., Al-Ahmad, H., Stewart, N. R., & Stewart, Jr. C. N., (2008). Plants to power: Bioenergy to fuel the future. *Trends Plant Sci., 13*(8), 421–429.

CHAPTER 4

Improving Plant Adaptation/Tolerance Under Stress Conditions: Future Prospects

RANJEET KAUR[1] and RUPAM KUMAR BHUNIA[2*]

[1]Department of Genetics, University of Delhi-South Campus, New Delhi – 110021, India

[2]National Agri-Food Biotechnology Institute (NABI), Plant Tissue Culture and Genetic Engineering, Sector-81 (Knowledge City), Mohali – 140306, Punjab, India

[*]Corresponding author. E-mail: rupamb2005@gmail.com

ABSTRACT

Plants are sessile beings and have to face numerous atrocities inflicted upon them throughout their lifespan by various adverse environmental facets comprising of biotic as well as abiotic factors that impose hostile conditions for their survival. This has led to an enormous loss in the quality and yield of the important food crops over the past and continues to pose as a serious threat to the future food security of the world. The conventional and modern plant breeding strategies have tried to tackle this problem but with limited success. With the advent of the genetic engineering mechanisms, generating tailor-made environment-ready plants that are better adapted as well as more tolerant towards these stresses has become strategic and faster. This chapter will encompass the recent developments in this field and discuss the strategies that can prove to be the breakthrough technologies of the future for developing the next-gen crops to cater the needs of the ever-increasing human population.

4.1 INTRODUCTION

Alteration in the climatic conditions has led to a dreadful situation in sessile plants, which have to face various stresses comprising of biotic and abiotic factors. Although plants try to adapt and acclimatize to these environmental cues, they fall short of the survival strategies in severe stress conditions. This is turn has adversely impacted the yield of major crops and harnessed the downfall of the food security of the teeming millions. The solution to this global problem lies in the genetic engineering of plants, particularly of crops, to make them climate-ready. The several strategies being employed to develop such tailored, smart crops revolve around the modulation of target downstream genes, whose products are directly involved in stress alleviation or the upstream transcription factor genes which are master regulators of the stress signaling pathway (Figure 4.1). The application of these two major strategies resulted in moderate to high-level success rates in combating stress along with the interplay of the signaling molecules. However, the future of the crop stress lies in the genome editing technology, which has the capacity to produce plants with broad-spectrum tolerance and minimal side effects. Although several

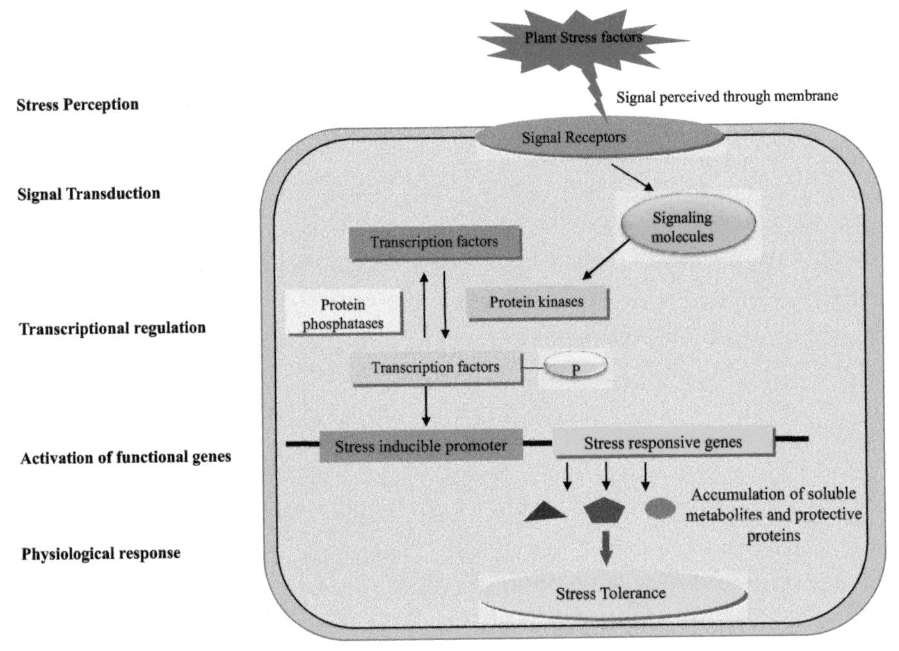

FIGURE 4.1 Schematic diagram showing the response mechanisms employed by a plant cell to improve stress tolerance/adaptation.

reviews have been published related to the production of stress-tolerant plants, no concrete step has been taken to overcome the problems that need to be tackled before the genetically engineered crops can become a part of our regular diet. In this chapter, we focus on the current strategies that are being undertaken to address the targeted goal of improving stress tolerance in plants, followed by a discussion on the limitations that are impeding the realization of these strategies and then we highlight the future prospects of achieving the goal and also suggest some possible research directions.

4.2 GENETIC ENGINEERING STRATEGIES FOR ACHIEVING PLANT STRESS TOLERANCE

The major strategies that govern the genetic engineering of plants for stress tolerance involve overexpression of three types of genes, namely the genes encoding the downstream effector molecules which are directly involved in mitigating stress, the signaling molecules which relay the stress signal to these effector molecules and the upstream regulatory genes which encrypt the genes involved in stress signaling pathways and control the expression of several other downstream genes.

4.2.1 OVEREXPRESSION OF DOWNSTREAM EFFECTOR MOLECULES

Single genes that encode for the end product of the stress-responsive pathways are the primary choice for overexpression in plants to improve stress tolerance. The products of these genes are directly involved in mitigating plant stress and mainly comprise of the enzymes/proteins that participate in the maintenance of cellular homeostasis. These are usually at the downstream or the end of a stress signaling pathway and are known as the effector genes and their products are called effector molecules. The effector molecules broadly fall into four different categories and include the antioxidants, ion transporters, molecular chaperones, and compatible osmolytes.

4.2.1.1 ANTIOXIDANTS

The toxic molecules that are rapidly generated in plants upon exposure to all types of stress are the reactive oxygen species (ROS). These comprise mostly of the superoxide radicals and hydrogen peroxide (H_2O_2), and are primarily responsible for the majority of the damage to biomolecules including DNA, proteins, and lipids. These are neutralized by special enzymatic molecules termed as antioxidants which include superoxide dismutase (SOD), catalase (CAT), ascorbate peroxidase (APX), aldehyde dehydrogenase (ALDH), and dehydroascorbate reductase (DHAR) among others. These play a crucial part in detoxifying the cellular environment from the harmful effects of the ROS. These ROS scavengers form an integral part of the stress tolerance mechanism in plants. Several studies have been performed on the overexpression of genes encoding the antioxidants and also have reported that the increased accumulation of antioxidants have significantly reduced the levels of ROS, thereby contributing in establishing cellular homeostasis (Table 4.1). However, the precise mechanism of action of these antioxidants is still unclear and needs more field trials related to stress tolerance on genetically engineered plants expressing these genes, before they can be truly exploited to the increase in stress adaptation and yield of major crop species.

4.2.1.2 WATER AND ION TRANSPORTERS

Maintenance of optimum water and ionic content is essential for plant cell survival under stress. This plant adaptation to stress is brought into effect with the help of genes functioning towards the synthesis of water and ion transport. The integral membrane proteins involved in the transport of water, solute, and CO_2 in plants is known as aquaporins (AQPs). Overexpression of aquaporin-encoding genes like AQP, plasma membrane intrinsic protein (PIP), nodulin 26-like intrinsic protein (NIP), and tonoplast intrinsic protein (TIP) have been genetically engineered in several plants undergoing water-related stresses. Similarly, ion channel transporters, like Na^+/H^+ antiporter and vacuolar Na^+/H^+ antiporter, are involved in maintaining the ionic balance under salinity induced stresses in plant cells. Major works related to these genes has been summarized in Table 4.2. However, overexpression of water channels may even lead to loss of cellular water molecules and lead to stressful conditions. Thus, the role of these water and ion transporters can be crucial in mitigating stress especially during osmotic stress, and hence their expression should be governed in a controlled manner.

4.2.1.3 COMPATIBLE OSMOLYTES

Exposure to any kind of stressful condition in plants leads to an imbalance in cellular homeostasis and accumulation of compatible osmolytes. The

TABLE 4.1 Summary of Overexpression of Genes Encoding Antioxidants

Antioxidant	Gene	Source Organism	Target Plant	Effect Seen in Transgenics	References
Superoxide dismutase (SOD)	*Cu/ZnSOD*	Mangrove	Tobacco	Enhanced salinity tolerance by the reduction of reactive oxygen species in chloroplast	Jing et al., 2015
	SOD1	Cotton	Cotton	Enhanced tolerance to methyl viologen and salt stresses	Luo et al., 2013
	Mn-SOD3.1	Wheat	Potato	Higher tolerance to abiotic stress and improved yield under field conditions	Waterer et al., 2010
	Mn-SOD3.1	Wheat	Canola	Enhanced tolerance to heat, drought, and cold stresses	Gusta et al., 2009
	Cu-ZnSOD	Tobacco	Tobacco	Higher tolerance to salt stress	Lee et al., 2007
Ascorbate peroxidase (APX)	*Cytosolic APX2 (cAPX2)*	Rice	Alfalfa	Increased salt tolerance	Zhang et al., 2014
	APX	*Camellia azalea*	Tobacco	Enhances cold and heat stress tolerance by ROS scavenging	Wang et al., 2017
	APX	-	Sweet potato	Enhanced salt stress tolerance	Yan et al., 2016
	APX	*Brassica campesteris*	*Arabidopsis*	Heat stress tolerance	Chiang et al., 2015
	cAPX	*Populus tomentosa*	Tobacco	Enhances abiotic stress tolerance	Cao et al., 2017
Catalase (CAT)	*CAT1*	Cotton	Cotton	Enhancing Tolerance to Methyl Viologen and Salt Stresses	Luo et al., 2013
	katE	*E. coli*	rice	improved growth and yield under salt stress	Nagamiya et al., 2007

TABLE 4.1 *(Continued)*

Glutathione S-transferase (GST)	*GST*	*Suaeda salsa*	rice	enhanced tolerance to salinity and oxidative stresses at the vegetative stage	Zhao et al., 2006
	Zeta GST	*Pyrus pyrifolia*	tobacco	enhanced tolerance to oxidative damage caused by drought, NaCl, and Cd stresses	Liu et al., 2013
	GST	*Suaeda salsa*	*Arabidopsis*	Salt stress tolerance	Qi et al., 2010
	$GSTU_4$	*Glycine max*	Tobacco	Salt stress tolerance	Kissoudis et al., 2015
	$GSTL_2$	Rice	*Arabidopsis*	Increased tolerance to multiple abiotic stresses like drought, cold, salinity, and heavy metal stresses	Kumar et al., 2013
Dehydroascorbate reductase (DHAR)	*DHAR*	Tobacco	Tobacco	higher tolerance to salt stress	Lee et al., 2007
	DHAR	*Jatropha curcas*	Tobacco	Oxidative stress tolerance	Chang et al., 2017
	DHAR1	*Rice*	Rice	Enhanced salt stress tolerance	Kim et al., 2014
	Cytosolic DHAR	*Arabidopsis*	tobacco	Aluminum stress tolerance	Yin et al., 2010

TABLE 4.2 Summary of Overexpression of Genes Encoding Water and Ion Transporters

Water/Ion Transporters	Gene	Source Plant	Target Plant	Effect Seen in Transgenics	References
Plasma membrane intrinsic protein	*PIP2;1, PIP2;5, PIP2;7*	Tomato	*Arabidopsis* and tobacco	Drought stress tolerance	Li et al., 2016
	PIP1;1	Maize	*Arabidopsis*	Improved salinity and dehydration tolerance	Zhou et al., 2018
	PIP2;9	Soybean	Soybean	Drought tolerance and seed development	Lu et al., 2018
	PIP-1 and PIP2-2	Rice	*Arabidopsis*	Improved salinity and dehydration tolerance	Guo et al., 2006
Tonoplast intrinsic protein	*TIP*	wild soybean (*Glycine soja*)	*Arabidopsis*	More sensitivity to salt and dehydration presumably due to enhanced water loss of the transgenic plants	Wang et al., 2011
	TIP2;2	Tomato	Tomato	Increased cell water permeability and whole-plant transpiration, which resulted in improved salt and drought tolerance under field conditions	Sade et al., 2009
Nodulin 26-like intrinsic protein	*NIP*	Wheat	*Arabidopsis*	Enhanced plants tolerance to salt stress	Gao et al., 2010
	NIP5;1	*Atriplexcanescens*	*Arabidopsis*	Salinity and Improves Drought Tolerance	Yu et al., 2015
Aquaporin	*AQP1*	Tobacco	Tomato	Improved salinity stress tolerance	Sade et al., 2010
	AQP8	Wheat	Tobacco	Salt stress tolerance	Hu et al., 2012

TABLE 4.2 *(Continued)*

Water/Ion Transporters	Gene	Source Plant	Target Plant	Effect Seen in Transgenics	References
	AQP7	Wheat	Tobacco	Drought stress tolerance	Zhou et al., 2012
Na^+/H^+ antiporter	*sodium2 (SOD2)*	Yeast (*Schizosaccharomyces pombe*)	Rice	Improved salinity tolerance, higher photosynthetic levels, and root proton transport capacity, whereas ROS generation was reduced.	Zhao et al., 2006
	SOS1	*Soybean*	*Arabidopsis*	Salt stress tolerance	Nie et al., 2015
vacuolar Na^+/H^+ antiporter	*NHX1*	*Arabidopsis*	Cotton	Improved salt stress tolerance	He et al., 2005
	NHX1	*Arabidopsis*	Beet (*Beta vulgaris*)	Improved salt stress tolerance	Yang et al., 2005
	NHX1	*Arabidopsis*	Tall fescue (*Festuca arundinacea*)	Improved salt stress tolerance	Zhao et al., 2007
	NHX1	Rice	Maize	Improved salt stress tolerance	Chen et al., 2007
	NHX3	*Arabidopsis*	Sugar beet	Salt stress tolerance	Liu et al., 2008
	NHX5	*Arabidopsis*	Rice seedlings	Enhanced salt and drought tolerance	Li et al., 2011a
	NHX5	*Arabidopsis*	paper mulberry (*Broussonetiapapyrifera* L. Vent)	Enhanced salt and drought tolerance	Li et al., 2011b

osmolytes are believed to work in re-establishing the normal conditions in the cell, such as maintenance of cellular water content and turgidity, so that it can carry out its regular functions in the advent of stress. At times of stress, these are also believed to play the role of chaperones as well as ROS scavengers in terms of function. The compatible osmolytes fall into four different categories viz., amino acids (proline), amino acid derivatives (polyamines and glycine betaine), sugars (trehalose and fructan), and sugar alcohols (galactinol and mannitol). Overexpression of genes that could escalate the levels of these compatible osmolytes in the stressed plant cell has been one of the preferred approaches to improve stress tolerance in plants over the years. Table 4.3 provides a summary of the recent works in this regard. Although plenty of research has been commenced with the targeted goal of increasing osmolyte content in the plant cell under stress, the real mechanism behind the action of these compatible osmolytes remains an untold mystery that needs to be unearthed by delving deep into the research on the genes that govern their biosynthesis. The multiple functions played by these genes suggest a systematic pathway that is behind the action of these genes and is universally the same across all plant species, which respond similarly under both stresses, biotic, and abiotic.

4.2.1.4 MOLECULAR CHAPERONES

Molecular chaperones play a significant role in the stabilization of protein structure and also ensure proper folding of proteins under normal conditions. Cellular environment is highly messed up when encountered by stress and proteins undergo aggregation due to decreased presence of these chaperones. In such a scenario, overexpression of genes encoding molecular chaperones leading to their fast accumulation is greatly encouraged. Some of the popular protein families that function as chaperones and chaperone-like molecules include the heat shock proteins (HSPs) and the late embryogenesis abundant (LEA) proteins. Although the HSPs are universal in their presence in stressed cells, they are majorly expressed under heat stress conditions and assist the molecular machinery, particularly the transcriptome, metabolome, and lipidome, to function normally and carry out the regular cellular activities. Thus, their primary role is to act directly as the molecular chaperones and have been regarded as master controllers of cellular homeostasis. On the other hand, the LEA proteins are small-sized proteins that are known to

TABLE 4.3 Summary of Overexpression of Genes Encoding Compatible Osmolytes

Compatible Osmolytes	Gene	Source Organism	Target Plant	Effect Seen in Transgenics	References
Polyamines	*Arginine decarboxylase (ADC)*	*Avena sativa*	*Solanum meloangena*	Tolerance to Salinity, drought, low, and high temperature, heavy metal, and fungus	Raman and Rajam, 2007
	Ornithine decarboxylase (ODC)	*Mus musculus*	Tobacco	Salt stress tolerance	Kumria and Rajam
	S-adenosylmethionine decarboxylase (SAMDC)	*Tritordeum*	Rice	Sodium chloride-stress tolerance	Roy and Wu, 2002
	SAMDC	*Dianthus caryophyllus*	Tobacco	Multiple abiotic stresses like stresses such as salt stress, cold stress, acidic stress, and abscisic acid treatment	Wi et al., 2006
	Spermidine synthase (SPDS)	*Cucurbita ficifolia*	*Arabidopsis*	Enhanced tolerance to various stresses including chilling, freezing, salinity, hyperostosis, drought, and paraquat toxicity.	Kasukabe et al., 2004
Glycinebetaine (GB)	*Betaine aldehyde dehydrogenase (BADH)*	*Atriplexhortensis*	Wheat	Increased tolerance to drought and heat	Wang et al., 2010
	betA (encoding choline dehydrogenase)	*E. coli*	Maize	Improved yield under stressful conditions in the field.	Quan et al., 2004
	betA (encoding choline dehydrogenase)	*E. coli*	Wheat	Improved yield under salt stress conditions in the field.	He et al., 2010
	Choline monooxygenase (CMO)	*Salicornicaeuropaea*	Tobacco	Higher yield production under saline field condition	Wu et al., 2010

TABLE 4.3 *(Continued)*

Compatible Osmolytes	Gene	Source Organism	Target Plant	Effect Seen in Transgenics	References
	ALDH10A8 and ALDH10A9	*Arabidopsis*	*Arabidopsis*	Salt stress tolerance	Missihoun et al., 2015
Proline	Δ1-pyrroline-5-carboxylate synthetase (P5CS)	*Vigna aconitifolia*	Rice	increased tolerance to stress	Su et al., 2004
	Δ1-pyrroline-5-carboxylate synthetase (P5CS)	*Vigna aconitifolia*	Wheat	Increased tolerance to water stress	Vendruscolo et al., 2007
	P5CSF129A	*Arabidopsis*	chickpea (*Cicer arietinum*)	Drought stress tolerance	Bhatnagar-Mathur et al., 2009
	Δ1-pyrroline-5-carboxylate reductase (P5CR)	*Arabidopsis*	Soybean	Drought stress tolerance	Ronde et al., 2004
Mannitol	*Mannitol-1-phosphate dehydrogenase (mtlD)*	*E. coli*	Wheat	Improved tolerance to drought and salinity	Abebe et al., 2003
Trehalose	*Trehalase*	*Arabidopsis*	*Arabidopsis*	Drought stress tolerance	Houtte et al., 2013
	Trehalose-6-phosphate synthase (TPS) and *trehalose-6-phosphate phosphatise (TPP)*	*E. coli*	Rice	Drought, salt, and cold stress tolerance	Jang et al., 2003
	TPS1 and TPS2	Yeast	Alfalfa	Drought, freezing, salinity, and heat stress tolerance	Suarez et al., 2009

play a significant role in cellular dehydration processes like seed desiccation and drought-related stresses. They are unusually intrinsically disordered proteins (IDPs), i.e., they remain unfolded in normal conditions but acquire a structured shape when the cell undergoes dehydration. Thus, LEA proteins function as chaperone-like molecules and are proposed to help in maintaining cellular shape as well as assist in protein/enzyme structure stabilization under stress. Several works have been done to date on overexpression of different classes of these chaperones and some of the major ones are discussed in Table 4.4. However, the exact mechanism of their action is still unknown and needs to be investigated in detail. This will not only help us in designing the proper strategies for improving stress tolerance in plants but also help us in understanding the subtle nuances of protein structure stability under normal and stress conditions.

4.2.2 MODULATION OF SIGNALING MOLECULES

The signaling molecules play a major role in the transduction of the stress signals from the receptors to the downstream signaling cascades. These include the various plant hormones and other molecular species like calcium (Ca^{2+}), nitric oxide (NO), and calmodulin (CaM). The important plant hormones that are known to play significant roles in stress-responsive signaling pathways are the abscisic acid (ABA), cytokinins (CKs), auxins, salicylic acid (SA), jasmonic acid (JA), brassinosteroids (BR), and strigolactones (SL) (Wani et al., 2016). The relay of the signal occurs in a fast and efficient manner, and is responsible for altered gene expression related to growth, development, and stress responses.

4.2.2.1 PLAYING WITH PLANT HORMONES

Vital processes of biochemical, physiological, and morphological importance are governed by the small yet significant plant molecules, also called phytohormones. The major strategy governing phytohormone modulation for stress tolerance is either by its exogenous application or by overexpression of the hormone biosynthetic genes leading to its endogenous accumulation. The first line of defense hormone is ABA which rapidly accumulates in the stressed plant tissues, especially in the dehydrated cells where it induces stomatal closure, thereby reducing transpirational losses (Wilkinson and Davies, 2010; Vishwakarma et al., 2017). Apart from

TABLE 4.4 Summary of Overexpression of Genes Encoding Molecular Chaperones

Molecular Chaperone	Gene	Source Plant	Target Plant	Effect Seen in Transgenics	References
Late embryogenesis abundant (LEA) protein, group 3	*HVA1*	Barley	Wheat	Drought stress tolerance and improved yield	Bahieldin et al., 2005
	HVA1	Barley	Rice	Drought stress tolerance and improved yield	Chen et al., 2015
	LEA3-1	Rice	Rice	Improved yields under drought stress, without yield penalties under control conditions	Xiao et al., 2007
	Wsi18	*Oryza nivara*	*Oryza sativa*	Improved drought stress tolerance without yield penalty	Kaur et al., 2018
Dehydrin (LEA protein, group 2)	*WCOR410*	Wheat	Strawberry	Improved leaf freezing tolerance	Houde et al., 2004
	Dhn1 and *Dhn2*	Sorghum	Tobacco	Oxidative stress tolerance	Halder et al., 2018
Heat shock protein	*sHSP17.7*	Rice	Rice	improved drought and osmotic stress	Sato et al., 2008
	HSP26	Wheat	*Arabidopsis*	Heat stress tolerance	Chauhan et al., 2012
	Hsf6A	Wheat	Wheat	Heat stress tolerance	Xue et al., 2015
	HsfC2a	Wheat	Wheat	Heat stress tolerance	Hu et al., 2018

its known function in regulating seed dormancy, seed germination, and stomatal closure, ABA is an essential plant stress hormone and promotes tolerance to a variety of stresses. Increased drought tolerance by virtue of improved cell membrane stability and leaf water content was observed when ABA, SA, and g-aminobutyric acid (GABA) were exogenously applied in creeping bentgrass (Li et al., 2017). Overexpression of 9-*cis*-epoxycarotenoid dioxygenase (NCED), the rate-limiting enzyme in ABA biosynthesis has shown to result in increased ABA accumulation and drought resistance in several plants including tobacco (Zhang et al., 2008), *Arabidopsis* (Wan et al., 2006), and petunia (Estrada-Melo et al., 2015). ABA has been shown to affect the stress signaling pathways in plants and influences the metabolome of the stressed cell which is then programmed to accumulate metabolites such as compatible osmolytes, osmoprotectants, antioxidants, and LEA proteins, involved in combating stress (Sah et al., 2016). It is also involved in improving root architecture and maintenance of root meristem as well as cell turgidity, both of which are important factors in determining the tolerance of plants towards water stress (Sah et al., 2016; Duan et al., 2013). Overexpression of genes encrypting for ABA receptors namely PYR/PYL/RCAR (Pyrabactin resistance/Pyrabactin-like/regulatory components of ABA receptor) have also shown to positively regulate a myriad of abiotic stresses and preferably water stress (Zhao et al., 2016; Yu et al., 2017; Yang et al., 2016; Li et al., 2018).

CK is another important phytohormone involved in modulation of stress responses by preventing leaf senescence and increasing the proline levels (Zwack and Rashotte, 2015). Overexpression of CK biosynthetic gene encoding for key enzyme isopentenyl transferase (*IPT*) has shown to enhance drought and salinity tolerance in several studies (Rivero et al., 2007, 2010; Peleg et al., 2011). However, recent studies have shown that IPT over-expression reduced SOD and CATactivities and negatively impacted drought and salinity tolerance in *Arabidopsis* (Prerostova et al., 2018; Xu et al., 2016). CKs are found to be ABA antagonists and an increased ABA/CK ratio is desirable to render water stress tolerance in plants (Wani et al., 2016). Thus, the role of CK in plant stress tolerance is complex and not well understood to date. More probing should be done in this regard to clearly understand the effect of CKs on stress tolerance in plants.

The most studied among auxins, indole acetic acid (IAA) is involved in evading salinity stress (Egamberdieva et al., 2009) and heavy metal stress (Potters et al., 2007) by boosting root and shoot elongation under stress conditions. Genes encoding the auxin response factors (ARFs) are also shown to play a role in heat stress responses in rice (Jain and Khurana, 2009) and sorghum (Wang et al., 2010). Accumulation of ET is observed in multiple stresses such as salinity, low as well as high-temperature stresses (Wani et al., 2016). Its function is modulated by the presence of JA and SA and together they help in plant defense against pest and pathogen attacks, thus indicating a phytohormone crosstalk to bring out cellular homeostasis under stress (Kazan, 2015). Recently, JA has been reported to regulate leaf senescence under cold stress by displaying interactive responses with other hormones such as auxin, gibberellin, and ET (Hu et al., 2017; Per et al., 2018). BR, the relatively newer entry to the phytohormone family, has emerged as a significant and potent player in stress response mechanisms in plants and has shown to impart tolerance under multiple stress including chilling, heat, drought, flooding, heavy metal, salinity, light, and oxidative stresses (Wani et al., 2016). On the other hand, SL has been found to act as a signaling molecule in plant-pathogen interactions and also induce root nodulation in leguminous plants (Foo and Davies, 2011). Thus, the interplay of various phytohormones is desirable for plant stress responses and cooperative crosstalk is essential to maintain harmony under stress conditions. However, the real mode of action is still not fully understood and more dedicated works in this regard can be a key in deciphering the basis of stress tolerance mechanisms in plants for formulating potential genetic engineering strategies for the future.

4.2.2.2 OTHER SIGNALING MOLECULES

Apart from phytohormones, the other potent messengers in plant stress responses are Ca^{2+}, CaM, and NO. These are responsible for a quick relay of stress signals from the membrane-bound receptors to the signaling cascades for providing stress tolerance in plants. Controlled expression and transport of cytosolic Ca^{2+} are essential because their excess as well as the deficit is harmful to plant survival (Robertson, 2013). Overexpression of cyclic nucleotide-gated calcium (CNGC) channels present in the cell membrane has provided tolerance to salinity (Kugler et al., 2009) as well as heat stresses (Finka et al., 2012). Transgenic expression of cation exchange (CAX) proteins also improved cadmium tolerance in petunia (Wu et al., 2011). The down-regulation of CAX in *Arabidopsis* resulted in the susceptibility of transgenic plants towards salinity (Cheng et al., 2003). CaMs act as the primary sensors of cytosolic calcium levels and participate in maintaining the Ca^{+} homeostasis in the cell. Heat stress-induced transgenic expression of the *CaM1-1* gene from rice showed improved the thermo-tolerance in transgenic *Arabidopsis* (Wu et al., 2012). Overexpression of CaM-like 44 (CML44) protein from wild tomato (*Solanum habrochaites*) conferred tolerance to multiple stresses like drought, salinity, and cold stresses, in transformed tomato plants (Munir et al., 2016). Ectopic expression of calcineurin B-like (CBL) protein, another calcium sensor, also provides immunity to plants against stresses like salinity and drought (Cheong et al., 2010). Endogenous NO levels were elevated in heat-stressed *Arabidopsis* seedlings followed by the accumulation of CaM3 molecules, thereby aiding in heat stress tolerance (Xuan et al., 2010). Thus, the role of these tiny molecules is of utmost importance in the successful and controlled transmission of stress signals, and ultimately leading to the acquisition of stress tolerance in plants by activation of several downstream components of the stress-responsive signaling cascades. Their elucidation and characterization is a major challenge in plant stress physiology.

4.2.3 OVEREXPRESSION OF REGULATORY GENES

Stress-responsive pathways in plants are complex and involve a simultaneous expression of several genes to combat the stress in a collective manner. Like for instance, when the growth temperatures are elevated above the normal, plants experience heat stress. However, heat stress is accompanied by dehydration and oxidative stresses, which play in tandem to upset the cellular homeostasis. Thus, manipulation of genes that are located upstream of the signaling cascade can control the concomitant expression of several downstream effector genes. Such genes are known as the regulatory genes and mainly encode for the transcription factors (TFs), kinases, and phosphatases.

4.2.3.1 TRANSCRIPTION FACTOR GENES

The major TF families that are involved in stress-responsive signaling pathways are ERF/AP2 (Ethylene response element-binding factor/Apetala2), NAC [NAM (no apical meristem), ATAF1-2, and CUC2 (Cup-shaped cotyledon)], DREB/CRT (Dehydration-responsive element-binding/C-repeat), WRKY (WRKY-domain containing), HSF (heat shock factor), bZIP (basic leucine zipper), MYC, MYB, and NF-Y (Nuclear Factor-Y). Table 4.5 summarizes a few of the studies conducted to obtain stress tolerance by modulating TF genes. Apart from controlling the downstream effector genes,

TFs have been shown to interact with the phytohormones and also control their expression levels. In a recent study, ABA receptor PYL9 has shown to be positively regulated by MYC2 in the presence of elevated ABA levels and stress conditions (Aleman et al., 2016). Thus, it points towards possible crosstalk that exists between the phytohormones and the TFs, which lead to activation of downstream effector molecules, thereby aiding in acquiring stress tolerance in plants. More such studies are essential to unearth the possible crosstalk in different TFs as well as signaling hormones.

4.2.3.2 REGULATION BY PROTEIN KINASES AND PHOSPHATASES

Protein kinases and phosphatases provide a key link between the various signal transduction pathways, where they govern the activation and deactivation of stress-responsive genes/TFs. A few of the major such proteins are CBL-interacting protein kinases (CIPKs), CaM-binding protein kinase (CBK), Protein phosphatase 2C (PP2C), SnRK2 (SNF1-related protein kinase 2), Protein phosphatase 7 (PP7), Calcium-Dependent Protein Kinase (CDPK). Overexpression of a CDPK gene *OsCPK4* resulted in salt and drought tolerance in rice (Campo et al., 2014). In *Arabidopsis*, the generation of the gain-of-function mutants for the *AtCBK3* gene displayed improved thermo-tolerance, whereas the loss-of-function mutants for *AtCBK3* gene were found to be thermo-sensitive (Liu et al., 2008). However, the mechanism of action of these phosphatases and kinases need to be studied in depth to get a clear picture of the stress responses *in-vivo*.

4.3 NEW SOLUTIONS TO OLD PROBLEMS

The strategy of overexpression of single genes whether upstream regulatory genes or the downstream effector genes has been extensively used to overcome stress in plants. The plants expressing these genes have shown successful tolerance towards stress but with certain limitations. These strategies can be augmented with the application of technologically advanced and futuristic ideas (Figure 4.2), which are discussed below:

4.3.1 GENE-STACKING

In nature, plants face a series of stresses mainly in combination. There is rarely any situation where the plant is subjected to a particular single type of stress in isolation (Mickelbart et al., 2015). Also, there is plenty of variation in the

TABLE 4.5 Summary of Overexpression of Genes Encoding Transcription Factors (TFs)

Transcription Factor	Gene	Source Plant	Target Plant	Effect Seen in Transgenics	References
Nuclear Factor Y (NF-Y)	*NF-YC*	*Arabidopsis*	*Arabidopsis*	ABA hypersensitivity	Bi et al., 2017
	NF-YB2	Maize	Maize	Enhanced tolerance to severe drought stress in field trials	Nelson et al., 2007
NAC	*SNAC1*	Upland rice *IRA109 (Oryza sativa L. ssp japonica)*	Rice	Drought stress tolerance and increased yield when grown under drought stress field conditions.	Hu et al., 2008
	NAC5 and NAC6	Rice	Rice	Enhanced abiotic stress tolerance	Takasaki et al., 2010
	NAC10	Rice	Rice	Yielded more grain in the field under drought conditions, with a larger root diameter	Jeong et al., 2010
DREB	*DREB1*	*Adonis amurensis*	Rice and *Arabidopsis*	Salinity, drought, and freezing tolerance	Zong et al., 2016
	DREB2A	Maize	Maize	Drought and heat stress tolerance	Qin et al., 2007
	DREB	Rice	Rice	Improved drought, salinity, and cold stress tolerance	Dubouzet et al., 2003
ERF/AP2	*HARDY*	*Arabidopsis*	Rice	Drought stress tolerance	Karaba et al., 2007
WRKY	*WRKY1 and WRKY33*	Wheat	*Arabidopsis*	Heat stress tolerance	He et al., 2016
	WRKY11	Rice	Rice	Pathogen defense and drought tolerance	Lee et al., 2018
bZIP	*bZIP46*	Rice	Rice	Drought and salt tolerance, ABA sensitivity	Xiang et al., 2008

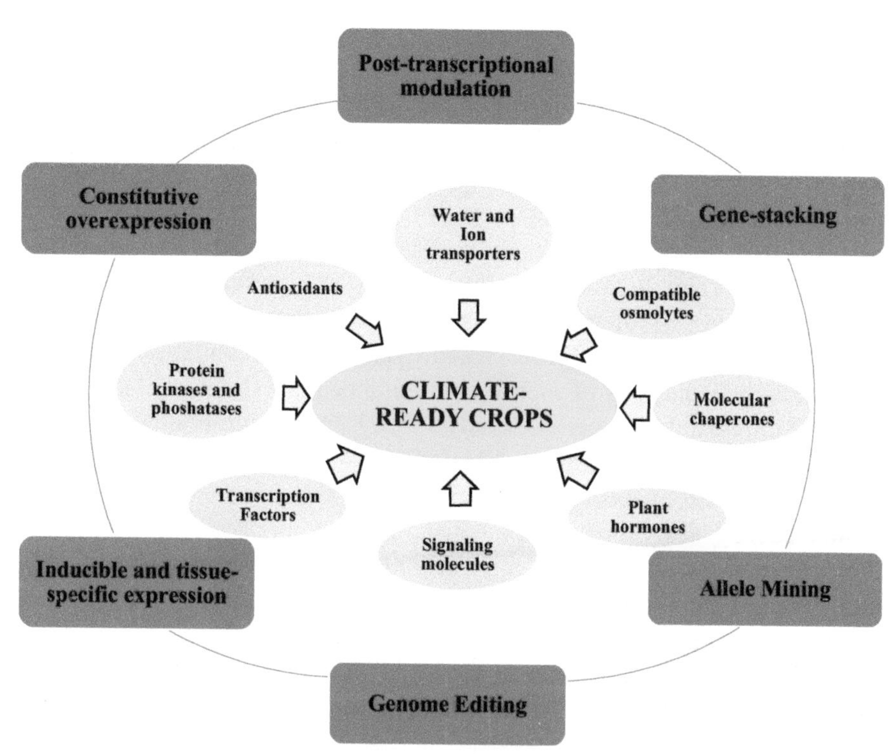

FIGURE 4.2 The various strategies that are currently being employed to improve stress tolerance/adaptation in plants.

duration of stress as well as the developmental stage of the plant which experiences stress (Reguera et al., 2012). Thus, there is a need to consider the genotype × environmental (G×E) effect on the experimental field plants. In such a scenario, the expression of a single gene cannot bring into effect the desired stress tolerance in the plant. These situations beckon the simultaneous expression of a multitude of genes for overcoming the multiple stresses as well as the environmental effect on plants. This strategy is known as gene stacking and has immense potential to render the plants tolerant towards multiple stresses and in varying environmental conditions in the natural field conditions.

Tobacco seeds suffering from oxidative stress and decreased germination rates due to long term storage and adverse environmental conditions were given a new lease of life by the simultaneous overexpression of genes encoding antioxidant enzymes viz., Cu/Zn-SOD and APX in plastids (Lee et al., 2010). In a rare strategy, proline levels were simultaneously increased and then decreased by co-expression of multiple genes, which ultimately enhanced tolerance of plants towards heat stress followed by an increased amount of the proline-rich proteins in the cell wall. This feat was achieved by overexpression of genes encoding for the biosynthetic enzymes gamma-glutamyl kinase 74 (GK74) and gamma-glutamyl phosphate reductase (GPR), along with down-regulation of proline dehydrogenase (*ProDH*) in tobacco and *Arabidopsis* (Stein et al., 2011). Simultaneous co-expression of a mixed combination of different antioxidant genes viz., DHAR, glutathione reductase (GR) and glutathione-S-transferase (GST) in tobacco chloroplasts resulted in improved tolerance of transgenic plants towards salinity and cold stress, whereas individual genes when over-expressed under same conditions, did not display any of such stress tolerance traits (Martret el al., 2011). Stacking of two different genes namely, *NCED*, and D-arabinono-1,4-lactone oxidase (*ALO*) resulted in improved tolerance to twin stresses of drought and chilling by increased accumulation of ABA and ascorbic acid levels in transgenic tobacco and stylo plants (Bao et al., 2016). Similarly, co-expression of TF genes namely, DREB, and phytochrome-interacting factor-like 1 (PIF) was done in rice where DREB expression contributed in enhanced drought tolerance and PIF expression increased plant height, thereby aiding to negate the DREB-induced retarded growth rate in transgenic plants (Kudo et al., 2017). In a recent study, successful pyramiding of ten different genes was done by marker-assisted selection in an elite indica rice variety Tapaswini and it resulted in improved tolerance traits towards multiple stresses (both biotic as well as abiotic) viz., submergence (*Sub1*), salinity (*Saltol*), gall midge (*Gm1*, *Gm4*), blast (*Pi2*, *Pi9*), and bacterial blight (*Xa 4*, *xa5*, *xa13*, *Xa21*) (Das and Rao, 2015; Das et al., 2018). However, a pyramiding of traits is a challenging strategy and it has to be ensured that the stacked genes offer a synergistic action towards a common target of stress tolerance. Any kind of antagonistic genes should not be stacked together and several trials should be performed *in-vitro* before making field trials.

4.3.2 ROLE OF PROMOTERS

Overexpression of genes for imparting stress tolerance is usually done under constitutive promoters, which leads to the accumulation of the gene products in large quantities even in the absence of stress. Their expression is not dependent on the developmental stage and environmental factors rather it is uniform throughout the plant (Singhal et al., 2016). The commonly used constitutive promoters for gene expression studies are the cauliflower mosaic virus 35S (CaMV35S), actin, and ubiquitin promoters (Singhal et al., 2016).

These are strong promoters and drive high levels of transgene expression constitutively, which in turn can govern the plant stress responses. However, this strategy is not successful in all plant types and in all situations. Plant stress is a complex trait and involves a myriad of genes that need to be regulated at different levels so that their action does not interfere with the other genes carrying out the essential functions. Several cases have been reported where the constitutive expression of the transgene negatively impacts the growth and development pattern of plants including smaller leaves, stunted growth, and delayed flowering (Pino et al., 2007; Kasuga et al., 2004). In the case of constitutive expression of genes conferring biotic stress resistance, constitutive expression of genes for one pathogen resistance can render the plant susceptible to attack by other pathogens (Berrocal-Lobo et al., 2002).

On the other hand, the deployment of inducible promoters provides liberty in terms of duration, stage, and amount of the expressed gene product. The gene expression levels are moderate and transgenic product is formed only when induced under proper conditions. The upstream promoter regions of the stress-responsive genes have been isolated, characterized, and used as the promoters to drive stress-inducible expression of the transgene. These include the *rd* (responsive to dehydration), *cor* (cold-regulated), *erd* (early response to dehydration), *kin* (cold-inducible), and *lti* (low-temperature induced). Among these, *rd29a* promoter has been successfully used to control the spatial and temporal expression of several genes specifically only under stress and without any kind of phenotypic defects (Wang et al., 2005; Qin et al., 2007; Jin et al., 2010; Das et al., 2010). The use of HSP101 promoter to drive the expression of *WRKY11* transgene in rice resulted in drought and heat tolerance along with improved survival rates of the green plant parts (Wu et al., 2009). Similarly, endogenous promoters of LEA genes *OsLEA3-1* and *WSI18* have been used to produce drought-tolerant transgenic rice plants without yield penalty (Xiao et al., 2007; Kaur et al., 2017, 2018). Thus, a strategically executed plan with careful use of stress-inducible promoters should be the basis of the controlled transgene expression under appropriate conditions. It is necessary from the point of view of increasing survival rate as well as the yield of stress-tolerant transgenic plants.

4.3.3 ALLELE MINING

Evolution and domestication of plants through the ages have led to the loss of numerous useful alleles from the gene pool of the crops. Intensive breeding and selection of plants for high yield and pest resistance have caused several immunities providing genes to be lost from the domesticated plants which have become more prone to the environmental stresses and pathogen attacks. This has narrowed the genetic diversity of cultivated crop varieties making them distinct from its wild relatives (Atwell et al., 2014; Callaway, 2014). These alterations govern morphological and physiological traits like leaf and sheath structure, grain shattering, canopy architecture, and in turn have left the cultivated crops more prone to disease and climate change (Atwell et al., 2014; Callaway, 2014). On the other hand, the wild progenitors of the elite cultivars provide a rich source for advanced alleles for crop protection and stress tolerance. Identification and isolation of such alleles from the ancestral varieties and then its introgression into elite cultivars is the new paradigm for generating stress-tolerant plants and is termed as allele mining. Allele mining refers to the dissection of naturally occurring variation at candidate genes/loci. The strategy of allele mining forms the basis of breeding technology where the molecular diversity of the crops has been exploited by breeders to create improved allelic combinations with enhanced performance, through sexual hybridization (Latha et al., 2004). This approach has helped in the production of high yielding and superior versions of the crops by introgression of useful alleles from vast plant genetic resources existing across the world. However, several incompatibility barriers inhibit the transfer of critical genes from wild species into cultivated species and this genetic bottleneck has been overcome with the help of genetic engineering approach (Sanchez et al., 2014). Thus, allele mining has now become an attractive tool for geneticists for tapping the genetic variation existing in wild relatives to make dramatic changes in trait expression in crop plants when moved to a suitable genetic background (Kumar et al., 2010).

The primary requirement to bring this strategy into effect is the identification of any existing polymorphism in the experimental plants (Kumar et al., 2010). For this, the whole genome sequence of the crop should be available so that a comparative analysis can be performed. Several studies have been conducted to this effect and brought forth the potentialities of allele mining in modern technology. Disease resistance was observed by mining of alleles for the locus *NBS-LRR* class R-genes in rice (Yang et al., 2008) and *Pto*gene in tomato (Rose et al., 2005). Expression of suitable alleles for the aluminum tolerance (*Alt*) locus conferred tolerance against aluminum toxicity in rye (Fontecha et al., 2007) and wheat (Raman et al., 2008). Tolerance towards drought stress was observed when water stress inducible 18 (*Wsi18*) gene from wild rice variety *Oryza nivara* was identified and transgenically expressed under its native promoter (Kaur et al., 2017) in elite rice cultivar

*O. sativa indica*cv. *IR20* (Kaur et al., 2018). Despite of several examples of introgressions of genes from the wild rice to the cultivated rice, little progress has been done on the transfer of genes for tolerance to stresses from wild species into elite cultivars. Thus, allele mining for complex stress tolerance traits can be a promising strategy of obtaining novel genes/alleles for the existing known genes from related genotypes and helping in creating elite cultivars with desirable traits. Improvement of staple crops in the future will, thus, involve targeted transfer of transgenes from landraces and wild germplasm into cultivated crop varieties, possibly working alongside conventional breeding programs aimed at identifying beneficial genes from quantitative trait locus (QTL) maps. Allele mining is a promising strategy and can retrieve the lost traits for beneficial means.

4.3.4 POST-TRANSCRIPTIONAL MODULATION

Among the recent strategies of imparting stress tolerance in plants is the one that is mediated by small, non-coding RNAs, such as short interfering RNAs (siRNAs) and micro RNAs (miRNAs), which are involved in post-transcription modulation of stress-responsive gene expression. MiRNAs act by silencing the target genes via the coordinated activity of the effector proteins such as Argonaute1 (AGO1) and their action has broad-spectrum effects that include plant growth, development, and stress-related responses. In *Arabidopsis*, the heat-inducible expression of *miR398* imparted thermo-tolerance by silencing of *Cu/ZnSOD* genes (*CSD1* and *CSD2*) along with their associated chaperone encoding *CCS* gene (Guan et al., 2013). Also, promoter regions of the three known *miR398* genes (*miR398a, miR398b,* and *miR398c*) were found to contain the heat shock elements (HSEs) to which the HSFa1b and HSFa7b were reported to bind under heat stress (Guan et al., 2013). In another study, heat stress application in *Arabidopsis* resulted in enhanced expression of *miR156* and its isoforms followed by thermal acclimatization of the experimental plants (Stief et al., 2014). Constitutive expression of *osa-miR396c* resulted in enhanced sensitivity to salinity stress in transgenic rice (Gao et al., 2010). Thus, post-transcription gene modulation can prove to be an effective strategy in regulating the stress-responsive genes.

4.3.5 THE GENOME EDITING PROMISE

The future of crop technology lies in genome editing for achieving better and improved versions of the existing varieties. The crop genome can be tailored with high precision and accuracy incorporating minimal alterations in the target genes and achieving functional modulation for the desired trait. The important tools for plant genome editing include the zinc finger nuclease (ZFN), transcriptional activator-like effector nuclease (TALEN), and clustered regularly interspaced short palindromic repeat (CRISPR)/Cas9 (CRISPR-associated nuclease 9), the latter being the most popular among all. In the CRISPR mediated techniques, a single guide RNA (sgRNA) recognizes the target DNA by complementation base pairing and signals the Cas-9 nuclease to cleave the sgRNA-bound target DNA. Thus, a properly designed profile of sgRNA holds the key to the successful genome editing of any desired plant genes involving CRISPR. Transgenic plants with targeted point mutations or gene-knockouts of heat stress-responsive factors can be of immense use in designing stress-tolerant crops.

4.4 CONCLUSIONS AND FUTURE PERSPECTIVES

In this chapter, we have discussed in detail the various strategies that are currently being employed to improve stress tolerance/adaptation in plants along with strategies that hold promise for the successful designing of climate-ready crops for the future (Figure 4.2). Apart from the enumerated approaches, there are independent novel researches that also favor the generation of stress-tolerant plants. These include the symbiont-based approaches (Coleman-Derr and Tringe, 2014), resurrection plants (Bechtold, 2018), and proteomics approach (Ahmad et al., 2016). However, there is a need to overcome the problem of acceptance of genetically modified (GM) crops among the common public so that the efforts do not go in vain. For these, researchers should aim to produce marker-free transgenic plants by use of different genetic engineering tools, such as transposons, ZFN, homologous recombination, *cre-loxP* (Chong-Pérez and Angenon, 2014). Often in nature, the stresses occur in combination, abiotic, and biotic factors together create a multi-facet attack on the plants (Pandey et al., 2017; Suzuki et al., 2014). Thus, in order to get a holistic view of the response of transgenic plants towards different stresses in field condition, one needs to design the experimental stress conditions closer to the field like conditions, so that the resultant stress-tolerant transgenic plant can cope with the environmental stresses in actuality. Also, transgenic plants should be checked routinely for any kind of transgene escape and loss of desired traits in the successive generations. More robust and strict measures should be adopted to rigorously check the adaptation of the designer crops towards natural stress conditions before releasing the variety for commercial use. Thus, the improvement of stress tolerance traits of crops/plants is an urgent need of the day and can be handled with the genetic engineering techniques. The future of the food security of the world lies in the production of designer crops that are immune to multiple stresses and also result in high yield.

KEYWORDS

- **stress tolerance**
- **signaling molecules**
- **aquaporins**
- **transcription factors**
- **molecular chaperones**
- **allele mining**

REFERENCES

Abebe, T., Arron, C. G., Bjorn, M., & John, C. C., (2003). Tolerance of mannitol-accumulating transgenic wheat to water stress and salinity. *Plant Physiology, 131*(4), 1748–1755. doi: 10.1104/pp.102.003616.

Ahmad, P., Arafat, A. H. A. L., Saiema, R., Nudrat, A. A., Muhammad, A., & Salih, G., (2016). Role of proteomics in crop stress tolerance. *Frontiers in Plant Science, 7*(9), 1336. doi: 10.3389/fpls.2016.01336.

Aleman, F., Junshi, Y., Melissa, L., Yohei, T., Alice, Y. K., Zixing, L., Toshinori, K., et al., (2016). An ABA-increased interaction of the PYL6 ABA receptor with MYC2 transcription factor : A putative link of ABA and JA signaling. *Scientific Report, 6*(3), 28941. doi: 10.1038/srep28941.

Atkinson, N. J., & Peter, E. U., (2012). The interaction of plant biotic and abiotic stresses: From genes to the field. *Journal of Experimental Botany, 63*(10), 3523–3544. doi: 10.1093/jxb/err313.

Atwell, B. J., Han, W., & Andrew, P. S., (2014). Could abiotic stress tolerance in wild relatives of rice be used to improve *Oryza sativa* ? *Plant Science, 215, 216*, 48–58. doi: 10.1016/j.plantsci.2013.10.007.

Bahieldin, A., Hesham, T. M., Hala, F. E., Osama, M. S., Ahmed, M. R., & Ismail, A. A., (2005). Field evaluation of transgenic wheat plants stably expressing the *HVA1* gene for drought tolerance. *PhysiologiaPlanatarum, 123*, 421–427. doi: 10.1111/j.1399-3054.2005.00470.x.

Bao, G., Chunliu, Z., Chunmei, Q., Ting, X., Zhenfei, G., & Shaoyun, L., (2016). Co-expression of *NCED* and *ALO* improves vitamin C level and tolerance to drought and chilling in transgenic tobacco and stylo plants. *Plant Biotechnology Journal, 14*, 206–214. doi: 10.1111/pbi.12374.

Bechtold, U., (2018). Plant life in extreme environments : How do you improve drought tolerance? *Frontiers in Plant Science, 9*(5), 543. doi: 10.3389/fpls.2018.00543.

Berrocal-lobo, M., Antonio, M., & Roberto, S., (2002). Constitutive expression of ethylene-response-factor1 in *Arabidopsis* confers resistance to several necrotrophic fungi. *The Plant Journal, 29*, 23–32. doi/abs/10.1046/j.1365–313x.2002.01191.x.

Bhatnagar-mathur, P., Vincent, V., Jyostna, D. M., Lavanya, M., Vani, G., & Kiran, K. S., (2009). Genetic engineering of chickpea (*Cicer arietinum* L.) with the *P5CSF129A* gene for osmoregulation with Implications on drought tolerance. *Molecular Breeding, 23*, 591–606. doi: 10.1007/s11032-009-9258-y.

Bi, C., Yu, M., Xiao, F., Wang, D., & Peng, Z., (2017). Over expression of the transcription factor NF-YC9 confers abscisic acid hypersensitivity in *Arabidopsis*. *Plant Molecular Biology, 95*(4), 425–439. doi: 10.1007/s11103-017-0661-1.

Callaway, E., (2014). The birth of rice. *Nature, 514*, S58–59. doi: Nature.com/articles/514S58a.pdf?origin=ppub.

Campo, S., Patricia, B., Joaquima, M., Eric, L., & María, C., (2014). Over expression of a calcium-dependent protein kinase confers salt and drought tolerance in rice by preventing membrane lipid peroxidation. *Plant Physiology, 165*(6), 688–704. doi: 10.1104/pp.113.230268.

Cao, S., Du, X. H., Li, L. H., Liu, Y. D., Zhang, L., Pan, X., Li, Y., Li, H., & Lu, H., (2017). Overexpression of *Populus tomentosa* cytosolic ascorbate peroxidase enhances abiotic stress tolerance in tobacco plants. *Russian Journal of Plant Physiology, 64*(2), 224–234. doi: 10.1134/S1021443717020029.

Chang, L., Huan, S., Hua, Y., Xuehua, W., Zhizhe, S., Fang, C., & Wei, W., (2017). Over-expression of dehydroascorbate reductase enhances oxidative stress tolerance in tobacco. *Electronic Journal of Biotechnology, 25*, 1–8. doi: 10.1016/j.ejbt.2016.10.009.

Chauhan, H., Neetika, K., Aashima, N., Jitendra, P. K., & Paramjit, K., (2012). The wheat chloroplastic small heat shock protein (SHSP26) is involved in seed maturation and germination and imparts tolerance to heat stress. *Plant, Cell and Environment, 35*(11), 1912–1931. doi: 10.1111/j.1365–3040.2012.02525.x.

Chen, M., Chen, Q., Niu, X., Zhang, R., Lin, H., Xu, C., Wang, X., Wang, G., & Chen, J., (2007). Expression of *OsNHX1* gene in maize confers salt tolerance and promotes plant growth in the field. *Plant, Soil and Environment, 53*(11), 490–498. doi: agriculturejournals.cz/publicFiles/00426.pdf.

Cheong, Y. H., Sun, J. S., Beom-Gi, K., Girdhar, K. P., Ju-Sik, C., Kyung-Nam, K., & Sheng, L., (2010). Constitutive overexpression of the calcium sensor cbl5 confers osmotic or drought stress tolerance in *Arabidopsis*. *Molecules and Cells, 29*, 159–165. doi: 10.1007/s10059–010–0025-z.

Chiang, C. M., Chien, H. L., Chen, L. F. O., Hsiung, T. C., Chiang, M. C., & Chen, S. P., (2003). Over expression of the genes, coding ascorbate peroxidase from *Brassica campestris* enhances heat tolerance in transgenic *Arabidopsis thaliana*. *Biologia Plantarum, 59*(2), 305–315. doi.org/10.1007/s10535-015-0489-y.

Chong-pérez, B., & Geert, A., (2014). Strategies for generating marker-free transgenic plants. In: Sithole-Niang, I., (ed.), *Genetic Engineering* (Vol. 10, pp. 17–48). IntechOpen.

Coleman-derr, D., & Susannah, G. T., (2014). Building the crops of tomorrow : Advantages of symbiont-based approaches to improving abiotic stress tolerance. *Frontiers in Microbiology5*, 1–6. doi: 10.3389/fmicb.2014.00283.

Das, G., & Gundimeda, J. N. R., (2015). Molecular marker assisted gene stacking for biotic and abiotic stress resistance genes in an elite rice cultivar. *Frontiers in Plant Science, 6*(9), 698. doi: 10.3389/fpls.2015.00698.

Das, G., Gundimeda, J. N. R., Varier, M., Prakash, A., & Dokku, P., (2018). Improved tapaswini having four BB resistance genes pyramided with six genes/QTLs, resistance/tolerance to biotic and abiotic stresses in rice. *Scientific Reports, 8*(1), (2413). doi: 10.1038/s41598-018-20495-x.

Das, M., Harsh, C., Anju, C., Qazi, M. R. H., & Paramjit, K., (2011). High-efficiency transformation and selective tolerance against biotic and abiotic stress in mulberry, *Morusindica* cv. K2, by constitutive and inducible expression of tobacco *Osmotin*. *Transgenic Research, 20*, 231–246. doi: 10.1007/s11248-010-9405-6.

Duan, L., Daniela, D., Han, N., Mei, Y., Rishikesh, B., Malcolm, J. B., & José, R. D., (2013). Endodermal ABA signaling promotes lateral root quiescence during salt stress in *Arabidopsis* seedlings. *The Plant Cell, 25*(1), 324–341. doi: 10.1105/tpc.112.107227.

Dubouzet, J. G., Yoh, S., Yusuke, I., Mie, K., Emilyn, G. D., Setsuko, M., Motoaki, S., et al., (2003). *OsDREB* genes in rice, *Oryza sativa* L., encode transcription activators that function in drought-, high-salt-and cold-responsive gene expression. *The Plant Journal, 33*, 751–763. doi/abs/10.1046/j.1365–313X.2003.01661.x.

Egamberdieva, D., (2009). Alleviation of salt stress by plant growth regulators and IAA producing bacteria in wheat. *Acta Physiologiae Plantarum, 31*, 861–864. doi: 10.1007/s11738-009-0297-0.

Estrada-melo, A. C., Chao, M., Michael, S. R., & Cai-zhong, J., (2015). Over-expression of an ABA biosynthesis gene using a stress-inducible promoter enhances drought resistance in petunia. *Horticulture Research, 2*(1), 15013. doi: 10.1038/hortres.2015.13.

Finka, A., America, F., Henriquez, C., Frans, J. M. M., Younousse, S., & Pierre, G., (2012). Plasma membrane cyclic nucleotide gated calcium channels control land plant thermal sensing and acquired thermo tolerance. *The Plant Cell, 24*, 3333–3348. doi: 10.1105/tpc.112.095844.

Fontecha, G., Silva-navas, J., Benito, C., Mestres, M. A., Espino, F. J., Hernández-riquer, M. V., & Gallego, F. J., (2007). Candidate gene identification of an aluminum-activated organic acid transporter gene at the *Alt4* locus for aluminum tolerance in rye (*Secale cereale*L.). *Theoretical and Applied Genetics, 114*, 249–260. doi: 10.1007/s00122-006-0427-7.

Foo, E., & Noel, W. D., (2011). Strigolactones promote nodulation in pea. *Planta, 234*, 1073–1081. doi: 10.1007/s00425-011-1516-7.

Gang-ping, X., Janneke, D., & Lynne, M. C., (2015). TaHsfA6f Is a transcriptional activator that regulates a suite of heat stress protection genes in wheat (*Triticum aestivum* L.) including previously unknown Hsf targets *Journal of Experimental Botany, 66*(3), 1025–1039. doi: 10.1093/jxb/eru462.

Gao, P., Xi, B., Liang, Y., Dekang, L., Yong, L., Hua, C., Wei, J., Dianjing, G., & Yanming, Z., (2010). Over-expression of *osa-MIR396c* decreases salt and alkali stress tolerance. *Planta, 231*, 991–1001. doi: 10.1007/s00425-010-1104-2.

Gao, Z., Xiaoliang, H., Baocun, Z., Chunjiang, Z., Yingzhu, L., Rongchao, G., Yinzhu, S., & Zhanjing, H., (2010). Over-expressing a putative aquaporin gene from wheat, *TaNIP*, enhances salt tolerance in transgenic *Arabidopsis*. *Plant and Cell Physiology, 51*(5), 767–775. doi: 10.1093/pcp/pcq036.

Guan, Q., Xiaoyan, L., Haitao, Z., Yanyan, Z., & Jianhua, Z., (2013). Heat stress induction of *MiR398* triggers a regulatory loop that is critical for thermo tolerance in *Arabidopsis*. *The Plant Journal, 74*, 840–851. doi: 10.1111/tpj.12169.

Guo, L., Lei, G., Zi, Y. W., Hong, L., Wei, E. C., Jun, C., Meihua, L., Zhang, L. C., Li, J. Q., & Hongya, G., (2006). Expression and functional analysis of the rice plasma-membrane intrinsic protein gene family. *Cell Research, 16*, 277–286. doi: 10.1038/sj.cr.7310035.

Gusta, L. V., Nicole, T. B., Guohai, W., Ximing, L., Xunjia, L., Michael, L. G., & Alan, M., (2009). Superoxide dismutase : An all-purpose gene for agri-biotechnology. *Molecular Breeding, 24*, 103–115. doi: 10.1007/s11032–009–9274-y.

Halder, T., Gouranga, U., Chandra, B., Arup, D., Chandrima, C., & Sudipta, R., (2018). Dehydrins impart protection against oxidative stress in transgenic tobacco plants. *Frontiers in Plant Science, 9*, 136. doi: 10.3389/fpls.2018.00136.

He, C., Aifang, Y., Weiwei, Z., Qiang, G., & Juren, Z., (2010). Improved Salt tolerance of transgenic wheat by introducing BetA gene for glycine betaine synthesis. *Plant Cell, Tissue and Organ Culture, 101*, 65–78. doi: 10.1007/s11240-009-9665-0.

He, C., Juqiang, Y., Guoxin, S., Lianhai, F., Scott, H. A., Dick, A., Eduardo, B., & Hong, Z., (2005). Expression of an *Arabidopsis* vacuolar sodium/proton antiporter gene in cotton improves photosynthetic performance under salt conditions and increases fiber yield in the field. *Plant and Cell Physiology, 46*(11), 1848–1854. doi: 10.1093/pcp/pci201.

He, G. H., Ji, Y. X., Yan, X. W., Jia, M. L., Pan, S. L., Ming, C., You, Z. M., & Zhao, S. X., (2016). Drought-responsive WRKY transcription factor genes *TaWRKY1* and *TaWRKY33* from wheat confer drought and/or heat resistance in *Arabidopsis*. *BMC Plant Biology, 16*(1), 116.

Hong-tao, L., Fei, G., Guo-liang, L., Jin-long, H., & Ren-gang, Z., (2008). The calmodulin-binding protein kinase 3 is part of heat-shock signal transduction in *Arabidopsis thaliana*. *The Plant Journal, 55*, 760–773. doi: 10.1111/j.1365-313X.2008.03544.x.

Houde, M., Sylvain, D., Daniel, N. D., & Fathey, S., (2004). Over-expression of the acidic dehydrin *WCOR410* improves freezing tolerance in transgenic strawberry leaves. *Plant Biotechnology Journal, 2*, 381–387. doi: 10.1111/j.1467-7652.2004.00082.x.

Houtte, H. V., Lies, V., López-galvis, L., Liesbeth, L., Ewaut, K., Sebastien, C., Regina, F., et al., (2013). Over-expression of the trehalase gene *AtTRE1* leads to increased drought stress tolerance in *Arabidopsis* and is involved in abscisic acid-induced stomatal closure. *Plant Physiology, 161*, 1158–1171. doi: 10.1104/pp.112.211391.

Hu, H., Jun, Y., Yujie, F., Xiaoyi, Z., Zhuyun, Q., & Lizhong, X., (2008). Characterization of transcription factor gene *SNAC2* conferring cold and salt tolerance in rice. *Plant Molecular Biology, 67*(1/2), 169–181. doi: 10.1007/s11103-008-9309-5.

Hu, W., Qianqian, Y., Yan, W., Rui, C., Xiaomin, D., Jie, W., Shiyi, Z., et al., (2012). Over-expression of a wheat aquaporin gene, *TaAQP8*, Enhances salt stress tolerance in transgenic tobacco. *Plant and Cell Physiology, 53*(12), 2127–2141. doi: 10.1093/pcp/pcs154.

Hu, Y., Yanjuan, J., Xiao, H., Houping, W., Jinjing, P., & Diqiu, Y., (2017). Jasmonate regulates leaf senescence and tolerance to cold stress : Crosstalk with other phytohormones. *Journal of Experimental Botany, 68*(6), 1361–1369. doi: 10.1093/jxb/erx004.

In-Cheol, J., Se-Jun, O., Ju-Seok, S., Won-Bin, C., Sang, I. S., Chung, H. K., Youn, S. K., et al., (2003). Expression of a bifunctional fusion of the *Escherichia coli* genes for trehalose-6-phosphate synthase and trehalose-6-phosphate phosphatase in transgenic rice plants increases trehalose accumulation and abiotic stress tolerance without stunting growth. *Plant Physiology, 131*, 516–524. doi: 10.1104/pp.007237.

Jain, M., & Jitendra, P. K., (2009). Transcript profiling reveals diverse roles of auxin-responsive genes during reproductive development and abiotic stress in rice *The FEBS Journal, 276*, 3148–3162. doi: 10.1111/j.1742-4658.2009.07033.x.

Jeong, J. S., Youn, S. K., Kwang, H. B., Harin, J., Sun-Hwa, H., Yang, D. C., Minkyun, K., et al., (2010). Root-specific expression of OsNAC10 improves drought tolerance and grain yield in rice under field drought conditions. *Plant Physiology, 153*, 185–197. doi: 10.1104/pp.110.154773.

Jin, T., Qing, C., Wangfeng, L., Dongxu, Y., Zhijian, L., Deli, W., Bao, L., & Lixia, L., (2010). Stress-inducible expression of *GmDREB1* conferred salt tolerance in transgenic Alfalfa. *Plant Cell, Tissue and Organ Culture, 100*, 219–227. doi: 10.1007/s11240-009-9628-5.

Jing, X., Peichen, H., Yanjun, L., Shurong, D., Niya, L., Rui, Z., Jian, S., et al., (2015). Overexpression of copper/zinc superoxide dismutase from mangrove *Kandeliacandel* in tobacco enhances salinity tolerance by the reduction of reactive oxygen species in chloroplast. *Frontiers in Plant Science, 6*, 23. doi: 10.3389/fpls.2015.00023.

Jun-Mei, Z., Xiao-Wei, L., Yuan-hang, Z., Fa-Wei, W., Nan, W., Yuan-Yuan, D., Yan-Xi, Y., Huan, C., Xiu-Ming, L., Na, Y., & Hai-Yan, L., (2016). The *AaDREB1* transcription factor from the cold-tolerant plant *Adonis amurensis* enhances abiotic stress tolerance in transgenic plant. *International Journal of Molecular Sciences, 17*, 611. doi: 10.3390/ijms17040611.

Karaba, A., Shital, D., Raffaella, G., Asaph, A., Kurniawan, R. T., Nayelli, M., Arjun, K., et al., (2007). Improvement of water use efficiency in rice by expression of HARDY an *Arabidopsis* drought and salt tolerance gene. *Proceedings of the National Academy of Sciences of the United States of America, 104*, 15270–15275.

Kasuga, M., Setsuko, M., Kazuo, S., & Yamaguchi-Shinozaki, K., (2004). A combination of the *ArabidopsisDREB1A* gene and stress-inducible *Rd29A* Promoter improved drought-and low-temperature stress tolerance in tobacco by gene transfer. *Plant and Cell Physiology, 45*(3), 346–350.

Kasukabe, Y., Lixiong, H., Kazuyoshi, N., Shuhei, M., Izumi, I., & Shoji, T., (2004). Over-expression of spermidine synthase enhances tolerance to multiple environmental stresses and up-regulates the expression of various stress-regulated genes in transgenic *Arabidopsis thaliana*. *Plant and Cell Physiology, 45*(6), 712–722.

Kaur, R., Anirban, C., Rupam, K. B., Jagannath, B., Asitava, B., Soumitra, K. S., & Ananta, K. G., (2017). Wsi18 promoter from wild rice genotype, *Oryza nivara*, shows enhanced expression under soil water stress in contrast to elite to elite cultivar, *IR20*. *Journal of Plant Biochemistry and Biotechnology, 26*(1), 14–26. doi: 10.1007/s13562-016-0355-9.

Kaur, R., Anirban, C., Rupam, K. B., Soumitra, K. S., & Ananta, K. G., (2018). Tolerance to soil water stress by an agronomically important rice cultivar (*Oryza sativa IR20*) is improved by the inducible expression of *Wsi18* gene locus from a wild rice (*Oryza nivara*). *Biologia Plantarum, 62*(1), 129–139. doi.org/10.1007/s10535-017-0742-7.

Kazan, K., (2015). Diverse roles of jasmonates and ethylene in abiotic stress tolerance. *Trends in Plant Science, 20*, 219–229. doi: 10.1016/j.tplants.2015.02.001.

Kim, Y. S., Kim, I. S., Shin, S. Y., Park, T. H., Park, H. M., Kim, Y. H., Lee, G. S., Kang, H. G., Lee, S. H., & Yoon, H. S., (2014). Overexpression of dehydroascorbate reductase confers enhanced tolerance to salt stress in rice plants (*Oryza sativa* L. *japonica*). *Journal of Agronomy and Crop Science, 200*, 444–456. doi: 10.1111/jac.12078.

Kissoudis, C., Clemens, V. D. W., Richard, V. G. F., & Gerard, V. D. L., (2014). Enhancing crop resilience to combined abiotic and biotic stress through the dissection of physiological and molecular crosstalk *Frontiers in Plant Science, 5*, 207. doi: 10.3389/fpls.2014.00207.

Kudo, M., Satoshi, K., Takuya, Y., Junya, M., Daisuke, T., Alisdair, R. F., Kazuo, S., et al., (2017). Double over-expression of DREB and PIF transcription factors improves drought stress tolerance and cell elongation in transgenic plants. *Plant Biotechnology Journal, 15*, 458–471. doi: 10.1111/pbi.12644.

Kugler, A., Barbara, K., Klaus, P., Patricia, W., & Petra, D., (2009). Salt-dependent regulation of a CNG channel subfamily in *Arabidopsis*. *BMC Plant Biology, 9*, 140. doi: 10.1186/1471–2229–9-140.

Kumar, G. R., Sakthivel, K., Sundaram, R. M., Neeraja, C. N., Balachandran, S. M., Shobha, R. N., Viraktamath, B. C., & Madhav, M. S., (2010). Allele mining in crops: Prospects and potentials. *Biotechnology Advances, 28*(4), 451–461. doi: 10.1016/j.biotechadv.2010.02.007.

Kumar, S., Mehar, H., Debasis, C., Rudra, D. T., Rama, S. D., & Prabodh, K. T., (2013). Expression of a rice lambda class of glutathione S-transferase, *OsGSTL2*, in *Arabidopsis* provides tolerance to heavy metal and other abiotic stresses. *Journal of Hazardous Materials, 248, 249*, 228–237. doi: 10.1016/j.jhazmat.2013.01.004.

Kumria, R., & Manchikatla, V. R., (2002). Ornithine decarboxylase transgene in tobacco affects polyamines, *in-vitro*-morphogenesis and response to salt stress *Journal of Plant Physiology, 159*, 983–990.

Latha, R., Rubia, L., Bennett, J., & Swaminathan, M. S., (2004). Allele mining for stress tolerance genes in *oryza* species and related germplasm. *Molecular Biotechnology, 27*(2), 101–108. doi: 10.1385/MB:27:2:101.

Lee, H., Jooyoung, C., Changhyun, C., Naeyoung, C., Hyun-so, J., Sang, R. P., Seungbum, L., & Duk-ju, H., (2018). Rice *WRKY11* plays a role in pathogen defense and drought tolerance. *Rice, 11*, 5. doi: 10.1186/s12284-018-0199-0.

Lee, Y. P., Kwang-Hyun, B., Haeng-Soon, L., Sang-Soo, K., Jae-Woog, B., & Suk-Yoon, K., (2010). Tobacco seeds simultaneously over-expressing Cu/Zn-superoxide dismutase and ascorbate peroxidase display enhanced seed longevity and germination rates under stress conditions. *Journal of Experimental Botany, 61*(9), 2499–2506. doi: 10.1093/jxb/erq085.

Li, M., Xiaojie, L., Hongqing, L., Xiaoping, P., & Guojiang, W., (2011). Over-expression of *AtNHX5* improves tolerance to both salt and water stress in rice (*Oryza sativa* L.). *Plant Cell, Tissue and Organ Culture, 107*, 283–293. doi: 10.1007/s11240-011-9979-6.

Li, M., Yan, L., Hongqing, L., & Guojiang, W., (2011). Over-expression of *AtNHX5* improves tolerance to both salt and drought stress in *Broussonetia papyrifera* (L.) vent. *Tree Physiology, 31*, 349–357. doi: 10.1093/treephys/tpr003.

Li, R., Jinfang, W., Shuangtao, L., Lei, Z., Chuandong, Q., Sarah, W., Bing, Z., et al., (2016). Plasma membrane intrinsic proteins *SlPIP2;1, SlPIP2;7* and *SlPIP2;5* conferring enhanced drought stress tolerance in tomato. *Scientific Reports, 6*, 31814. doi: 10.1038/srep31814.

Li, X., Gaoming, L., Ying, L., Xiangge, K., Liang, Z., Jianmei, W., Xufeng, L., & Yi, Y., (2018). ABA receptor subfamily III enhances abscisic acid sensitivity and improves the drought tolerance of *Arabidopsis*. *International Journal of Molecular Sciences, 19*, 1938. doi: 10.3390/ijms19071938.

Li, Z., Jingjin, Y., Yan, P., & Bingru, H., (2016). Metabolic pathways regulated by abscisic acid, salicylic acid, and γ-aminobutyric acid in association with improved drought tolerance in creeping bentgrass (*Agrostis stolonifera*). *Physiologia plantarum, 159*, 42–58. doi: 10.1111/ppl.12483.

Liu, D., Liu, Y., Rao, J., Wang, G., Li, H., Ge, F., & Chen, C., (2013). Over-expression of the glutathione S-transferase gene from *Pyrus pyrifolia* fruit improves tolerance to abiotic stress in transgenic tobacco plants. *Molecular Biology* (*Moscow), 47*(4), 591–601.

Liu, H., Qiuqing, W., Mengmeng, Y., Yanyan, Z., Yingbao, W., & Hongxia, Z., (2008). Transgenic salt-tolerant sugar beet (*Beta vulgaris* L.) constitutively expressing an *Arabidopsis thaliana* vacuolar Na^+/ H^+ antiporter gene, *AtNHX3*, Accumulates more soluble sugar but less salt in storage roots. *Plant, Cell and Environment, 31*, 1325–1334. doi: 10.1111/j.1365-3040.2008.01838.x.

Lu, L., Changhe, D., Ruifang, L., Bin, Z., Chuang, W., & Huixia, S., (2018). Roles of soybean plasma membrane intrinsic protein GmPIP2;9 in drought tolerance and seed development *Frontiers in Plant Science, 9*, 530. doi: 10.3389/fpls.2018.00530.

María-teresa, P., Jeffrey, S. S., Eung-jun, P., Zoran, J., Patrick, M. H., Michael, F. T., & Tony, H. H. C., (2007). Use of a stress inducible promoter to drive ectopic *AtCBF* expression improves potato freezing tolerance while minimizing negative effects on tuber yield. *Plant Biotechnology Journal, 5*, 591–604. doi: 10.1111/j.1467-7652.2007.00269.x.

Martret, L., Miranda, P., Karen, S., Gregory, D. N., & Philip, J. D., (2011). tobacco chloroplast transformants expressing genes encoding dehydroascorbate reductase, glutathione reductase, and glutathione-S-transferase, exhibit altered anti-oxidant metabolism and improved abiotic stress tolerance. *Plant Biotechnology Journal, 9*, 661–673. doi: 10.1111/j.1467-7652.2011.00611.x.

Mickelbart, M. V., Paul, M. H., & Bailey-Serres, J., (2015). Genetic mechanisms of abiotic stress tolerance that translate to crop yield stability. *Nature Reviews Genetics, 16*(4), 237–251. doi: 10.1038/nrg3901.

Missihoun, T. D., Eva, W., Jean-paul, G., Solenne, B., Muhammad, R. S., Alain, B., & Dorothea, B., (2015). Over-expression of *ALDH10A8* and *ALDH10A9* genes provides insight into their role in glycine betaine synthesis and affects primary metabolism in *Arabidopsis thaliana.Plant and Cell Physiology, 56*(9), 1798–1807. doi: 10.1093/pcp/pcv105.

Mittler, R., & Eduardo, B., (2010). Genetic engineering for modern agriculture: Challenges and perspectives. *Annual Review of Plant Biology, 61*, 443–462. doi: 10.1146/annurev-arplant-042809-112116.

Munir, S., Hui, L., Yali, X., Saddam, H., Bo, O., Yuyang, Z., & Hanxia, L., (2016). Over-expression of calmodulin-enhances tolerance to multiple abiotic stresses. *Scientific Reports, 6*, 31772. doi: 10.1038/srep31772.

Nagamiya, K., Tsuyoshi, M., Kimiko, N., Shamsul, H., Eriko, H., Sakiko, H., Kenjiro, O., et al., (2007). Enhancement of salt tolerance in transgenic rice expressing an *Escherichia Coli* catalase gene, *Kat E. Plant Biotechnology Reports, 1*, 49–55. doi: 10.1007/s11816-007-0007-6.

Nelson, D. E., Peter, P. R., Tom, R. A., Robert, A. C., Jingrui, W., David, C. W., Don, C. A., et al., (2007). Plant nuclear factor Y (NF-Y) B subunits confer drought tolerance and lead to improved corn yields on water-limited acres. *Proceedings of the National Academy of Sciences of the United States of America* (Vol. 104, p. 42). doi: 10.1073_pnas.0707193104.

Ning-hui, C., Jon, K. P., Bronwyn, J. B., Toshiro, S., & Kendal, D. H., (2003). The *Arabidopsis* Cax1 mutant exhibits impaired ion homeostasis, development, and hormonal responses and reveals interplay among vacuolar transporters. *The Plant Cell, 15*(2), 347–364. doi: 10.1105/tpc.007385.348.

Pandey, P., Vadivelmurugan, I., Muthukumar, V. B., & Luis, E. S., (2017). Impact of combined abiotic and biotic stresses on plant growth and avenues for crop improvement by exploiting physio-morphological traits. *Frontiers in Plant Science, 8*, 537. doi: 10.3389/fpls.2017.00537.

Peleg, Z., Maria, R., Ellen, T., Harkamal, W., & Eduardo, B., (2011). Cytokinin-mediated source/sink modifications improve drought tolerance and increase grain yield in rice under water-stress. *Plant Biotechnology Journal, 9*(7), 747–758. doi: 10.1111/j.1467-7652.2010.00584.x.

Per, T. S., Iqbal, R. K. M., Naser, A. A., Asim, M., & Hussain, J., (2018). Jasmonates in plants under abiotic stresses : Crosstalk with other phytohormones matters. *Environmental and Experimental Botany, 145*, 104–120. doi: 10.1016/j.envexpbot.2017.11.004.

Potters, G., Taras, P. P., Yves, G., Klaus, J. P., & Marcel, A. K. J., (2007). Stress-induced morphogenic responses : Growing out of trouble ? *Trends in Plant Science, 12*(3), 98–105. doi: 10.1016/j.tplants.2007.01.004.

Prabhavathi, V. R., & Manchikatla, V. R., (2007). Polyamine accumulation in transgenic eggplant enhances tolerance to multiple abiotic stresses and fungal resistance. *Plant Biotechnology, 24*, 273–282.

Prerostova, S., Petre, I. D., Alena, G., Vojtech, K., Niklas, K., & Martin, C., (2018). Cytokinins : Their Impact on molecular and growth responses to drought stress and recovery in *Arabidopsis.Frontiers in Plant Science, 9*, 655. doi: 10.3389/fpls.2018.00655.

Qi, Y. C., Liu, W. Q., Qiu, L. Y., Zhang, S. M., Ma, L., & Zhang, H., (2010). Over-expression of glutathione s transferase gene increases salt tolerance of *Arabidopsis. Russian Journal of Plant Physiology, 57*(2), 233–240. doi: 10.1134/S102144371002010X.

Qin, F., Masayuki, K., Yoh, S., Kyonoshin, M., Yuriko, O., Lam-son, P. T., Kazuo, S., & Yamaguchi-Shinozaki, K., (2007). Regulation and functional analysis of *ZmDREB2A* in response to drought and heat stresses in *Zea mays* L. *The Plant Journal, 50*, 54–69. doi: 10.1111/j.1365-313X.2007.03034.x.

Quan, R., Mei, S., Hui, Z., Yanxiu, Z., & Juren, Z., (2004). Engineering of enhanced glycine Betaine synthesis improves drought tolerance in maize. *Plant Biotechnology Journal, 2*, 477–486. doi: 10.1111/j.1467-7652.2004.00093.x.

Raman, H., Peter, R. R., Rosy, R., Benjamin, J. S., Kerong, Z., Peter, M., Rachel, W., et al., (2008). Analysis of *TaALMT1* Traces the transmission of aluminum resistance in cultivated common wheat (*Triticum aestivum* L.). *Theoretical and Applied Genetics, 116*, 343–354. doi: 10.1007/s00122-007-0672-4.

Reguera, M., Zvi, P., & Eduardo, B., (2015). Biochimica et biophysica acta targeting metabolic pathways for genetic engineering abiotic stress-tolerance in crops. *Biochimica et Biophysica Acta, 1819*(2), 186–194. doi: 10.1016/j.bbagrm.2011.08.005.

Rivero, R. M., Jacinta, G., Allen, V. D., Harkamal, W., & Eduardo, B., (2010). Enhanced cytokinin synthesis in tobacco plants expressing PSARK: IPT prevents the degradation of photosynthetic protein complexes during drought. *Plant and Cell Physiology, 51*(11), 1929–1941. doi: 10.1093/pcp/pcq143.

Rivero, R. M., Mikiko, K., Amira, G., Hitoshi, S., Ron, M., Shimon, G., & Eduardo, B., (2007). Delayed leaf senescence induces extreme drought tolerance in a flowering plant. *Proceedings of the National Academy of Sciences of the United States of America, 104*(49), 19631–19636.

Robertson, D. N., (2013). *Modulating Plant Calcium for Better Nutrition and Stress Tolerance* (p. 952043).Hindawi Publishing Corporation.

Ronde, D. J. A., Laurie, R. N., Caetano, T., Greyling, M. M., & Kerepesi, I., (2004). Comparative study between transgenic and non-transgenic soybean lines proved transgenic lines to be more drought tolerant. *Euphytica, 138*, 123–132123–32.

Rose, L. E., Charles, H. L., Adriana, J. B., & Richard, W. M., (2005). Natural variation in the *Pto*pathogen resistance gene within species of wild tomato (*Lycopersicon*). I. Functional analysis of pto alleles. *Genetics, 171*, 345–357. doi: 10.1534/genetics.104.039339.

Roy, M., & Ray, W., (2002). Over-expression of S-adenosylmethionine decarboxylase gene in rice increases polyamine level and enhances sodium chloride-stress tolerance. *Plant Science, 163*, 987–992.

Sade, N., Basia, J. V., Alex, D., Arava, S., Gil, R., Hagit, N., Rony, W., Hagai, K., & Menachem, M., (2009). Improving plant stress tolerance and yield production: Is the tonoplast aquaporin SlTIP2;2 a key to isohydric to anisohydric conversion? *New Phytologist, 181*, 651–661. doi: 10.1111/j.1469-8137.2008.02689.x.

Sade, N., Michaele, G., Ron, S., Amnon, S., Rony, W., & Menachem, M., (2010). the role of tobacco aquaporin1 in improving water use efficiency, hydraulic conductivity, and yield production under salt stress. *Plant Physiology, 152*, 245–254. doi: 10.1104/pp.109.145854.

Sah, S. K., Kambham, R. R., & Jiaxu, L., (2016). Abscisic acid and abiotic stress tolerance in crop plants. *Frontiers in Plant Science, 7*, 571. doi: 10.3389/fpls.2016.00571.

Sanchez, P. L., Rod, A. W., & Darshan, S. B., (2014). The wild relative of rice : Genomes and genomics the genus *Oryza* : Broadening the gene pool species into rice. In: Zhang, Q., & Wing, R. A., (eds.), *Genetics and Genomics of Rice* (Vol. 5, pp. 9–25). Springer: New York. doi: 10.1007/978-1-4614-7903-1.

Sato, Y., & Sakiko, Y., (2008). Enhanced tolerance to drought stress in transgenic rice plants over-expressing a small heat-shock protein, sHSP17.7. *Plant Cell Reports, 27*, 329–334. doi: 10.1007/s00299-007-0470-0.

Singhal, P., Arif, T. J., Mudsser, A., Qazi, M., & Rizwanul, H., (2016). Plant abiotic stress : A prospective strategy of exploiting promoters as alternative to overcome the escalating burden. *Frontiers in Life Science, 9*(1), 52–63. doi: 10.1080/21553769.2015.1077478.

Stein, H., Arik, H., Gad, M., Oran, E., Haviva, E., Laszlo, N. C., László, S., Csaba, K., & Aviah, Z., (2011). Elevation of free proline and proline-rich protein levels by simultaneous manipulations of proline biosynthesis and degradation in plants. *Plant Science, 181*(2), 140-150. doi: 10.1016/j.plantsci.2011.04.013.

Stief, A., Simone, A., Karen, H., Bikram, D. P., Wolf-Rüdiger, S., & Isabel, B., (2014). *Arabidopsis MiR156* regulates tolerance to recurring environmental stress through *SPL* transcription factors. *The Plant Cell, 26*, 1792–1807. doi: 10.1105/tpc.114.123851.

Su, J., & Ray, W., (2004). Stress-inducible synthesis of proline in transgenic rice confers faster growth under stress conditions than that with constitutive synthesis. *Plant Science, 166*, 941–948. doi: 10.1016/j.plantsci.2003.12.004.

Suárez, R., Cecilia, C., & Gabriel, I., (2013). Enhanced tolerance to multiple abiotic stresses in transgenic alfalfa accumulating trehalose. *Crop Science, 49*, 1791–1799. doi: 10.2135/cropsci2008.09.0573.

Suzuki, N., Rosa, M. R., Vladimir, S., Eduardo, B., & Ron, M., (2014). Abiotic and biotic stress combinations. *New Phytologist, 203*(1), 32–43. doi: 10.1111/nph.12797.

Takasaki, H., Kyonoshin, M., Satoshi, K., Yusuke, I., Yasunari, F., Kazuo, S., Yamaguchi-Shinozaki, K., & Kazuo, N., (2010). The abiotic stress-responsive NAC-type transcription factor OsNAC5 regulates stress-inducible genes and stress tolerance in rice. *Molecular Genetics and Genomics, 284*, 173–183. doi: 10.1007/s00438-010-0557-0.

Vendruscolo, E. C. G., Ivan, S., Marcos, P., Carlos, A., Hugo, B., Correa, M., Celso, J., Luiz, G., & Esteves, V., (2007). Stress-induced synthesis of proline confers tolerance to water deficit in transgenic wheat. *Journal of Plant Physiology, 164*, 1367–1376. doi: 10.1016/j.jplph.2007.05.001.

Vishwakarma, K., Neha, U., Nitin, K., Gaurav, Y., Jaspreet, S., Rohit, K. M., Vivek, K., et al., (2017). Abscisic acid signaling and abiotic stress tolerance in plants : A review on current knowledge and future prospects. *Frontiers in Plant Science, 8*, 161. doi: 10.3389/fpls.2017.00161.

Wang, G. P., Zhang, X. Y., Li, F., Luo, Y., & Wang, W., (2010). Over accumulation of glycine betaine enhances tolerance to drought and heat stress in wheat leaves in the protection of photosynthesis. *Photosynthetica, 48*(1), 117–126. doi.org/10.1007/s11099-010-0016-5.

Wang, J., Bin, W., Hengfu, Y., Zhengqi, F., Xinlei, L., Sui, N., Libo, H., & Jiyuan, L., (2017). Over-expression of *CaAPX* induces orchestrated reactive oxygen scavenging and enhances cold and heat tolerances in tobacco *Hindawi Biomed. Research International*, 4049534. doi.org/10.1155/2017/4049534.

Wang, J., Qian, L., Xinguo, M., Ang, L., & Ruilian, J., (2016). Wheat transcription factor TaAREB3 participates in drought and freezing tolerances in *Arabidopsis. International Journal of Biological Sciences, 12*, 257–269. doi: 10.7150/ijbs.13538.

Wang, S., Youhuang, B., Chenjia, S., Yunrong, W., Saina, Z., Dean, J., Tom, J. G., Ming, C., & Yan, H. Q., (2010). Auxin-related gene families in abiotic stress response in *Sorghum bicolor. Functional and Integrative Genomics, 10*, 533–546. doi: 10.1007/s10142-010-0174-3.

Wang, X., Yong, L., Wei, J., Xi, B., Hua, C., Dan, Z., Xiao-Li, S., Lian-Jiang, C., & Yan-Ming, Z., (2011). A novel *glycine soja* tonoplast intrinsic protein gene responds to abiotic stress and depresses salt and dehydration tolerance in transgenic *Arabidopsis thaliana. Journal of Plant Physiology, 168*(11), 1241–1248. doi: 10.1016/j.jplph.2011.01.016.

Wang, Y., Jifeng, Y., Monika, K., Maryse, C., Angela, S., Charlene, M., Tina, U., et al., (2005). Molecular tailoring of farnesylation for plant drought tolerance and yield protection. *The Plant Journal, 43*, 413–424. doi: 10.1111/j.1365-313X.2005.02463.x.

Wang-Xing, N., Lin, X., & Bing-Jun, Y., (2014). A putative soybean *GmsSOS1* confers enhanced salt tolerance to transgenic *Arabidopsis sos1–1* mutant. *Protoplasma, 252*, 127–134. doi: 10.1007/s00709-014-0663-7.

Wani, S. H., Vinay, K., Varsha, S., & Saroj, K., (2016). Science direct phytohormones and their metabolic engineering for abiotic stress tolerance in crop plants. *The Crop Journal, 4*(3), 162–176. doi: 10.1016/j.cj.2016.01.010.

Waterer, D., Benning, G. T., Ximing, L., Xunjia, L., Michael, G., Alan, M., & Lawrence, V. G., (2010). Evaluation of abiotic stress tolerance of genetically modified potatoes (*Solanum tuberosum* cv. *Desiree*). *Molecular Breeding, 25*, 527–540. doi: 10.1007/s11032-009-9351-2.

Wi, S., Kim, W., & Park, K., (2006) Over-expression of carnation S-adenosylmethionine decarboxylase gene generates a broad-spectrum tolerance to abiotic stresses in transgenic tobacco plants. *Plant Cell Reports, 25*, 1111–1121. doi: 10.1007/s00299-006-0160-3.

Wilkinson, S., & William, J. D., (2010). Drought, ozone, ABA, and ethylene : New insights from cell to plant to community. *Functional Plant Biology, 33*, 510–525. doi: 10.1111/j.1365-3040.2009.02052.x.

Wu, H. C., Luo, D. L., Vignols, F., & Jinn, T. L., (2012). Heat shock-induced biphasic Ca^{2+} signature and *OsCaM1-1* nuclear localization mediate downstream signaling in acquisition of thermotolerance in rice (*Oryza sativa* L.). *Plant, Cell and Environment, 35*, 1543–1557. doi: 10.1111/j.1365-3040.2012.02508.x.

Wu, Q., Toshiro, S., Kimberly, A. W., Jeung-sul, H., Chang, K., Kendal, D. H., & Sunghun, P., (2011). Expression of an *Arabidopsis* Ca^{2+}/H^{+} antiporter *CAX1* variant in *Petunia* enhances cadmium tolerance and accumulation. *Journal of Plant Physiology, 168*(2), 167–173. doi: 10.1016/j.jplph.2010.06.005.

Wu, S., Su, Q., & An, L. J., (2010). Isolation of choline monooxygenase (CMO) gene from *Salicornia europaea* and enhanced salt tolerance of transgenic tobacco with *CMO* genes. *Indian Journal of Biochemistry and Biophysics, 47*, 298–305.

Wu, X., Yoko, S., Sachie, K., Yukihiro, I., & Kinya, T., (2009). Enhanced heat and drought tolerance in transgenic rice seedlings overexpressing *OsWRKY11* under the control of *HSP101* promoter. *Plant Cell Reports, 28*(1), 21–30. doi: 10.1007/s00299–008–0614-x.

Xiang, Y., Tang, N., Du, H., Ye, H., Xiong, L., Yong, X., Ning, T., Hao, D., Haiyan, Y., & Lizhong, X., (2008). Characterization of *OsbZIP23* as a key player of the basic leucine zipper transcription factor family for conferring abscisic acid sensitivity and salinity and drought tolerance in rice. *Plant Physiology, 148*, 1938–1952. doi: 10.1104/pp.108.128199.

Xiao, B., Yuemin, H., Ning, T., & Lizhong, X., (2007). Over-expression of a *LEA* gene in rice improves drought resistance under the field conditions *Theoretical and Applied Genetics, 115*(1), 35–46. doi: 10.1007/s00122-007-0538-9.

Xiao-jun, H., Dandan, C., Lynne, M. C., Fernanda, D. M., Zheng-Bin, Z., Janneke, D., Sundaravelpandian, K., et al., (2018). Heat Shock Factor C2a Serves as a proactive mechanism for heat protection in developing grains in wheat via an ABA-mediated regulatory pathway. *Plant Cell and Environment, 41*, 79–98. doi: 10.1111/pce.12957.

Xiao-rong, W., & Ling, L., (2006). Regulation of ABA level and water-stress tolerance of *Arabidopsis* by ectopic expression of a peanut 9-Cis-epoxycarotenoid dioxygenase gene. *Biochemical and Biophysical Research Communications, 347*, 1030–1038. doi: 10.1016/j.bbrc.2006.07.026.

Xu, Y., Patrick, B., Xunzhong, Z., & Bingru, H., (2016). Enhancing cytokinin synthesis by over-expressing Ipt alleviated drought inhibition of root growth through activating ROS-scavenging systems in *Agrostis stolonifera. Journal of Experimental Botany, 67*(6), 1979–1992. doi: 10.1093/jxb/erw019.

Xuan, Y., Shuo, Z., Lei, W., Yudou, C., & Liqun, Z., (2010). Nitric oxide functions as a signal and acts upstream of *AtCaM3* in thermo tolerance in *Arabidopsis* seedlings. *Plant Physiology, 153*, 1895–1906. doi: 10.1104/pp.110.160424.

Yan, H., Qiang, L., Sung-Chul, P., Xin, W., Ya-Ju, L., Yun-Gang, Z., Wei, T., Meng, K., & Dai-Fu, M., (2016). Over-expression of *CuZnSOD* and *APX* Enhance salt stress tolerance in sweet potato. *Plant Physiology et Biochemistry, 109*, 20–27. doi: 10.1016/j.plaphy.2016.09.003.

Yang, A. F., Duan, X. G., Gu, X. F., Gao, F., & Zhang, J. R., (2005). Efficient transformation of beet (*Beta vulgaris*) and production of plants with improved salt-tolerance. *Plant Cell, Tissue and Organ Culture, 83*, 259–270. doi: 10.1007/s11240-005-6670-9.

Yang, S., Tingting, G., Chunyu, P., Zhumei, F., Jing, D., Yueyu, H., Jian-Qun, C., & Dacheng, T., (2008). Genetic variation of NBS-LRR class resistance genes in rice lines. *Theoretical and Applied Genetics, 116*, 165–177. doi: 10.1007/s00122-007-0656-4.

Yang, Z., Jinghui, L., Stefanie, V. T., Alexander, C., Wilhelm, W., & Hans, S., (2016). Leveraging abscisic acid receptors for efficient water use in *Arabidopsis. Proceedings of the National Academy of Sciences of the United States of America, 113*(24), 6791–6796. doi: 10.1073/pnas.1601954113.

Yin, L., Shiwen, W., Amin, E., Imtiaz, U., Yoko, Y., Wataru, T., Yuichi, T., & Kiyoshi, T., (2010). Overexpression of dehydroascorbate reductase, but not monodehydroascorbate reductase, confers tolerance to aluminum stress in transgenic tobacco. *Planta, 231*, 609–621. doi: 10.1007/s00425-009-1075-3.

Yi-Shih, C., Shuen-Fang, L., Peng-Kai, S., Chung-An, L., Tuan-Hua, D. H., & Su-May, Y., (2015). A late embryogenesis abundant protein HVA1 regulated by an inducible promoter Enhances root growth and abiotic stress tolerance in rice without yields Penalty. *Plant Biotechnology Journal, 13*(1), 105–116. doi: 10.1111/pbi.12241.

Young-Pyo, L., Sun-Hyung, K., Jae-Wook, B., Haeng-Soon, L., Sang-Soo, K., & Suk-Yoon, K., (2007). Enhanced tolerance to oxidative stress in transgenic tobacco plants expressing three antioxidant enzymes in chloroplasts. *Plant Cell Reports, 26*, 591–598. doi: 10.1007/s00299-006-0253-z.

Yu, G., Jingtao, L., Xinhua, S., Xianghui, Z., Jinliang, L., & Hongyu, P., (2015). Over-expression of AcNIP5;1, a novel nodulin-like intrinsic protein from halophyte *Atriplexcanescens*, enhances sensitivity to salinity and improves drought tolerance in *Arabidopsis*. *Plant Molecular Biology Reporter, 33*, 1864–1875. doi: 10.1007/s11105-015-0881-y.

Yu, J., Haiman, G., Xiaokun, W., Renjie, T., Yuan, W., & Fugeng, Z., (2017). Overexpression of pyrabactin resistance-like abscisic acid receptors enhances drought, osmotic, and cold tolerance in transgenic poplars. *Frontiers in Plant Science, 8*, 1752. doi: 10.3389/fpls.2017.01752.

Zhang, L., Guangyao, Z., Chuan, X., Jizeng, J., Xu, L., & Xiuying, K., (2018). A wheat R2R3-MYB gene, *TaMYB30-B*, improves drought methylation and chromatin patterning stress tolerance in transgenic *Arabidopsis*. *Journal of Experimental Botany, 63*(16), 5873–5885. doi: 10.1093/jxb/err313.

Zhang, L., Xiaoli, G., Haiyan, Z., Chunlei, Z., Aiju, Z., Fei, W., Yue, Z., & Xuejun, T., (2017). Isolation and characterization of heat-responsive gene *TaGASR1* from wheat (*Triticum aestivum* L.). *Journal of Plant Biology, 60*, 57–65. doi: 10.1007/s12374-016-0484-7.

Zhang, P., Wen-Quan, W., Gen-Liang, Z., Miroslav, K., Petre, D., Jia, X., & Wilhelm, G., (2010). Senescence-inducible expression of isopentenyl transferase extends leaf life, increases drought stress resistance, and alters cytokinin metabolism in cassava. *Journal of Integrative Plant Biology, 52*(7), 653–669. doi: 10.1111/j.1744-7909.2010.00956.x.

Zhang, Q., Cui, M., Xin, X., Ming, X., Jing, L., & Jin-Xia, W., (2014). Over-expression of a cytosolic ascorbate peroxidase gene, *OsAPX2*, Increases salt tolerance in transgenic alfalfa. *Journal of Integrative Agriculture, 13*(11), 2500–2507. doi: 10.1016/S2095-3119(13)60691-7.

Zhang, Y., Jinfen, Y., Shaoyun, L., Jiongliang, C., & Zhenfei, G., (2008). Over-expressing *SgNCED1* in tobacco increases ABA Level, antioxidant enzyme activities, and stress tolerance. *Journal of Plant Growth Regulation, 27*, 151–158. doi: 10.1007/s00344-008-9041-z.

Zhao, F., & Hui, Z., (2006). Salt and paraquat stress tolerance results from co-expression of the *Suaeda salsa* glutathione s-transferase and catalase in transgenic rice. *Plant Cell, Tissue, and Organ Culture, 86*, 349–158. doi: 10.1007/s11240-006-9133-z.

Zhao, F., Shanli, G., Hui, Z., & Yanxiu, Z., (2006). Expression of yeast SOD2 in Transgenic rice results in increased salt tolerance. *Plant Science, 170*(2), 216–224. doi: 10.1016/j.plantsci.2005.08.017.

Zhao, J., Bronwyn, J. B., Joy, M., Jon, K. P., & Kendal, D. H., (2008). The *ArabidopsisCax3* mutants display altered salt tolerance, pH sensitivity and reduced plasma membrane H^+-ATPase activity. *Planta, 227*, 659–669. doi: 10.1007/s00425-007-0648-2.

Zhao, J., Daying, Z., Zheyong, X., Heng, L., & Guangmin, X. Ã., (2007). Enhanced salt tolerance of transgenic progeny of tall fescue (*Festuca arundinacea*) expressing a vacuolar Na^+/H^+ antiporter gene from *Arabidopsis.Journal of Plant Physiology, 164*, 1377–1383. doi: 10.1016/j.jplph.2007.04.001.

Zhao, Y., Zhulong, C., Jinghui, G., Lu, X., Minjie, C., Chunmei, Y., Yuanlei, H., & Jun, Y., (2016). ABA receptor PYL9 promotes drought resistance and leaf senescence. *Proceedings of the National Academy of Sciences of the United States of America, 113*(7), 1949–1954. doi: 10.1073/pnas.1522840113.

Zhou, L., Jing, Z., Yuhan, X., Chaoxian, L., Jiuguang, W., Guoqiang, W., & Yilin, C., (2018). Over-expression of a maize plasma membrane intrinsic protein mPIP1; 1 confers drought and salt tolerance in *Arabidopsis*. *PLoS One, 13*(6), e0198639. doi.org/10.1371/journal.pone.0198639.

Zhou, S., Wei, H., Xiaomin, D., Zhanbing, M., Lihong, C., Chao, H., Chen, W., Jie, W., Yanzhen, H., Guangxiao, Y., & Guangyuan, H., (2012). Over-expression of the wheat aquaporin gene, *TaAQP7*, enhances drought tolerance in transgenic tobacco. *PLoS One, 7*(12), e52439. doi: 10.1371/journal.pone.0052439.

Zwack, P. J., & Aaron, M. R., (2015). Interactions between cytokinin signaling and abiotic stress responses. *Journal of Experimental Botany, 66*(16), 4863–4871. doi: 10.1093/jxb/erv172.

CHAPTER 5

MicroRNAs as a Potential Tool to Improve Abiotic Stress Tolerance in Plants

CHANDRA PAL SINGH

Department of Botany, University of Rajasthan, Jawahar Lal Nehru Marg, Jaipur – 302004, Rajasthan, India, E-mail: chandrapal203@gmail.com

ABSTRACT

MicroRNAs (miRNAs) are small non-coding RNAs that regulate the expression of target genes in a sequence-specific manner. miRNAs have been shown to play a vital role in almost all biological processes. In plants, several agricultural traits are significantly influenced by miRNAs including growth, reproduction, biomass production, grain yield, and biotic and abiotic stresses. Abiotic stress negatively affects the plant productivity, hence the molecular basis of it needs to be investigated. Altered expression patterns of miRNAs in response to a wide range of abiotic stress conditions such as extreme temperatures, drought, salinity, oxidative stress, hypoxia, and nutrient deficiency, have been reported in several plants. These studies suggest the involvement of miRNA in combating abiotic stress responses and shows the potential of miRNAs, to use them as targets for developing abiotic stress-tolerant crop plants. Therefore, here, we describe the current status and future perspective of developing abiotic stress resistance crop plants using miRNA as key targets.

5.1 INTRODUCTION

Globally, abiotic stress is the major cause of crop loss, dipping average productivity for most of the major crop plants by more than 40%. Because of their sessile nature, plants encounter to various adverse environmental conditions (Ausubel, 2005). These adverse environmental conditions comprise both biotic (living organisms causing stress) and abiotic components (non-living factors causing stress). Abiotic components are important factors and essentially required for the growth and development of plants at optimum levels. Alteration in the intensity of these factors possesses stress to the plants that result in suppression of growth and development (Lobell, Schlenker, and Costa-Roberts, 2011). Many studies have demonstrated that abiotic stresses inhibit seed germination, seedling development, root development, photosynthesis, and chlorophyll biosynthesis (Cramer et al., 2011; Fernandez, 2014). Abiotic stresses also enhance the accumulation of reactive oxygen species (ROS) and that further damage plant growth and development (Fernandez, 2014; Mathur, Agrawal, and Jajoo, 2014; Suzuki et al., 2014; Wang, Vinocur, and Altman, 2003). Plants have evolved intricate sensing and responsive mechanisms to withstand the adverse environmental conditions (Wani et al., 2016). Understanding plant responses at the molecular level to different abiotic stresses is important for the improvement of crop productivity.

Plant responses to abiotic stresses involve complex integrated networks branching in multiple cellular pathways (Wani et al., 2016). The number of genes has been identified for the role in abiotic stresses in plants (Seki et al., 2002; Zhu, 2002). These genes can be categorized into three classes; first, upregulated genes, second downregulated genes, and third unaltered genes. Generally, upregulated genes are considered as a positive regulator, however, downregulated genes are taken as negative regulators (Zhang and Wang, 2015). Abiotic stress-induced gene expression can be regulated at different levels such as chromatin, transcription, post-transcriptional, and translation and post-translational (Zhang and Wang, 2015). Researchers are exploring novel genes and their regulators for developing genetic engineering-based abiotic stress-tolerant crops. In the last two decades, microRNAs (miRNAs) have emerged important gene regulators and revolutionized our understanding towards gene regulation (Bartel, 2004). miRNAs are the 19–28 nucleotide long small non-coding RNAs, which regulate gene expression at the post-transcriptional level through complementary binding to the target mRNA sequence (Bartel, 2004). These tiny RNA molecules are pivotal for plants in maintaining homeostasis and restoring biological functions under adverse environmental conditions (Sunkar et al., 2007).

miRNAs are the part of the RNAi mechanism which expresses endogenously in the cell (Riedmann and Schwentner, 2010). There are multiple ways by which miRNAs play a key role in gene regulation in a wide range of organisms including viruses, algae, fungi, insects, plants, and animals (Jones-Rhoades, Bartel, and Bartel, 2006; Ambros, 2004). miRNAs can operate at the chromatin level, transcription level, and post-transcription level. Most of the miRNAs are functionally conserved in plants and animals and encoded by the nuclear genome. The advancement of molecular biology and genomics field has revealed that plants respond to different types of abiotic stresses not only at the transcription or translation level but also at the post-transcriptional level (Phillips, Dalmay, and Bartels, 2007). miRNAs are widespread and cover almost all biological processes of plants mainly including germination, development, growth, flowering, reproduction, cellular metabolism, immunity, abiotic, and biotic stresses response, etc., (Cramer et al., 2011; Fernandez, 2014). In this chapter, we will provide an overview about the biogenesis and functions of plant miRNAs and their responses to abiotic stresses. Moreover, studies done for the functional characterization of miRNAs through overexpressing them in transgenic plants under various abiotic stresses will also be discussed.

5.2 PLANT MIRNA: BIOGENESIS AND FUNCTIONS

Since the discovery of the first microRNA, lin-4 in *Caenorhabditis elegans* numerous miRNAs have been discovered in the number of species (Lee, Feinbaum, and Ambros, 1993). But it took almost ten years to identify the first miRNA in plants (Reinhart et al., 2002). Advancement in sequencing technology specifically next-generation small RNA sequencing is one of the major factors in the astonishing growth of miRNA entries in the miRBase database (Kozomara and Griffiths-Jones, 2014). Unlike animals, most of the miRNAs in plants are produced from their independent miRNA-encoding loci that are non-protein-coding transcriptional units (Axtell, Westholm, and Lai, 2011). Though, in plants, few miRNAs are derived from precursors situated in mRNA regulatory regions and also precursors located in tandem (Axtell, Westholm, and Lai, 2011). These miRNA-encoding transcriptional units are transcribed by RNA polymerase II. The transcription involves the recruitment of RNA Pol-II and several transcriptional activators to promoters and regulatory motifs of miRNA genes, respectively (Rogers and Chen, 2013). Transcripts within having complementary nucleotide sequences when read in opposite directions, form stem-loop structures. That contains single or many loops and single stem which remain unpaired at the ends. Similarly, transcripts which produce miRNAs have characteristic stem loop structures and are referred to as primary-miRNAs (pri-miRNAs), ranging from hundreds to thousands of nucleotides in length. Pri-miRNAs like mRNAs, are protected by the addition of a 7-methylguanosine cap at 5' (Xie et al., 2005a) and a polyadenylated tail at 3' (Baohong et al., 2006). Once primary miRNAs are formed, they are recognized by processing machinery mainly involving dicer like (DCL) enzymes. Most of the MIR genes in plants are processed by DCL1 (Reinhart et al., 2002; Park et al., 2005). The number of RNA binding proteins assists DCL1 in the formation of mature miRNA duplex (miRNA/miRNA*) from pri-miRNA. The processing comprises two crucial steps pri-miRNA to precursor-miRNA (pre-miRNA) generation then pre-miRNA to mature miRNA generation. Major RNA binding proteins contributors are the zinc finger protein SERRATE (SE) (Yu et al., 2005; Grigg et al., 2005; Dajana et al., 2006), the G-patch domain protein TOUGH (TGH) (Ren et al., 2012), and hyponastic leaves1 (HYL1) (Han et al., 2004; Vazquez et al., 2004; Kurihara et al., 2006). HYL1 and TOUGH are the double-stranded RNA (dsRNA) and single-stranded RNA (ssRNA) binding proteins, respectively; however, SE putatively binds at the junction of ssRNA/dsRNA (Figure 5.1) (Han et al., 2004; Vazquez et al., 2004; Kurihara et al., 2006). In the plant system, pri-miRNA to mature miRNA processing machinery operates in the nucleus. In the last step of processing in nucleus, mature miRNA duplexes get methylated at the 3′ terminus by HUA enhancer 1 (HEN 1), a RNA methyltransferase enzyme. Subsequently, mature miRNA duplexes are exported by HASTY (HST) to the cytoplasm (MY et al., 2005). This finally produces a mature functional miRNA duplex in the cytoplasm. Each mature miRNA consists of two-strand, generally referred to as guide and passenger strand. The stand loaded on the RNA-induced silencing complex (RISC) is called a guide strand that has relatively lower stability of base-pairing at its 5'end. The other miRNA strand is generally discarded and is known as the passenger strand. As soon as miRNA-loaded on to the RISC, the miRNA guide strand directs RISC to find a complementary target mRNA sequence (Rogers and Chen, 2013). Argonaute1 (AGO1) is the main effector molecule of RISC, particularly in the miRNA pathway. Gene regulation through miRNA-mediated mechanism may result in cleavage of target mRNA or translational repression based on sequence complementarity (Figure 5.1) (Vaucheret et al., 2004; Baumberger and Baulcombe, 2005; Qi, Denli, and Hannon, 2005; Tang et al., 2003). If there is complete complementarity between the miRNA and target mRNA sequence, AGO1 generally cleaves the mRNA and leads to target degradation. In the case of partial complementarity between miRNA and target mRNA, translation repression occurred. Plant miRNA usually have a perfect or near-perfect pairing with their target mRNA and induce gene repression through degradation of their target mRNA (Vaucheret et al., 2004; Baumberger and Baulcombe, 2005; Qi, Denli, and Hannon, 2005; Tang et al., 2003). Beside to miRNAs main mode of posttranscriptional gene regulation, plant miRNAs are also reported to act at the level of transcription (Wu et al., 2010).

In the current perspective, miRNAs are having a broad functional spectrum. Genes of almost all cellular pathways are predicted to be regulated by miRNAs in higher organisms including plants. miRNAs role in the regulation of following processes have been functionally characterized; plant growth, shoot, and root development, vascular development, morphogenesis, transcription factor, signal transduction, RNAi components, hormone response, abiotic, and biotic stresses response regulation, etc., (Baohong et al., 2006). Primary targets of plant miRNAs are transcription factors (TFs), which serve as the regulators of important biological processes.

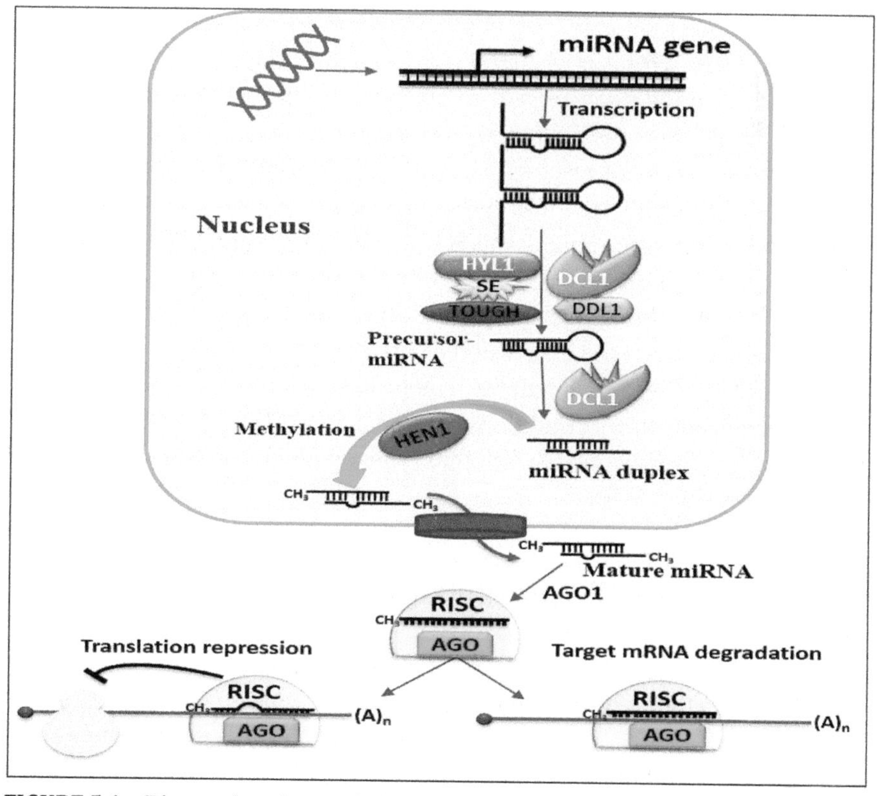

FIGURE 5.1 Biogenesis pathway of miRNAs in plants.

5.3 MIRNAS ROLE IN PLANT ABIOTIC STRESS RESPONSES

Plants, being sessile organisms, have to deal with various environmental abiotic stress conditions such as drought, salt, extreme temperatures, oxidative stress, hypoxia, salinity, and nutrient deficiency (Wani et al., 2016). Abiotic stress causes various damaging effects to the plants at the cellular and molecular levels. Their harmful effect induces accumulation of osmolytes and oxidative damage to major metabolites of the cell including lipids, proteins, and DNA (Zhu, 2002; Scandalios, 2005; Yamaguchi-Shinozaki and Shinozaki, 2006). Plants molecular networks, which maintain the homeostasis during normal conditions get misbalanced by abiotic stresses and require quick reprogramming of molecular switches. These molecular networks are operated at different levels of gene regulation which mainly includes chromatin, transcription, post-transcription, translation, and post-translation. By modulating cellular factors, plants pose a counter defense to the adverse abiotic stresses. A large number of genes and miRNAs are affected by different types of abiotic stresses (Khraiwesh et al., 2012; Jian-Kang, 2016; Grant et al., 2011). Alteration in the expression levels of miRNAs has opened a new area of investigation for gene regulation in plants. miRNAs can regulate the expression of plant genes which are either negative or positive regulators of responses against abiotic stresses. Altered miRNAs can regulate the gene which is harming plants upon abiotic stress. Small RNA sequencing, microarray, and RT-qPCR based studies have provided miRNA expression data for a number of plant species under abiotic stresses (Li et al., 2017). More importantly, phenotypic analysis of mutants or transgenic plants, in which the expression of either stress-responsive miRNAs or their target genes was manipulated, has proven the role of miRNAs during the corresponding stress conditions (Zhang, 2015). Various miRNAs found differentially expressed under different types of abiotic stresses are summarized in Table 5.1.

5.4 POTENTIAL ABIOTIC STRESS-RESPONSIVE MIRNAS IN PLANTS

5.4.1 DROUGHT AND SALINITY STRESS

Drought is the most frequent and serious problem across the word (Shukla, Chinnusamy, and Sunkar, 2008). Uneven patterns of annual rainfall and

TABLE 5.1 Differentially Expressed miRNAs Under Various Abiotic Stresses and Their Predicted Targets with Putative Functions

Type of Abiotic Stress	miRNAs	Targets	Target Function
Drought stress	miR159	MYB	ABA signaling pathway
	miR164	OMTN/NAC	Lateral root development and drought response genes
	miR168	AGO	miRNA biogenesis-pathway
	miR169	NFYA5	Stomatal opening
	miR319	TCP	Enhancement of drought tolerance
	miR394	LCR	ABA-dependent signaling
	miR396	GRF	Modifying plant morphology
	miR408	Plantacyanin and laccase	Promoting electron transport
	miR474	PDH	Proline metabolism
	miR156	SBP	family of transcription factors—promote phase transitions, flowering time
	miR171	GRAS	GRAS transcription factors—response to abiotic stresses and floral development
	miR1444	Polyphenol oxydase	Probable role for improving plant water stress
	miR2118	TIR-NBS-LRR domain protein	Response to salinity, drought, cold, and ABA stress
Salt stress	miR158	NFY/MtHAP2-1	Meristem development
	miR162	DCL	miRNA processing pathways
	miR167	ARF	Auxin response regulation
	miR168	AGO	miRNA biogenesis-pathway
	miR169	NFY/MtHAP2-1	Meristem development
	miR171	SCL	Chlorophyll biosynthesis
	miR319	TCP	Responsible for Na+ uptake
	miR393	TIR1, AFB2	Auxin signaling
	miR394	LCR	Inhibition of ABA-mediated signaling
	miR397	Laccases	Regulating cell wall function
	miR398	CSD	Conversion of superoxide radicals to H_2O_2 and O_2
	miR528	AsAAO/AsCBP1	Regulation of oxidation homeostasis and electron transfer
	miRNVL5	GhCHR	Sodium accumulation control

TABLE 5.1 *(Continued)*

Type of Abiotic Stress	miRNAs	Targets	Target Function
Heat stress	miR156	SPLs	Promoting heating memory
	miR158	NFY/MtHAP2-1	Meristem development
	miR162	DCL	miRNA processing pathways
	miR167	ARF	Auxin response regulation
	miR168	AGO	miRNA biogenesis-pathway
	miR398	CSD1, CSD2, CCS	Regulation of heat shock transcription factor (HSFs) heat shock proteins (HSPs)
	miR173	HTT	HSP activity enhancement
	miR159	GaMYB	Starch metabolism regulation
	miR172	AP2-like genes	Regulation of TARGET OF EAT1 (TOE1), TOE2, and SCHLAFMUTZE (SMZ)
	miR827	SPX-MFS	Regulation of vacuolar SPX-MFS transporters
	miR399	PHO2	Phosphate metabolism regulation
	miR408	Plastocyanin	Electron transport regulation
	miR160	ARF	Auxin response regulation
Cold Stress	miR156	SPLs	Phase transition
	miR160	ARF10, ARF16, ARF17	Seed germination and management of abiotic stress response
	miR164	NAC domain transcription factors	Developmental regulation of root and leaf
	miR166	HD-ZIP transcription factors	Plant development
	miR167	ARF6 and ARF8	Reproductive development
	miR172	AP2	Flowering timing; floral organ identity
	miR390	ARF	Auxin responsive signaling pathway
	miR395	Sulfate transporters	Response to sulfate deficiency
	miR399	Phosphate transporter	Response to phosphate starvation
	miR402	DML3	DNA methylation regulation
	miR408	Laccase; TaCLP1	Regulation of abiotic stress responses
ABA Stress	miR159	MYB	Flower development
	miR160	ARF	Auxin response regulation
	miR167	ARF	ABA-response regulation
	miR169	NFYA	Regulation of stomatal opening and ABA response

TABLE 5.1 *(Continued)*

Type of Abiotic Stress	miRNAs	Targets	Target Function
	miR319	TCP	Responsible for Na+ uptake
	miR393	TIR1/AFB	Auxin response regulation
	miR397	Laccase	Regulating cell wall function
	miR398	CSD	Conversion of superoxide radicals to H_2O_2 and O_2
	miR402	HhH-GPD	DNA methylation
	miR417	RDRP	Secondary miRNA generation
Nutrient-Stress	miR167	ARF7	Stimulating lateral root growth
	miR169	NFYA5	Expression regulation of N transporter
	miR393	ARF3	Primary and lateral root growth induction
	miR395	APS1/APS4	Sulfur transport
	miR397	Laccase	Regulating cell wall function
	miR398	CSD1/CSD2	Regulation of antioxidant responses
	miR399	PHO2	Regulation of Phosphate transport between root and shoot
	miR408	PCL	Regulation of antioxidant responses
	miR778	SUVH6	Regulation of Phosphate uptake
	miR826	AOP2	Glucosinolate synthesis pathway
	miR827	NLA	Regulation of Phosphate transport
	miR5090	AOP2	Glucosinolate synthesis pathway
Oxidative Stress	miR398	CSD1/CSD2	Regulation of antioxidant responses
	miR408	PCL	Regulation of antioxidant responses

Abbreviations: AGO: Argonaute protein; ARF: auxin response factor; COX: cytochrome C oxidase; CSD: Cu/Zn-superoxide dismutase; CCS: copper chaperon of CSD; GRF: growth-regulating factor; HD-Zip: homeodomain-leucine zipper; NFY: nuclear factor Y; PDH: proline dehydrogenase; PLD: phospholipase D; PPR: protein kinase, kinesin, LRR: leucine-rich repeat; POD: peroxidase; SCL: scarecrow-like proteins; TIR1: transport inhibitor response 1; SPL: Squamosa promoter binding protein-like; AP: Apetala; TFs: Transcription factors; TCP: Teosinte branched cycloidea and PCF family; LCR: leaf curling responsiveness; OMTN: Oryza miR164-targeted NAC genes; NAC: NAM, ATAF1/2, and CUC2 proteins; SBP: Squamosa promoter binding protein; GRAS: gibberellic-acid insensitive (GAI), repressor of GAI (RGA) and SCARECROW (SCR) protein family; MtHAP2-1: Medicagotruncatula NF-YA subunit; DCL: dicer-like gene; AsCBP1: AsAAO and copper ion binding protein 1; AsAAO: *Agrostis stolonifera* ascorbic acid oxidase; GhCHR: *Gossypium hirsutum* cys/His-rich; DC1/PHD domains-containing protein; AP2: Apetala2-like genes; PHO2: Pi-responsive genes; SPX-MFS: SYG1/PHO81/XPR1 domain containing Pi transporters proteins-major Facility Superfamily; DML3: demeter-like3; RDRP: RNA-dependent RNA polymerase; HhH-GPD: helix-hairpin-helix and glycine/proline/aspartate (D) containing protein; APS: adenosine 5′-phosphosulfate; PCL: plastocyaninlike; SUVH6: Histone H3 lysine 6 methyltransferase-SUV(R) SET domain protein; AOP2: Glucosinolate-ALK-OHP-AOP locus 2 protein; NLA: nitrogen limitation adaptation gene.

inadequate rainfall generally results in less soil water that leads to drought condition. Plants tolerate drought conditions by enhanced water uptake, reduced water loss, and by other mechanisms (Farooq et al., 2009). Drought stress has been revealed to alter the expression of many genes and metabolites, which are directly or indirectly involved in maintaining homeostasis of water intake (Farooq et al., 2009). These genes and metabolites mainly include abscisic acid (ABA)-induced genes, auxin response factors (ARFs), dehydrins, functional ATPase domains, TFs, invertase, osmotic stress tolerance pathway genes, glutathione S-transferase (GST), transmembrane proteins, helicase, proline osmolytes, and carbohydrate-related metabolites, etc., (Arash, Zakaria and Golam, 2013).

Recent studies were done on *Arabidopsis thaliana* (Liu et al., 2008), *Oryza sativa* (Zhao et al., 2007), *Triticum species* (Kantar, Lucas, and Budak, 2011), *Saccharum officinarum* (Agustina et al., 2015),*Sorghum bicolor* (Amit et al., 2015), *Populus trichocarpa* (Lu, Sun, and Chiang, 2008),*Zea mays* (Wei et al., 2009), *Vigna unguiculata* (Barrera-Figueroa et al., 2011), *Glycine max* (Kulcheski et al., 2011), *Solanum lycopersicum* (Liu et al., 2018),*Phaseolus vulgaris* (Arenas-Huertero et al., 2009), and *Nicotiana tabacum* (Frazier et al., 2011) have found differential expression patterns of miRNAs in relation to drought stress. In the model plant, *Arabidopsis*, miR156, miR158, miR159, miR165, miR167, miR168, miR169, miR171, miR319, miR393, miR394, miR396, and miR397 were reported for altered expression upon drought stress in seedlings (Liu et al., 2008). A link has been established between upregulation of miR393 and downregulation of Transport inhibitor response1 (TIR1) gene, to determine the miR393 role in drought response. TIR1 gene is a putative target of miR393 (Hao et al., 2012). TIR1 is an inducer of the growth and development, therefore its downregulation by miR393 negatively affects plant growth (Hao et al., 2012).

A study done on rice using microarray identified drought-responsive miRNAs (Zhou et al., 2010). Eight miRNAs were upregulated (miR395, miR474, miR845, miR851, miR854, miR901, miR903, and miR1125) and 11 miRNAs were found downregulated (miR170, miR172, miR397, miR408, miR529, miR896, miR1030, miR1035, miR1050, miR1088, and miR1126), under drought stress (Zhou et al., 2010). Surprisingly nine miRNAs (miR156, miR168, miR170, miR171, miR172, miR319, miR396, miR397, and miR408) of rice showed a contrary response to drought stress as reported in *Arabidopsis* (Zhou et al., 2010). Another study done in Dongxiang wild rice found 33 differentially expressed miRNAs under drought stress including miR160f-5p, miR164c, miR319b, miR444, miR166h-5p, miR172d-5p, miR167h-3p and miR160f-5p (Zhang et al., 2017). In sorghum expression of drought, stress-responsive miRNAs were determined by a high-throughput sequencing analysis. That resulted in identifying higher levels of miR156g*, miR166f*, miR166g*, miR167g*, miR167h*, miR169e*, miR169h*, miR-383*, and miR398* in drought-stressed samples (Amit et al., 2015). Thirteen differential expressing miRNAs were identified in leaf and root tissues of *Triticum turgidum ssp. dicoccoides* under drought-stressed conditions (Kantar, Lucas, and Budak, 2011). miRNAs, miR166-5p, miR169-3p, miR397a, miR397b, and miR1513c showed a differential pattern upon drought stress conditions in soybean (Kulcheski et al., 2011). A study done in greenhouse sugarcane plants reported that miR397 levels first increases (post 2 days) then subsequently decreases (post 4 days) under drought stress (Kulcheski et al., 2011). However, another miRNA, miR399 remains repressed for during the same duration. Sugarcane plants grown in fields showed upregulation of miR160 and miR399 and downregulation of miR166, miR171, and miR396 in rainfed plants compared to watered plants (Kulcheski et al., 2011). Enhancement was reported in the levels of miRS1, miR159, miR393, miR1514a, miR2118, and miR2119 upon drought stress-condition in *P. vulgaris* (Arenas-Huertero et al., 2009). In *Medicago truncatula*, miR398a, miR398b, and miR408 were significantly induced while miR169 was suppressed under drought treatments (Trindade et al., 2010). Moreover, in Populus, miR1711-n, miR1445, miR1446a, miR1446b, miR1446c, miR1446d, miR1446e, and miR1447 showed aberrant expression upon drought-stress (Lu, Sun, and Chiang, 2008). In maize, miR474 was reported upregulated under drought stress, and that targets the proline dehydrogenase (PDH) gene (Wei, Zhang, Xiang, and Zhang, 2009).

Salinity stress inhibits plant growth and development process by inducing changes in different physiological and metabolic pathways (Zhu, 2002, 2007). Several studies have determined plant responses at the level of genes and miRNAs to the salt stress (Shiozaki, Yamada, and Yoshiba, 2005). Most of the experiments conducted to establish salinity effect on miRNA expressions are done using NaCl. In *Arabidopsis*, miR156, miR158, miR159, miR165, miR167, miR168, miR169, miR171, miR319, miR393, miR394, miR396, and miR397 were upregulated in response to salt stress, while the accumulation of miR398 was decreased (Liu et al., 2008). However, in salt-stressed sorghum plants, levels of miR156, miR156a, miR156h, miR156i, miR159, miR166a, miR166d, miR171 and miR1435 were found repressed and levels of miR160f, miR166, miR167, miR168, miR393, and miR399 accumulated higher (El Sanousi et al.,

2016). A study done in *Raphanus sativus* reported 71 stress-responsive miRNAs, including 49 known and 22 novel miRNAs (Sun et al., 2015). In *P. trichocarpa*, salt stress decreased the accumulation of miR530a, miR1445, miR1446a-e, miR1447, and miR1711-n, although levels of miR482.2 and miR1450 were positively induced in same conditions (Lu and Huang, 2008). A microarray-based analysis performed in roots of *Z. mays* revealed a higher abundance of miR156, miR164, miR167, and miR396 family member miRNAs, whereas lower levels were detected for miR162, miR168, miR395, and miR474 family member miRNAs, under salt treatment (Mohammad et al., 2017). Arenas-Huertero et al. reported the upregulation of miRS1 and miR159.2 upon NaCl treatment in *P. vulgaris* (Arenas-Huertero et al., 2009).

5.4.2 HEAT AND COLD STRESS

Due to their sessile nature, plants experience fluctuations in environmental temperature conditions. Change in the environmental temperature is a very crucial factor for plants grown in the specific area, because it affects their survival through causing an alteration in their physiology and metabolism (Mohammad et al., 2017). A comprehensive study explored the miRNAs response to heat stress (40°C) in wheat leaves using Solexa platform-based small RNA sequencing (Xin et al., 2010). In this analysis, 153 miRNAs were identified in total, belonging to 51 known and 81 novel miRNA families. Nine conserved wheat miRNAs showed differential expression upon heat stress. Eight miRNAs involving miR156, miR159, miR160, miR166, miR168, miR169, miR393, and miR827 were found upregulated, while miR172 showed downregulation under heat stress (Xin et al., 2010). In rice, miR1423a-5p, miR1427, miR1863a, miR2055, and miR5072 accumulated at higher levels under heat treatment compared to control samples in shoots of cultivar N22, however levels of miR166n-5p, miR2863b, and miR396f-3p found repressed (Satendra et al., 2017). In roots of the same N22 cultivar, miR1427 was induced while miR1878 was repressed after heat treatment. Furthermore, in shoots of Vandana rice cultivar, expression levels of miR394, miR408-5p, miR440, miR444a-5p, miR444f, miR531a, miR1879, and miR3979-3p exhibited upregulation upon heat treatment (Satendra et al., 2017). Though heat treatment in roots of the Vandana cultivar showed enhancement for the levels of miR396c-5p, miR528-5p, miR169i-5p, miR5072, and miR5082, whereas reduction in the levels of miR1878 was observed (Satendra et al., 2017).

The miRNAs response in cold stress has been evaluated in *Arabidopsis* (Liu et al., 2008), rice (Lv et al., 2010), tomato (*Solanum habrochaites*) (Cao et al., 2014), Populus (Lu, Sun, and Chiang, 2008), tea (*Camellia sinensis*) (Zhang et al., 2014a), and Brachypodium (Zhang et al., 2009). In *Arabidopsis*, miRNAs, miR165, miR166, miR169, miR172,miR393, miR396, miR397 showed upregulation upon cold stress treatment (Liu et al., 2008). In Populus, higher accumulation of miR168a, miR168b, and miR477a, miR477b was observed however lower levels of miR156g-j; miR475a, miR475b, and miR476a were reported upon cold shock (Lu, Sun, and Chiang, 2008). In wild tomato (*Solanum habrochaites*), 192 miRNAs reported for increased expression, and 205 miRNAs showed downregulation in the expression upon cold stress (Cao et al., 2014). Likewise, a study conducted by Zhang et al. in 2014, in tea plants, revealed a decrease in expression levels of 43 miRNAs and enhancement in the levels of 31 miRNAs under cold stress (Zhang et al., 2014a). In rice, Lv et al. (2010) determined the effect of cold stress on the expression of miRNAs and found that eighteen rice miRNAs were downregulated under cold treatment; hence, these miRNAs are most likely cold-responsive miRNAs (Lv et al., 2010).

5.4.3 NUTRITIONAL STRESS

Mineral nutrients are essential elements for the growth and development of plants. Optimum concentrations of these mineral nutrients promote the growth of the plants, however, fluctuations in optimum concentration impose stress to plants. Recent evidences suggests the role of miRNAs in maintaining nutrient homeostasis in plants (Liang, Ai, and Yu, 2015). Most of the miRNA study has centered on the phosphorus (P), nitrogen (N), potassium (K), sulfur (S), iron (Fe), and copper (Cu) (Khraiwesh, Zhu, and Zhu, 2012). In *Arabidopsis*, miR395, miR398, and miR399 were enhanced in the sulfate, copper, and phosphate-deficient conditions, respectively (Jones-Rhoades and Bartel, 2004; Yamasaki et al., 2007; Aung et al., 2006; Bari et al., 2006). miR399 of *Arabidopsis* targets the PHO_2 gene which encodes the E2-UBC24 enzyme that controls phosphate uptake and root-to-shoot distribution, negatively (Aung et al., 2006; Bari et al., 2006). Regulatory mechanism of phosphate uptake has revealed miR399 regulation by phosphate starvation response 1 (PHR1) gene, which itself induced by phosphate deprived condition and regulates positively GNATATNC cis-elements-containing phosphate-responsive genes (Aung et al., 2006; Bari, Datt, and Stitt, 2006; Pant et al., 2008; Rubio et al., 2001). Cu is another essential micronutrient required for photosynthesis and plant growth (Marschner, 1995). In many plant species, miR398 is reported as a master regulator of Cu homeostasis (Yamasaki et al., 2007). In the Cu deficient condition, the level of miR398 rises, and that slows down the allocation of Cu to CSDs (Cu-superoxide dismutases) (Abdel-Ghany and Pilon, 2008). In Brassica, miR398 was found upregulated in all parts of the plants upon Cu deficiency (Buhtz et al., 2010). In sulfur deprived conditions role of miR395 has been implicated in plants (Jones-Rhoades, Bartel, and Bartel, 2006; Jones-Rhoades and Bartel, 2004). miR395 expression levels were induced by the sulfur deprived condition. ATP sulfurylases (APSs) and the sulfate transporter AST68 both are targeted by the miR395 (Jones-Rhoades, Bartel, and Bartel, 2006; Jones-Rhoades and Bartel, 2004). A link between increasing levels of miR395 and decreased levels of APSs and AST68 target genes were observed under sulfur deprived condition (Jones-Rhoades, Bartel, and Bartel, 2006; Jones-Rhoades and Bartel, 2004). In Brassica, higher accumulation of miR395 was observed in the phloem sap compared to other tissues including root, stem, or leaves upon sulfur deficiency (Buhtz et al., 2008, 2010). Levels of miR164b and miR394a-c were induced by sulfur deficiency, while miR160 was repressed under the same condition in Brassica (Huang et al., 2010). Nitrogen starvation was found to repress the expression of miR169 in *Arabidopsis*; however, its predicted target gene, NFYA showed higher expression under nitrogen deficiency (Zhao et al., 2011). Precursor analysis of miR169 revealed that the expression of miR169a was significantly reduced in both roots and shoots under N deficiency (Zhao et al., 2011). Another *Arabidopsis* miRNA, miR167 showed lesser expression in the presence of N and its predicted target, ARF8, showed higher levels in the pericycle region (Gifford et al., 2008). In *Arabidopsis*, Fe-deprived condition induced expression of miR169c, miR172c, miR172d, miR173, miR394a, and miR394b in shoots and miR169b, miR169c, miR172c, miR172d, miR173 and miR394b in roots (Kong and Yang, 2010). Initially, the expression of these miRNAs were induced and subsequently decreased with time upon Fe-deficiency (Kong and Yang, 2010).

5.4.4 OXIDATIVE STRESS AND HYPOXIA

ROS are mainly produced in the chloroplasts, mitochondria, and peroxisomes and are linked with different types of abiotic stresses such as UV radiation, heavy metals, high-intensity light, salinity, extreme temperatures, and drought stress

(Mittler et al., 2004). ROS inflicts oxidative destruction to the cells, includes superoxide radicals (O_2^-), hydrogen peroxide (H_2O_2), and hydroxyl radicals (OH^-). SODs are responsible for the conversion of the very toxic superoxide radicals (O_2^-) into slightly toxic hydrogen peroxide (H_2O_2) (Fridovich, 1995). In *Arabidopsis* Cu-Zn-related CSDs, CSD1, CSD2, and CSD3 are encoded by CSD genes (Sunkar, Kapoor, and Zhu, 2006). Sunkar et al. (2006) reported that miR398 levels are reduced under oxidative stress, however, the expression of its predicted targets, CSD1 and CSD2 were found upregulated under the same stress conditions (Sunkar, Kapoor, and Zhu, 2006). This suggests under oxidative stress miR398 is repressed which results in relieved higher expression of CSD1 and CSD2, which neutralizes the toxic effects of superoxide free radicals (Sunkar, Kapoor, and Zhu, 2006). In rice, the effect of oxidative stress using H_2O_2 was studied in the seedlings. Levels of miR169, miR397, miR827, and miR1425 were noticed higher, whereas levels of miR528 were found reduced upon H_2O_2-mediated oxidative stress (Li et al., 2010).

Oxygen deprived condition (hypoxia) causes stress and also obstructs cellular respiration (Agarwal and Grover, 2006). In *Zea mays,*miR166, miR167, miR171, and miR399 were detected upregulated while miR159 was found decreased under hypoxia condition (Zhang et al., 2008). Similarly, miR396 and miR528 of rice showed increased and decreased expression upon hypoxia, respectively (Zhang et al., 2008). In *Arabidopsis*, 19 miRNAs were determined as hypoxia-responsive and out of that miR156g, miR157d, miR158a, miR159a, miR172a, miR172b, miR391, and miR775 showed upregulation under hypoxia (Moldovan et al., 2009).

5.4.5 UV RADIATION

High levels of UV-B radiation cause a type of abiotic stress. Wang et al. (2013), determined differential expression of novel miRNA, miR6000 in wheat upon UV-B treatment (Wang et al., 2013). Among known miRNAs, miR159, miR167a, and miR171 were found induced and miR156, miR164, and miR395 showed downregulation under UV-B exposure (Wang et al., 2013). An *in silico* study predicted the role of miRNAs in response to UV-B radiation. Upregulated miRNAs were miR156, miR157, miR159, miR160, miR165, miR166, miR167, miR169, miR170, miR171, miR172, miR319, miR393, miR398, and miR401 under UV-B-mediated stress (Zhou, Wang, and Zhang, 2007). Most of these miRNAs (miR156, miR160, miR165/166, miR167, miR398, and miR168) were downregulated in *Populus tremula* under the same stress (Jia et al., 2009). These results indicate species-specific regulation of miRNAs under UV-B stress.

5.4.6 ABSCISIC ACID (ABA)-INDUCED STRESS

Abscisic acid (ABA), an important phytohormone, plays a key role in stomatal opening, plant adaptation, germination, and development in plants (Saroj et al., 2016). Therefore, identifying ABA-responsive miRNAs in plants is of significance, because it would provide evidence of operated regulatory mechanisms under ABA-stress. In *Arabidopsis*, miR159 was accumulated in higher concentration upon ABA treatment in seedlings of germinating seeds, (Reyes and Chua, 2007). A comprehensive study done by Sunkar and Zhu et al. (2004) reported that the expression of miR393, miR397b, and miR402 were induced by ABA treatment, while miR389a showed downregulation (Sunkar and Zhu, 2004). Moreover, separate studies conducted in *Arabidopsis* found that miR160 (Liu et al., 2007) and miR417 (Jung and Kang, 2007) are positively induced and miR169 (Li et al., 2008) and miR398 (Jia et al., 2009) are repressed upon ABA treatment. ABA-induced the expression of miR159.2, miR393, and miR2118 in *P. vulgaris*, while miRS1, miR1514, and miR2119 showed subtle upregulation under ABA treatment (Arenas-Huertero et al., 2009). In rice, ABA-treated plants showed induction in the levels of miR319 and repression in the levels of miR167 and miR169 (Liu et al., 2009). A recent report has shown that an ABA-responsive miRNA regulates DNA methylation in *Physcomitrella patens* (Khraiwesh et al., 2010). miR1026 showed higher expression upon ABA treatment while transcript levels of putative target, PpbHLH was reported downregulated and its gene was found hypermethylated (Khraiwesh et al., 2010). This suggested role of miRNAs in epigenetic regulation of stress-responsive genes in plants.

5.5 miRNAs AS IMPORTANT TOOL TO IMPROVE ABIOTIC STRESS TOLERANCE IN PLANTS

Sustainable agriculture is needed to address the food and energy requirement of a rapidly increasing global population. However, sustainable crop production of important crops requires focused efforts to cope with the changing environment. Advance approaches are more focused on the manipulation of gene expression to acquire traits-of-interest such as crop yield, quality, resistance against pathogens and abiotic stress-tolerance, etc. Recent evidences suggest that miRNAs are a key factor in determining various traits in plants, including abiotic stress tolerance, pathogen-caused disease resistance, germination, biomass yields, flowering time, and other cellular pathways. These facts advocate about the significance of miRNAs in crop improvement. Therefore, miRNAs can be used as a potential tool for engineering abiotic-stress tolerant plants. Numerous methods or combinations of them can be applied for miRNA manipulations in plants. These methods include overexpression or knockdown of specific miRNA(s), similarly overexpression, or knockdown of their validated target genes, overexpression of artificial miRNA (amiRNA) against specific target genes, overexpression of the target-mimics, and overexpression of miRNA-resistant target genes (Zhou and Luo, 2013).

A number of studies have been done involving the overexpression of miRNAs in many plant species. Transgenic plants engineered by overexpressing specific miRNA(s) showed increased tolerance or sensitivity to a specific type of abiotic stress (Varsha et al. 2016). Obtained miRNA-based transgenic plants displayed the potential to engineer abiotic stress-tolerant plants using modern genetic engineering approaches. Here we have discussed some important examples, where tolerance against abiotic stresses was acquired through miRNA manipulations in plants. Furthermore, an account of transgenic plants engineered though miRNA-overexpression and their tolerance and sensitivity under various abiotic stresses are given in Table 5.2.

The first discovered plant miRNA, miR156 has now been shown to play a key role in various abiotic stresses and flowering phase transition. *In silico,* prediction tools predicted the target binding sites for miR156 on the squamosa-promoter binding-like (SPL) TFs that play a vital role indevelopmental transitions. Transgenic plants for miR156 were produced by overexpressing precursor of miR156 in *Arabidopsis*. The results of this study suggested the role of miR156 in managing the memory of heat stress and miR156-transgenic plants also showed greater tolerance to heat stress (Stief et al., 2014). However, overexpression of miR156 exhibited decreased cold tolerance in rice (Cui et al., 2015). Another transgenic study was done by overexpressing miR156 in switchgrass resulted in increased biomass, 58–101%, compared to non-transgenic control plants (Fu et al., 2012).

TABLE 5.2 A List of miRNA Overexpression Studies Done for Conferring Abiotic Stress Tolerance in Plants

Targeted miRNA	Source Plant Species of Targeted miRNA Gene	Transgenic Plants	Targeted Stress Responses	References
miR156	*Oryza sativa*	*O. sativa*	Heat stress tolerance	Stief et al., 2014
miR156	*Arabidopsis thaliana*	*A. thaliana*	Decrease in cold tolerance	Cui et al., 2015
miR156	*O. sativa*	*Panicum virgatum*	Increase in biomass	Fu et al., 2012
miR159	*O. sativa*	*O. sativa*	Sensitivity to heat stress	Wang et al., 2012
miR169	*Solanum lycopersicum*	*S. lycopersicum*	Enhancement of tolerance against drought	Zhang et al., 2011
miR169	*A. thaliana*	*A. thaliana*	Sensitive to nitrogen deficiency	Zhao et al., 2011
miR172	*Glycine max*	*A. thaliana*	Drought and salt tolerance	Li et al., 2016
miR173	*A. thaliana*	*A. thaliana*	Tolerance to high temperature	Li et al., 2014
miR319	*O. sativa*	*Agrostis stolonifera*	Enhanced drought and salt tolerance	Zhou et al., 2013
miR319	*O. sativa*	*O. sativa*	Enhanced cold tolerance	Yang et al., 2013
miR390	*O. sativa*	*O. sativa*	Decreased Cd tolerance and increase in Cd accumulation	Ding et al., 2016
miR393 Rice	*O. sativa*	*O. sativa*	Increase in sensitivity to salinity and alkalinity	Gao et al., 2011
miR393	*A. thaliana*	*A. thaliana*	Increase in sensitivity to salinity and alkalinity	Gao et al., 2011
miR394	*G. max*	*A. thaliana*	Drought Tolerance	Ni et al., 2012
miR394	*A. thaliana*	*A. thaliana*	Improved cold tolerance	Song et al., 2016
miR395	*A. thaliana*	*A. thaliana*	Drought and salinity stress	Kim et al., 2010

TABLE 5.2 *(Continued)*

Targeted miRNA	Source Plant Species of Targeted miRNA Gene	Transgenic Plants	Targeted Stress Responses	References
miR395	*A. thaliana*	*Brassica napus*	Higher tolerance to oxidative stress and heavy metal stress	Huang et al., 2010
miR396	*O. sativa*	*O. sativa*	Increase in sensitivity to salinity and alkalinity	Gao et al., 2010
miR396	*A. thaliana*	*A. thaliana*	Increase in sensitivity to salinity and alkalinity	Gao et al., 2010
miR398	*A. thaliana*	*A. thaliana*	Improved thermo-tolerance	Guan et al., 2013
miR399	*A. thaliana*	*S. lycopersicum*	Enhanced growth under phosphorus deprived condition and low temperature	Gao et al., 2015
miR402	*A. thaliana*	*A. thaliana*	Enhanced tolerance to drought, salinity, and cold stress	Kim et al., 2010b
miR408	*A. thaliana*	*Cicer arietinum*	Higher drought tolerance	Hajyzadeh et al., 2015
miR417	*A. thaliana*	*A. thaliana*	Increase in sensitivity to salinity and ABA	Jung and Kang, 2007

Studies have shown miR159 upregulation upon ABA treatment, which is a master switch in regulating osmotic stress responses in plants. Reyes and Chua (2007) carried out the functional analysis of miR159 in *Arabidopsis* by overexpressing it. They demonstrated that miR159 conquers the expression of MYB33 and MYB101 genes through RNA cleavage that are predicted targets of miR159 and makes plants less sensitive to ABA. While overexpression of its cleavage-resistant targets resulted in hypersensitivity to ABA. Furthermore, in rice overexpression of miR159 made plants more sensitive to heat stress that indicates knockdown of miR159 may provide resistance to heat stress in rice (Wang et al., 2012).

The role of another big miRNA family, miR169, conserved in almost all plant species, is determined in abiotic stress tolerance. In *Arabidopsis* overexpression of miR169 resulted in hypersensitive plants to nitrogen deficiency (Zhao et al., 2011). However, the major role of miR169 in drought tolerance was determined in transgenic tomato. Transgenic tomato overexpressing miR169 showed considerable improvement in tolerance to drought stress post one week of drought treatment (Zhang et al., 2011). Transgenic plants showed no symptom of drought whereas non-transgenic control plants exhibited noticeable symptoms of drought stress such as wilting and turgor loss and dehydration of leaves, etc. miR169 targets three nuclear factor Y (NFY) subunit genes (SlNF-YA1/2/3), which are suppressed under drought stress in tomato. Results also suggested overexpression of miR169 controls the stomatal opening and reduces the transpiration rate, which contributes to tolerance against drought stress (Zhang et al., 2011). The miR172 of soybean was able to improve water deficit and salinity tolerance, upon overexpression in *Arabidopsis* (Li, Wang, Zhang, and Li, 2016).

The expression of miR319 was reported induced under multiple abiotic stresses (Zhou et al., 2010). When miR319a of rice overexpressed in creeping bentgrass (*Agrostis stolonifera*), it substantially enhanced the drought and salinity tolerance of transgenic plants compared to wild type non-transgenic plants (Zhou et al., 2013). However, overexpression of miR319 in rice increased the cold tolerance of transgenic plants (Yang et al., 2013). Developed transgenic rice plants also showed a higher survival rate (~50%) in comparison to non-transgenic control plants (13% of survival rate) (Yang et al., 2013). The miR319 has target binding sites on teosinte branched 1, cycloidea, PCF1 (TCP) TFs, and the resulted outcome may be due to the inhibition of these target genes. These findings suggest that miR319 can be used as a potential target miRNA to engineer transgenic plants with improved tolerance against multiple abiotic stresses.

The expression of miR393 is reported responsive to salinity and alkalinity abiotic stresses. Transgenic studies done in *Arabidopsis* and rice by Gao and colleagues (2011) revealed that the overexpression of miR393 in transgenic *Arabidopsis* and rice plants made them more sensitive to salinity and alkalinity stresses (Gao et al., 2011). Which resulted in inhibition of seedling growth and root development in both the plant species (Gao et al., 2011). These outcomes indicate that suppression of miR393 can be useful in improving the tolerance of plants against different abiotic stresses.

miR394 is another highly conserved miRNAs and it was found differentially expressed under various abiotic stress conditions including drought, salinity, and iron and sulfate deficient conditions (Liu et al., 2008). Transgenic *Arabidopsis* plants overexpressing miR394 of soybean showed tolerance to drought stress (Ni, Hu, Jiang, and Zhang, 2012). Moreover, the leaf curling responsiveness (LCR) gene is a putative target of miR394 in *Arabidopsis*. A study was done in mutant *Arabidopsis*, which was the loss of function mutant for LCR, overexpression of miR394 resulted in improved tolerance to cold stress (Song et al., 2016).

Similarly, miR395 has been also reported for differential expression upon treatment of multiple abiotic stresses. Members of the miR395 family exhibited a change in the extent of target cleavage, due to a single nucleotide difference in their mature miRNA sequences. Elucidation of miR395 done by overexpressing it in *Arabidopsis* resulted in reduced tolerance to drought and salt stress in transgenic plants (Kim et al., 2010a). Another overexpression study performed in rapeseed established the role of miR395 in enhancing tolerance to cadmium stress (Zhang et al., 2013).

The miR396 was observed for its aberrant expression under salinity and alkalinity stress. Transgenic plants overexpressing miR396 showed diminished growth and development in *Arabidopsis* under salinity and alkalinity stresses (Gao et al., 2010). However, there was no change observed in phenotypes of these miR396-transgenic plants compared to control non-transgenic plants. Observations indicate that miR396 may be a negative regulator of plant genes which are responsible for neutralizing effects of salinity and alkalinity stresses.

In *Arabidopsis* levels of miR402 was determined higher under dehydration, salinity, and cold stresses compared to unstressed plants (Sunkar and Zhu, 2004; Planta et al., 2010b). In transgenic *Arabidopsis* plants, overexpression of miR402 enhanced the seed germination under dehydration, salinity, and cold stresses. Whereas enhancement in the plant growth was observed only under salinity stress in miR402-transgenic plants (Planta et al., 2010b).

Similarly, the expression of miR417 was found altered under salt and dehydration stresses (Jung and Kang, 2007). Transgenic *Arabidopsis* plants overexpressing miR417 reduced the seed germination and plant growth under high salt treatment given in the presence of ABA (Jung and Kang, 2007).

5.6 CONCLUSION AND FUTURE PERSPECTIVES

Sustainable development is required to manage the plant-based resources. However, changes in the environmental conditions impose abiotic stress, which significantly reduces the plant production. Major abiotic stresses include drought, salinity, heat, cold, nutrient deficiency, oxidative stress, oxygen deficiency, and hormone concentration induced-stress, etc. Moreover, not only reduction in the impact of abiotic stresses on crop production is needed but simultaneously improvement in the plant production is essential to fulfill the supply of increasing population. In order to mitigate the deleterious effects of abiotic stresses on crop productivity and improve the crop production by producing stress-tolerance plants, strategies beyond conventional and molecular plant breeding are required. To achieve those effective approaches can be designed for developing more tolerant plant varieties to sustain effects of abiotic stresses. Tolerant plant varieties can be engineered by manipulating expression of protein-coding genes or non-protein coding genes responsive to abiotic stresses. Increasing evidence suggests that miRNAs, a class of small non-protein coding RNAs, regulates plethora of genes involving in vital cellular pathways including abiotic stress-responsive pathways (Sanan-Mishra et al., 2009; Sonali and Jolly, 2014; Arnaud et al., 2017). Number of studies has shown altered expression of miRNAs in various plant species under multiple abiotic stresses. High throughput sequencing technologies have facilitated the enrichment of the miRNome expressed under different abiotic stress conditions (Khraiwesh, Zhu, and Zhu, 2012). *In silico,* prediction of putative targets for abiotic stress-responsive miRNAs has aided pace to their functional characterization. Additionally, transgenic approaches have also played crucial role in determining functions of the stress-responsive miRNAs in model as well as crop plants (Zhang, 2015; Baohong, Zhanga, and Qinglian, 2016). But still identified stress responsive-miRNAs are much more in numbers compared to functionally validated, therefore researchers should focus more on functionally establishing their role in regulating responses under different abiotic stress conditions rather than their profiling.

Gene-manipulation for engineering improved plant varieties for desired traits is on the rise across the globe and producing more qualitative and quantitative products (Zhang and Wang, 2015; Arnaud et al., 2017). Owing to the key role of miRNAs in the maintaining homeostasis in cellular pathways under abiotic stresses by titrating the expression of pathway components genes, make them potential candidates for engineering stress-tolerant crop plants using transgenic technologies. Transgenic plants developed by overexpressing specific miRNA(s) showed tolerance against specific types of abiotic stress. These findings suggest that miRNAs can serve as a suitable target for engineering abiotic stress-tolerant plants. However, unintended side effects should be considered before designing miRNA-based transgenic overexpression to obtain desired traits.

KEYWORDS

- **abiotic stress**
- **crop plants**
- **double-stranded RNA**
- **gene regulation**
- **MicroRNA**
- **non-coding RNA**

REFERENCES

Abdel-Ghany, S. E., & Pilon, M., (2008). MicroRNA-mediated systemic down-regulation of copper protein expression in response to low copper availability in *Arabidopsis*. *J. Biol. Chem., 283*, 15932–15945.

Agarwal, S., & Grover, A., (2006). Molecular biology. Biotechnology and genomics of flooding-associated low O_2 stress response in plants. *Crit. Rev. Plant Sci., 25*, 1–21.

Agustina, G., Lara, I. D., Raphael, S. M., Thaís, H. F., & Marcelo, M., (2015). MicroRNAs and drought responses in sugarcane. *Front Plant Sci., 6*, 58.

Ambros, V., (2004). The functions of animal microRNAs. *Nature, 431*(7006), 350–355.

Amit, K., Shuchi, S., Senthilkumar, K. M., Viswanathan, C., Dev, M. P., & Kailash, C. B., (2015). Identification of novel drought-responsive microRNAs and trans-acting siRNAs from *Sorghum* bicolor (L.) Moench by high-throughput sequencing analysis. *Front. Plant Sci.*, https://doi.org/10.3389/fpls.2015.00506 (accessed on 13 June 2020).

Arash, N., Zakaria, H. P., & Golam, F., (2013). Drought tolerance in wheat. *The Scientific World Journal.* Article ID 610721.

Arenas-Huertero, C., Pérez, B., Rabanal, F., Blanco-Melo, D., De La Rosa, C., Estrada-Navarrete, G., Sanchez, F., et al., (2009). Conserved and novel miRNAs in the legume *Phaseolus vulgaris* in response to stress. *Plant Molecular Biology, 70*, 385–401.

Aung, K., Lin, S. I., Wu, C. C., Huang, Y. T., Su, C. L., & Chiou, T. J., (2006). pho2, a phosphate over accumulator, is caused by a nonsense mutation in a microRNA399 target gene. *Plant Physiol., 141,* 1000–1011.

Ausubel, F. M., (2005). Are innate immune signaling pathways in plants and animals conserved? *Nat. Immunol., 6*(10), 973–979.

Axtell, M. J., Westholm, J. O., & Lai, E. C., (2011). Vive la difference: Biogenesis and evolution of microRNAs in plants and animals. *Genome Biol., 12*(4), 221.

Baohong, Z., & Qinglian, W., (2016). MicroRNA, a new target for engineering new crop cultivars. *Bioengineered, 7*(1), 7–10.

Baohong, Z., Xiaoping, P., George, P. C., & Todd, A. A., (2006). *Plant microRNA: A Small Regulatory Molecule with Big Impact* (Vol. 289, No. 1, pp. 3–16).

Bari, R., Datt, P. B., Stitt, M., & Scheible, W. R., (2006). PHO2 MicroRNA399 and PHR1 define a phosphate signaling pathway in plants. *Plant Physiol., 141*, 988–999.

Barrera-Figueroa, B. E., Gao, L., Diop, N. N., Wu, Z., Ehlers, J. D., Roberts, P. A., Close, T. J., Zhu, J. K., & Liu, R. Y., (2011). Identification and comparative analysis of drought-associated MicroRNAs in two cowpea genotypes. *BMC Plant Biology, 11*, 127.

Bartel, D. P., (2004). microRNAs: Genomics, biogenesis, mechanism, and function. *Cell, 116*(2), 281–297.

Baumberger, N., & Baulcombe, D. C., (2005). *Arabidopsis* ARGONAUTE1 is an RNA Slicer that selectively recruits microRNAs and short interfering RNAs. *Proc. Natl. Acad. Sci. US., 102*(33), 11928–11933.

Buhtz, A., Pieritz, J., Springer, F., & Kehr, J., (2010). Phloem small RNAs, nutrient stress responses, and systemic mobility. *BMC Plant Biol., 10*, 64.

Buhtz, A., Springer, F., Chappell, L., Baulcombe, D. C., & Kehr, J., (2008). Identification and characterization of small RNAs from the phloem of *Brassica napus. Plant J., 53*, 739–749.

Cao, X., Wu, Z., Jiang, F., Zhou, R., & Yang, Z., (2014). Identification of chilling stress-responsive tomato microRNAs and their target genes by high throughput sequencing and degradome analysis. *BMC Genomics, 15*, 1130.

Cramer, G. R., Urano, K., Delrot, S., Pezzotti, M., & Shinozaki, K., (2011). Effects of abiotic stress on plants: A systems biology perspective. *BMC Plant Biol., 11*, 163.

Cui, N., Sun, X., Sun, M., Jia, B., Duanmu, H., Lv, D., et al., (2015). Overexpression of OsmiR156k leads to reduced tolerance to cold stress in rice (*Oryza sativa*). *Mol. Breed., 35*, 214. doi: 10.1007/s11032-015-0402-6.

Dajana, L., Ghanasyam, R., Dominik, D. S., Cathie, M., & Jonathan, C., (2006). Serrate: A new player on the plant microRNA scene. *EMBO Rep., 7*(10), 1052–1058.

Ding, D., Zhang, L., Wang, H., Liu, Z., Zhang, Z., & Zheng, Y., (2009). Differential expression of miRNAs in response to salt stress in maize roots. *Annals of Botany, 103*, 29–38.

Djami-Tchatchou, A. T., Sanan-Mishra, N., Khayalethu, N., & Ian, A. D., (2017). Functional roles of microRNAs in agronomically important plants-potential as targets for crop improvement and protection. *Front Plant Sci., 8*, 378.

El Sanousi, R. S., Hamza, N. B., Abdelmula, A. A., Mohammed, I. A., Gasim, S. M., & Sanan-Mishra, N., (2016). Differential expression of miRNAs in *Sorghum* bicolor under drought and salt stress. *American Journal of Plant Sciences, 7*, 870–878.

Farooq, M., Wahid, A., Kobayashi, N., Fujita, D., & Basra, S. M. A., (2009). Plant drought stress: Effects, mechanisms and management. *Agronomy for Sustainable Development, 29*(1).

Fernandez, J. E., (2014). Understanding olive adaptation to abiotic stresses as a tool to increase crop performance. *Environmental and Experimental Botany, 103*, 158–179.

Frazier, T. P., Sun, G., Burklew, C. E., & Zhang, B., (2011). Salt and drought stresses induce the aberrant expression of microRNA genes in tobacco. *Molecular Biotechnology, 49*, 159–165.

Fridovich, I., (1995). Superoxide radical and superoxide dismutases. *Annual Review of Biochemistry, 64*, 97–112.

Fu, C., Sunkar, R., Zhou, C., et al., (2012). Over-expression of miR156 in switch grass (*Panicum virgatum* L.) results in various morphological alterations and leads to improved biomass production. *Plant Biotechnology Journal, 10*, 443–452.

Gao, P., Bai, X., Yang, L., Lv, D., Li, Y., Cai, H., Ji, W., Guo, D., & Zhu, Y., (2010). Overexpression of osa-MIR396c decreases salt and alkali stress tolerance. *Planta, 231*, 991–1001.

Gao, P., Bai, X., Yang, L., Lv, D., Pan, X., Li, Y., Cai, H., Ji, W., Chen, Q., & Zhu, Y., (2011). Osa-MIR393: A salinity-and alkaline stress-related microRNA gene. *Molecular Biology Reports, 38*, 237–242.

Gifford, M. L., Dean, A., Gutierrez, R. A., Coruzzi, G. M., & Birnbaum, K. D., (2008). Cell-specific nitrogen responses mediate developmental plasticity. *Proc. Natl. Acad. Sci. U.S.A., 105*, 803–808.

Grant, R. C., Kaoru, U., Serge, D., Mario, P., & Kazuo, S., (2011). Effects of abiotic stress on plants: A systems biology perspective. *BMC Plant Biol., 11*, 163.

Grigg, S. P., Canales, C., Hay, A., & Tsiantis, M., (2005). SERRATE coordinates shoot meristem function and leaf axial patterning in *Arabidopsis*. *Nature, 437*, 1022–1026.

Han M H., Goud, S., Song, L., & Fedoroff, N., (2004). The *Arabidopsis* double-stranded RNA-binding protein HYL1 plays a role in microRNA-mediated gene regulation. *Proc. Natl. Acad. Sci. USA, 101*, 1093–1098.

Hao, C., Zhuofu, L., & Liming, X., (2012). *A Plant microRNA Regulates the Adaptation of Roots to Drought Stress* (Vol. 586, No. 12).

Huang, S. Q., Xiang, A. L., Che, L. L., Chen, S., Li, H., Song, J. B., et al., (2010). Aset of miRNAs from *Brassica napus* in response to sulfate deficiency and cadmium stress. *Plant Biotechnol. J., 8*, 887–899.

Jia, X., Ren, L., Chen, Q. J., Li, R., & Tang, G., (2009). UV-B-responsive microRNAs in *Populus tremula*. *J. Plant Physiol., 166*, 2046–2057.

Jia, X., Wang, W. X., Ren, L., Chen, Q. J., Mendu, V., Willcut, B., Dinkins, R., Tang, X., & Tang, G., (2009). Differential and dynamic regulation of miR398 in response to ABA and salt stress in *Populus tremula* and *Arabidopsis thaliana*. *Plant Mol. Biol., 71*, 51–59.

Jian-Kang, Z., (2016). Abiotic stress signaling and responses in plants. *Cell*, 167.

Jones-Rhoades, M. W., & Bartel, D. P., (2004). Computational identification of plant microRNAs their targets, including a stress-induced miRNA. *Mol. Cell, 14*, 787–799.

Jones-Rhoades, M. W., Bartel, D. P., & Bartel, B., (2006). MicroRNAS and their regulatory roles in plants. *Annu. Rev. Plant Biol., 57*, 19–53.

Jung, H. J., & Kang, H., (2007). Expression and functional analyses of microRNA417 in *Arabidopsis thaliana* under stress conditions. *Plant Physiol. Biochem., 45*, 805–811.

Kantar, M., Lucas, S. J., & Budak, H., (2011). miRNA expression patterns of *Triticum dicoccoides* in response to shock drought stress. *Planta, 233*(3), 471–484.

Khraiwesh, B., Arif, M. A., Seumel, G. I., Ossowski, S., Weigel, D., Reski, R., & Frank, W., (2010). Transcriptional control of gene expression by microRNAs. *Cell, 140*, 111–122.

Khraiwesh, B., Zhu, J. K., & Zhu, J., (2012). Role of miRNAs and siRNAs in biotic and abiotic stress responses of plants. *Biochim. Biophys. Acta, 1819*(2), 137–148.

Kim, J. Y., Kwak, K. J., Jung, H. J., Lee, H. J., & Kang, H., (2010). MicroRNA402 affects seed germination of *Arabidopsis thaliana* under stress conditions via targeting demeter-like Protein3 mRNA. *Plant and Cell Physiology, 51*, 1079–1083.

Kim, J., Lee, H., Jung, H., Maruyama, K., Suzuki, N., & Kang, H., (2010). Overexpression of microRNA395c or 395e affects differently the seed germination of *Arabidopsis thaliana* under stress conditions. *Planta, 232*, 1447–1454.

Kong, W. W., & Yang, Z. M., (2010). Identification of iron-deficiency responsive microRNA genes and cis-elements in *Arabidopsis*. *Plant Physiology and Biochemistry, 48*, 153–159.

Kozomara, A., & Griffiths-Jones, S., (2014). miRBase: Annotating high confidence microRNAs using deep sequencing data. *Nucleic Acids Res., 42* (Database issue), D68–73.

Kulcheski, F. R., De Oliveira, L. F., Molina, L. G., Almerao, M. P., Rodrigues, F. A., Marcolino, J., et al., (2011). Identification of novel soybean microRNAs involved in abiotic and biotic stresses. *BMC Genomics, 12*, 307.

Kulcheski, F. R., De Oliveira, L. F., Molina, L. G., Almerão, M. P., Rodrigues, F. A., Marcolino, J., Barbosa, J. F., et al., (2011). Identification of novel soybean microRNAs involved in abiotic and biotic stresses. *BMC Genomics, 12*, 307.

Kurihara, Y., Takashi, Y., & Watanabe, Y., (2006). The interaction between DCL1 and HYL1 is important for efficient and precise processing of pri-miRNA in plant microRNA biogenesis. *RNA, 12*(2), 206–212. Park.

Lee, R. C., Feinbaum, R. L., & Ambros, V., (1993). The *C. elegans* heterochronic gene lin-4 encodes small RNAs with antisense complementarity to lin-14. *Cell, 75*(5), 843–854.

Li, S., Castillo-González, C., Yu, B., & Zhang, X., (2017). The functions of plant small RNAs in development and in stress responses. *Plant J., 90*(4), 654–670. doi: 10.1111/tpj.13444. Epub 20th February 2017.

Li, T., Li, H., Zhang, Y. X., & Liu, J. Y., (2010). Identification and analysis of seven H_2O_2-responsive miRNAs and 32 new miRNAs in the seedlings of rice (*Oryza sativa* L. ssp. indica). *Nucleic Acids Res.*

Li, W. X., Oono, Y., Zhu, J., He, X. J., Wu, J. M., Iida, K., Lu, X. Y., Cui, X., Jin, H., & Zhu, J. K., (2008). The *Arabidopsis* NFYA5 transcription factor is regulated transcriptionally and posttranscriptionally to promote drought resistance. *Plant Cell, 20*, 2238–2251.

Li, W., Wang, T., Zhang, Y., & Li, Y., (2016). Over-expression of soybean miR172c confers tolerance to water deficit and salt stress, but increases ABA sensitivity in transgenic *Arabidopsis thaliana. J. Exp. Bot., 67*, 175–194.

Liang, G., Ai, Q., & Yu, D., (2015). Uncovering miRNAs involved in crosstalk between nutrient deficiencies in *Arabidopsis*. *Sci. Rep., 5*, 11813.

Liu, H. H., Tian, X., Li, Y. J., Wu, C. A., & Zheng, C. C., (2008). Microarray-based analysis of stress-regulated MicroRNAs in *Arabidopsis thaliana. RNA, 14*, 836–843.

Liu, M., Yu, H., Zhao, G., Huang, Q., Lu, Y., & Ouyang, B., (2018). Identification of drought-responsive microRNAs in tomato using high-throughput sequencing. *Funct. Integr. Genomics, 18*(1), 67–78.

Liu, P. P., Montgomery, T. A., Fahlgren, N., Kasschau, K. D., Nonogaki, H., & Carrington, J. C., (2007). Repression of auxin response factor10 by microRNA160 is critical for seed germination and post germination stages. *Plant J., 52*, 133–146.

Liu, Q., Zhang, Y. C., Wang, C. Y., Luo, Y. C., Huang, Q. J., Chen, S. Y., Zhou, H., Qu, L. H., & Chen, Y. Q., (2009). Expression analysis of phytohormone-regulated microRNAs in rice, implying their regulation roles in plant hormone signaling. *FEBS Lett., 583*, 723–728.

Lobell, D. B., Schlenker, W., & Costa-Roberts, J., (2011). Climate trends and global crop production since 1980. *Science, 333*(6042), 616–620.

Lu, S. F., Sun, Y. H., & Chiang, V. L., (2008). Stress-responsive microRNAs in populous. *Plant Journal, 55*, 131–151.

Lu, X., Y., & Huang, X. L., (2008). Plant miRNAs and abiotic stress responses. *Biochemical and Biophysical Research Communications, 368*, 458–462.

Lv, D. K., Bai, X., Li, Y., Ding, X. D., Ge, Y., Cai, H., Ji, W., Wu, N., & Zhu, Y. M., (2010). Profiling of cold-stress-responsive miRNAs in rice by microarrays. *Gene, 459*(1/2), 39–47.

Marschner, H., (1995). *Mineral Nutrition of Higher Plants* (2nd edn.). London: Academic Press.

Mathur, S., Agrawal, D., & Jajoo, A., (2014). Photosynthesis: Response to high temperature stress. *Journal of Photochemistry and Photobiology B Biology, 137*, 116–126.

Mittler, R., Vanderauwera, S., Gollery, M., & Van, B. F., (2004). Reactive oxygen gene network of plants. *Trends Plant Sci., 9*, 490–498.

Mohammad, A. A., Nudrat, A. A., Muhammad, A., Mohammed, N. A., Leonard, W., & Parvaiz, A., (2017). Plant responses to environmental stresses—from gene to biotechnology. *AoB Plants* (Vol. 9, No. 4).

Moldovan, D., Spriggs, A., Yang, J., Pogson, B. J., Dennis, E. S., & Wilson, I. W., (2009). Hypoxia-responsive microRNAs and trans-acting small interfering RNAs in *Arabidopsis. J. Exp. Bot., 61*, 165–177.

Ni, Z., Hu, Z., Jiang, Q., & Zhang, H., (2012). Overexpression of gma-MIR394a confers tolerance to drought in transgenic *Arabidopsis thaliana. Biochemical and Biophysical Research Communications, 427*, 330–335.

Pant, B. D., Buhtz, A., Kehr, J., & Scheible, W. R., (2008). MicroRNA399 is a long-distance signal for the regulation of plant phosphate homeostasis. *Plant J., 53*, 731–738.

Park, M. Y., Wu, G., Gonzalez-Sulser, A., Vaucheret, H., & Poethig, R. S., (2005). Nuclear processing and export of microRNAs in *Arabidopsis. Proc. Natl. Acad. Sci. USA., 102*(10), 3691–3696.

Park, M. Y., Wu, G., Gonzalez-Sulser, A., Vaucheret, H., & Poethig, R. S., (2005). Nuclear processing and export of microRNAs in *Arabidopsis. Proc. Natl. Acad. Sci. US., 102*(10), 3691–3696.

Phillips, J. R., Dalmay, T., & Bartels, D., (2007). The role of small RNAs in abiotic stress. *FEBS Letters, 581*, 3592–3597.

Qi, Y., Denli, A. M., & Hannon, G. J., (2005). Biochemical specialization within *Arabidopsis* RNA silencing pathways. *Mol. Cell, 19*(3), 421–428.

Reinhart, B. J., Weinstein, E. G., Rhoades, M. W., Bartel, B., & Bartel, D. P., (2002). microRNAs in plants. *Genes Dev., 16*(13), 1616–1626.

Ren, G., Xie, M., Dou, Y., Zhang, S., Zhang, C., & Yu, B., (2012). Regulation of miRNA abundance by RNA binding protein TOUGH in *Arabidopsis. Proc. Natl. Acad. Sci. USA., 109*(31), 12817–12821.

Reyes, J. L., & Chua, N. H., (2007). ABA induction of miR159 controls transcript levels of two MYB factors during *Arabidopsis* seed germination. *Plant J., 49*, 592–606.

Riedmann, L. T., & Schwentner, R., (2010). miRNA, siRNA, piRNA and argonautes: News in small matters. *RNA Biol., 7*(2), 133–139.

Rogers, K., & Chen, X., (2013). Biogenesis, turnover, and mode of action of plant microRNAs. *Plant Cell, 25*(7), 2383–2399.

Rubio, V., Linhares, F., Solano, R., Martin, A. C., Iglesias, J., Leyva, A., & Paz-Ares, J., (2001). A conserved MYB transcription factor involved in phosphate starvation signaling both in vascular plants and in unicellular algae. *Genes Dev., 15*, 2122–2133.

Sanan-Mishra, N., Kumar, V., Sopory, S. K., & Mukherjee, S. K., (2009). Cloning and validation of novel miRNA from basmati rice indicates cross talk between abiotic and biotic stresses. *Mol. Genet. Genomics, 282*(5), 463–474.

Saroj, K. S., Kambham, R. R., & Jiaxu, L., (2016). Abscisic acid and abiotic stress tolerance in crop plants. *Front Plant Sci., 7*, 571.

Satendra, K. M., Sailaja, B., Surekha, A., Vishnu, V. P., Voleti, S. R., Sarla, N., & Desiraju, S., (2017). Genome-wide changes in microRNA expression during short and prolonged heat stress and recovery in contrasting rice cultivars. *J. Exp. Bot., 68*(9), 2399–2412.

Scandalios, J. G B., (2005). Oxidative stress: Molecular perception and transduction of signals triggering antioxidant gene defenses. *J. Med. Biol. Res., 38*(7), 995–1014.

Seki, M., Narusaka, M., Ishida, J., Nanjo, T., Fujita, M., Oono, Y., Kamiya, A., et al., (2002). Monitoring the expression profiles of 7000 *Arabidopsis* genes under drought, cold, and high-salinity stresses using a full length cDNA microarray. *Plant J., 31*, 279–292.

Shiozaki, N., Yamada, M., & Yoshiba, Y., (2005). Analysis of salt-stress inducible ESTs isolated by PCR subtraction in salt-tolerant rice. *Theor. Appl. Genet., 110*, 1177–1186.

Shukla, L. I., Chinnusamy, V., & Sunkar, R., (2008). The role of MicroRNAs and other endogenous small RNAs in plant stress responses. *Biochimica etBiophysica Acta, 1779*, 743–748.

Sonali, B., & Jolly, B., (2014). MicroRNAs: The potential biomarkers in plant stress response. *American Journal of Plant Sciences, 5*, 748–759.

Song, J. B., Gao, S., Wang, Y., Li, B. W., Zhang, Y. L., & Yang, Z. M., (2016). miR394 and its target gene LCR are involved in cold stress response in *Arabidopsis*. *Plant Gene, 5*, 56–64. doi: 10.1016/j.plgene.2015.12.001.

Stief, A., Altmann, S., Hoffmann, K., Pant, B. D., Scheible, W. R., & Bäurle, I., (2014). *Arabidopsis* miR156 regulates tolerance to recurring environmental stress through SPL transcription factors. *The Plant Cell, 26*, 1792–1807.

Sun, X., Xu, L., Wang, Y., Yu, R., Zhu, X., Luo, X., et al., (2015). Identification of novel and salt-responsive miRNAs to explore miRNA-mediated regulatory network of salt stress response in radish (*Raphanus sativus* L.). *BMC Genomics, 16*, 197. doi: 10.1186/s12864-015-1416-5.

Sunkar, R., & Zhu, J. K., (2004). Novel and stress-regulated microRNAs and other small RNAs from *Arabidopsis*. *Plant Cell, 16*, 2001–2019.

Sunkar, R., Chinnusamy, V., Zhu, J., & Zhu, J. K., (2007). Small RNAs as big players in plant abiotic stress responses and nutrient deprivation. *Trends Plant Sci., 12*(7), 301–309.

Sunkar, R., Kapoor, A., & Zhu, J. K., (2006). Posttranscriptional induction of two Cu/Zn superoxide dismutase genes in *Arabidopsis* is mediated by down regulation of miR398 and important for oxidative stress tolerance. *Plant Cell, 18*, 2051–2065.

Suzuki, N., Rivero, R. M., Shulaev, V., Blumwald, E., & Mittler, R., (2014). Abiotic and biotic stress combinations. *New Phytologist, 203*, 32–43.

Tang, G., Reinhart, B. J., Bartel, D. P., & Zamore, P. D., (2003). A biochemical framework for RNA silencing in plants. *Genes Dev., 17*(1), 49–63.

Trindade, I., Capitão, C., Dalmay, T., Fevereiro, M. P., & Santos, D. M., (2010). miR398 and miR408 are up-regulated in response to water deficit in *Medicago truncatula*. *Planta, 231*, 705–716.

Varsha, S., (2016). Vinay, K., Rachayya, M. D., Tushar, S. K., & Shabir, H. W. *MicroRNAs As Potential Targets for Abiotic Stress Tolerance in Plants*.

Vaucheret, H., Vazquez, F., Crete, P., & Bartel, D. P., (2004). The action of argonaute1 in the miRNA pathway and its regulation by the miRNA pathway are crucial for plant development. *Genes Dev., 18*(10), 1187–1197.

Vazquez, F., Gasciolli, V., Crété, P., & Vaucheret, H., (2004). The nuclear dsRNA binding protein HYL1 is required for microRNA accumulation and plant development, but not posttranscriptional transgene silencing. *Curr. Biol., 14*, 346–351.

Wang, B., Sun, Y. F., Song, N., Wang, X. J., Feng, H., Huang, L. L., et al., (2013). Identification of UV-B-induced microRNAs in wheat. *Genet. Mol. Res., 12*, 4213–4221.

Wang, W., Vinocur, B., & Altman, A., (2003). Plant responses to drought, salinity, and extreme temperatures: Towards genetic engineering for stress tolerance. *Planta, 218*, 1–14.

Wang, Y., Sun, F., Cao, H., Peng, H., Ni, Z., Sun, Q., & Yao, Y., (2012). TamiR159 directed wheat TaGAMYB cleavage and its involvement in another development and heat response. *PLoS One, 7*, e48445.

Wani, S. H., Sah, S. K., Hussain, M. A., Kumar, V., & Balachandra, S. M., (2016). Transgenic approaches for abiotic stress tolerance in crop plants. In: Al-Khayri, J. M., Jain, S. M., & Johnson, D. V., (eds.), *Advances in Plant Breeding Strategies: Agronomic, Abioticand Biotic Stress Traits* (Vol. 2, pp. 345–396). Gewerbestrasse: Springer International.

Wei, L., Zhang, D., Xiang, F., & Zhang, Z., (2009). Differentially expressed miRNAs potentially involved in the regulation of defense mechanism to drought stress in maize Seedlings. *Int. J. Plant Sci., 170*, 979–989.

Wu, L., Zhou, H., Zhang, Q., Zhang, J., Ni, F., Liu, C., & Qi, Y., (2010). DNA methylation mediated by a microRNA pathway. *Mol Cell, 38*(3), 465–475.

Xin, M., Wang, Y., Yao, Y., Xie, C., Peng, H., Ni, Z., & Sun, Q., (2010). Diverse set of microRNAs are responsive to powdery mildew infection and heat stress in wheat (*Triticum aestivum* L.). *BMC Plant Biology, 10*, 123.

Yamaguchi-Shinozaki, K., & Shinozaki, K., (2006). Transcriptional regulatory networks in cellular responses and tolerance to dehydration and cold stresses. *Annu. Rev. Plant. Biol., 57*, 781–803.

Yamasaki, H., Abdel-Ghany, S. E., Cohu, C. M., Kobayashi, Y., Shikanai, T., & Pilon, M., (2007). Regulation of copper homeostasis by microRNA in *Arabidopsis*. *J. Biol. Chem., 282*, 16369–16378.

Yang, C. H., Li, D. Y., Mao, D. H., Liu, X., Ji, C. J., et al., (2013). Over-expression of microRNA319 impacts leaf morphogenesis and leads to enhanced cold tolerance in rice (*Oryza sativa* L.). *Plant Cell Environ., 36*, 2207–2218.

Yu, B., Yang, Z., Li, J., Minakhina, S., Yang, M., Padgett, R. W., et al., (2005). Methylation as a crucial step in plant microRNA biogenesis. *Science, 307*(5711), 932–935.

Zhang, B., & Wang, Q., (2015). MicroRNA-based biotechnology for plant improvement. *J. Cell Physiol., 230*, 1–15. doi: 10.1002/jcp.24685.

Zhang, B., (2015). MicroRNA: A new target for improving plant tolerance to abiotic stress. *J. Exp. Bot., 66*(7), 1749–1761.

Zhang, J. W., Long, Y., Xue, M. D., Xiao, X. G., & Pei, X. W., (2017). Identification of microRNAs in response to drought in common wild rice (*Oryza rufipogon* Griff.) shoots and roots. *PLoS One, 12*(1), e0170330. https://doi.org/10.1371/journal.pone.0170330 (accessed on 13 June 2020).

Zhang, J., Xu, Y., Huan, Q., & Chong, K., (2009). Deep sequencing of *Brachypodium* small RNAs at the global genome level identifies microRNAs involved in cold stress response. *BMC Genomics, 10*, 449.

Zhang, L. W., Song, J. B., Shu, X. X., Zhang, Y., & Yang, Z. M., (2013). miR395 is involved in detoxification of cadmium in *Brassica napus*. *Journal of Hazardous Materials, 250, 251*, 204–211.

Zhang, X., Zou, Z., Gong, P., Zhang, J., Ziaf, K., Li, H., Xiao, F., & Ye, Z., (2011). Over-expression of microRNA 169 confers enhanced drought tolerance to tomato. *Biotechnology Letters, 33*, 403–409.

Zhang, Y., Zhu, X., Chen, X., Song, C., Zou, Z., Wang, Y., et al., (2014a). Identification and characterization of cold-responsive microRNAs in tea plant (*Camellia sinensis*) and their targets using high-throughput sequencing and degradome analysis. *BMC Plant Biol., 14*, 271.

Zhang, Z., Wei, L., Zou, X., Tao, Y., Liu, Z., & Zheng, Y., (2008). Submergence-responsive MicroRNAs are potentially involved in the regulation of morphological and metabolic adaptations in maize root cells. *Ann. Bot., 102*, 509–519.

Zhao, B. T., Liang, R. Q., Ge, L. F., Li, W., Xiao, H. S., Lin, H. X., Ruan, K. C., & Jin, Y. X., (2007). Identification of drought-induced MicroRNAs in rice. *Biochemical and Biophysical Research Communications, 354*, 585–590.

Zhao, M., Ding, H., Zhu, J. K., Zhang, F., & Li, W. X., (2011). Involvement of miR169 in the nitrogen-starvation responses in *Arabidopsis*. *New Phytol., 190*(4), 906–915.

Zhou, L., Liu, Y., Liu, Z., Kong, D., Duan, M., & Luo, L., (2010). Genome-wide identification and analysis of drought-responsive microRNAs in *Oryza sativa*. *Journal of Experimental Botany, 61*, 4157–4168.

Zhou, M., & Luo, H., (2013). MicroRNA-mediated gene regulation: Potential application for plant genetic engineering. *Plant Mol. Biol., 83*, 59–75.

Zhou, M., Li, D., Li, Z., Hu, Q., Yang, C., Zhu, L., & Luo, H., (2013). Constitutive expression of a miR319 gene alters plant development and enhances salt and drought tolerance in transgenic creeping bent grass. *Plant Physiology, 161*, 1375–1391.

Zhou, X., Wang, G., & Zhang, W., (2007). UV-B responsive microRNA genes in *Arabidopsis thaliana*. *Mol. Syst. Biol., 3*, 103.

Zhu, J. K., (2002). Salt and drought stress signal transduction in plants. *Annu. Rev. Plant Biol., 53*, 247–273.

Zhu, J. K., (2007). *Plant Salt Stress*. Encyclopedia of life sciences. doi: 10.1002/9780470015902.

CHAPTER 6

Molecular Approaches for Engineering Plants Resistant to Viruses

NIRBHAY KUMAR KUSHWAHA[1], SHWETA ROY[2], MANSI[3], HASTHI RAM[4], and PRAVEEN SONI[5*]

[1]*Department of Plant Biology, Swedish University of Agricultural Sciences, Uppsala-75007, Sweden*

[2]*Department of Cell and Molecular Biology, Uppsala University, Uppsala-75236, Sweden*

[3]*ICAR-National Institute for Plant Biotechnology, Indian Agricultural Research Institute, New Delhi-110012, India*

[4]*National Agri-Food Biotechnology Institute, Mohali-140306, India*

[5]*Department of Botany, University of Rajasthan, Jaipur-302004, India*

[*]*Corresponding author. E-mail: praveen.soni15@gmail.com, praveensoni@uniraj.ac.in*

ABSTRACT

Plant viruses are one of the devastating pathogens and major constraints to several crops. They account for huge economic losses worldwide. The conventional method of breeding has limited success in diminishing the impact of plant viruses. Because of the vast variation in the genetic compositions of different viruses, it is difficult to develop a plant resistant to all viruses. An effective and efficient antiviral strategy can be designed only by understanding the molecular mechanism of pathogenesis of viruses. Efforts have been done to raise virus-resistant crops by genetic engineering. Several virus-derived genes such as those encoding mutated forms of viral movement protein, replicase protein, coat protein, and interfering RNA have been utilized in this regard. Moreover, host-derived genes such as R-genes, microRNAs, and other defense-related genes have also been employed successfully. In recent years, extensive research has generated massive information about host-virus interaction and host factors involved in viral replication, transcription, and pathogenesis. New viral strains have also been identified and their genomes have also been sequenced using advanced tools of next generation sequencing (NGS) and metagenomics. The emergence of the CRISPR/Cas9 method has also revolutionized the approaches for developing virus-resistant plants. The high sensitivity of NGS coupled with the specificity of the CRISPR/Cas9 based method has emerged as one of the most successful and popular biotechnological tools to engineer virus-resistant plants. Because of its tremendous potential, the CRISPR/Cas9 method will have extensive applications in the field of plant virology. In this chapter, we have explained the molecular mechanisms of plant defense against viruses. The application of different approaches to raise virus-resistant plants has also been described.

6.1 INTRODUCTION

Plant viruses are known to infect several crops and cause huge economic loss each year. Viruses encode limited proteins by their small genomes and therefore, heavily depend on host proteins and intracellular machinery for their replication, transcription, and movement to establish a successful infection. Plants are equipped with several lines of defense mechanisms that are specialized in the inhibition of virus multiplication. Similarly, for a counter-attack, viruses have evolved proteins that are capable of suppressing plant defense mechanisms. Thus, the development of the disease is a complex interaction between the array of proteins encoded by hosts and viruses. In fields, crops are usually targeted by several viruses at a time. The cumulative effect of suppressor proteins encoded by viruses may dominate over the defense mechanism of the crop that ultimately leads to the development of disease in the infected crop. With time, plants have also evolved themselves to cope with multiple virus attacks. Plants employ different defense mechanisms for viruses and target several steps of viral entry and multiplication. Plants have evolved virus-specific R-genes. Each R-gene encoded protein recognizes a specific virus and inhibits its propagation. Furthermore, transcriptional and post-transcriptional gene silencing mechanisms operative against viruses increase the plant fortification. The details of the molecular mechanisms of plant defense are described below.

6.1.1 MOLECULAR MECHANISM OF PLANT DEFENSE AGAINST VIRUSES

Plants have evolved either RNA or protein-mediated defense mechanisms against viruses (Baulcombe 2004; Meister and Tuschl 2004; Jones and Dangl, 2006; Eamens et al., 2008; Bhattacharjee et al., 2009; Dodds and Rathjen, 2010; Schwessinger and Ronald, 2012). The first mechanism involves the generation of small interfering RNAs (siRNAs) of 21–24 nucleotides that trigger either the degradation of viral genome RNA (post-transcriptional gene silencing-PTGS) or methylation of viral DNA (transcriptional gene silencing-TGS) (Brodersen et al., 2008; Matzke et al., 2009; Haag and Pikaard 2011; Figure 6.1). During post-transcriptional gene silencing, the RNA genome of plant viruses is copied into complementary dsRNA by viral RNA-dependent RNA polymerases (RDR) (Figure 6.1A). The dsRNA is subsequently cleaved into siRNAs by a host ribonuclease III-like protein, known as Dicer-like (DCL) (Hamilton and Baulcombe, 1999; Hutvagner et al., 2001; Borsani et al., 2005). The siRNAs are directed to an RNA-included silencing complex (RISC) containing Argonaute (AGO) protein. Subsequently, RISC complex with siRNA and Ago protein is guided to target RNAs. They are consequently degraded by the RISC complex (Sontheimer and Carthew, 2005; Fabian et al., 2010).

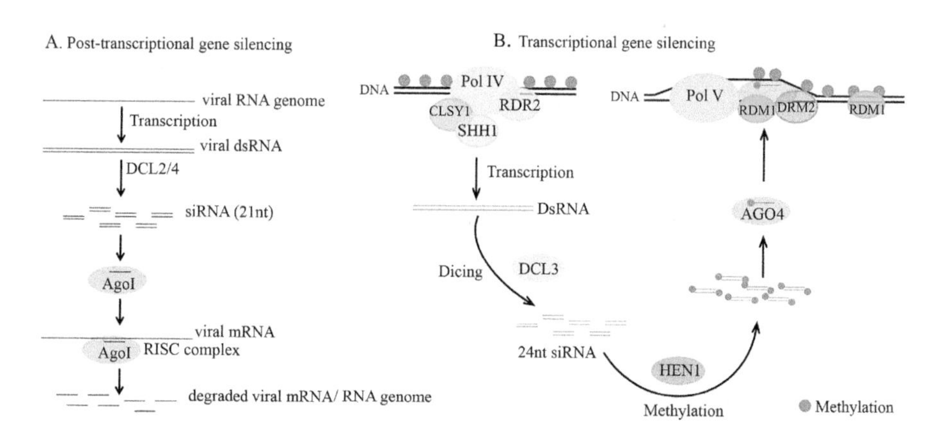

FIGURE 6.1 Mechanism of (A) post-transcriptional and (B) transcriptional gene silencing.

In the case of transcriptional gene silencing (TGS), DNA methylation acts as an epigenetic repressive mark that negatively regulates gene expression (Figure 6.1B). Plant DNA viruses such as geminiviruses are victims of the host transcriptional gene silencing machinery (Raja et al., 2008). During TGS, methylated DNA is transcribed by RNA Pol IV and V in coordination with RNA Pol II to produce non-coding RNA (Herr et al., 2005; Kanno et al., 2005; Onodera et al., 2005; Pikaard et al., 2008; Ream et al., 2009; Haag et al., 2011). In the next step RDR2 converts non-coding RNAs into dsRNAs, which are subsequently processed into 24 nucleotides siRNAs by the DCL3 protein (Figure 6.1B). The siRNAs thus produced are recruited by RISC complex containing AGO4 or AGO6. The resultant complex along with RNA Pol V is paired with the target DNA region and eventually recruits Domains Rearranged Methyltransferase 2 (DRM2) to set off DNA methylation. The symmetric methylation is maintained throughout DNA replication by Methyltransferase 1 (Liu et al., 2012).

Plants have evolved specific R-genes to countermeasure the attack of plant viruses. Most of the R-genes are dominant but a few recessive ones have also been reported. Proteins encoded by R-genes generally contain leucine-rich repeat domains and a binding site for nucleotide. Their N-termini may contain putative coiled-coil or toll-interleukin-1 receptor-like domains (Collier et al., 2011). Several R-genes encoded proteins recognize the Avr protein encoded by viral pathogen and the interaction triggers a defense response to limit the viral attack. Another gene called the 'N' gene is also related to defense against viruses. The protein encoded the 'N' gene in tobacco interacts with the replicase protein of *Tobacco mosaic virus* (TMV) and restricts infection. In *Solanum tuberosum*, Rx1 protein recognizes the CP of *Potato virus X* (PVX) and activates defense response against it. The HRT1 protein encoded by *Arabidopsis thaliana* confers resistance against *Cucumber mosaic virus* (CMV) through coat protein recognition. Recessive R-genes encode translation initiation factors (eIF4E/eIF4G) that confer resistant against *Turnip crinkle virus* (TCV) (Yoshii et al., 1998), *Barley yellow mosaic virus* (BYMV) (Kanyuka et al., 2005; Pellio et al., 2005), CMV (Yoshii et al., 2004), *Melon necrotic spot virus* (MNSV) (Nieto et al., 2006), and *Potyviruses* (Kang et al., 2005; Truniger and Aranda, 2009).

Apart from R-genes, TGS, and PTGS, several other defense mechanisms have also been reported in plants against viruses. In one of the studies, *Solanum lycopersicum* remorin protein interacted with a movement protein of PVX named Triple Gene Block Protein1, and the interaction led to the prevention of virus movement (Raffaele, et al., 2009). A hexose transporter LeHT1 conferred natural protection to *Tomato yellow leaf curl virus* (TYLCV) (Eybishtz et al., 2010).

6.2 MOLECULAR APPROACHES USED TO ENGINEER VIRUS-RESISTANT CROPS

Understanding the molecular mechanism of resistance provides crucial information which can be further utilized for engineering virus-resistant crops. In the past few decades, several approaches have been attempted to develop virus-resistant crops. Plant biologists have employed RNAi and CRISPR/Cas9 based strategies to engineer virus-resistant plants.

6.2.1 NON-VIRAL GENES USED TO ENGINEER VIRUS-RESISTANT CROPS

As mentioned earlier several R-genes confer natural defense against viruses (Albar et al., 2006; Bhattacharjee et al., 2009). They are a prominent source of defense genes for engineering virus-resistant plants using biotechnological approaches. Since R-genes are naturally present in plants, they are considered as the safest candidate genes for improving the immunity of susceptible plants against virus attack without causing any abnormal effects on plant growth and development and human health. R-gene encoded protein recognizes the Avr protein of the pathogen and the Avr/R protein interaction triggers activation of defense response involving salicylic acid, jasmonic acid, reactive oxygen species, and nitric oxide. In most of the cases, the activated defense response is manifested as a hypersensitive response that restricts the virus spread (Culver and Padmanabhan, 2007; Carr and Loebenstein, 2010; Pallas and García, 2011). The identification and characterization of N-gene from *Nicotiana glutinosa* opened a new direction to develop R-gene based virus-resistant crop. The reaction between the N-gene product and TMV replicase results in a hypersensitive response that restricts the TMV effect at the infected site. The N-gene has been used to develop transgenic tomato (*Lycopersicum esculentum*) and tobacco (*N. tabacum*) plants to improve resistance against TMV (Holmes, 1937; Holmes, 1954; Whitham et al., 1994; Whitham et al., 1996). In another case, transgenic tomato plants expressing peptide aptamer showed reduced infection of TYLCV or *Tomato mottle virus*. Enhanced resistance is achieved by the interaction between peptide aptamer and viral Rep protein (Reyes et al., 2013).

Bacterial genes have also been used to develop virus-resistant crops. *Bacillus amyloliquefaciens* genes encoding barnase and barstar proteins were introduced under the promoter of *African cassava mosaic virus*

(ACMV). During ACMV infection, the expression of barnase and barstar genes was induced that increased the active RNase and subsequently caused cell death at infected sites, which restricted the viral spread (Zhang and Gruissem, 2003).

6.2.2 PATHOGEN-DERIVED RESISTANCE

This approach depends on engineering plants expressing viral proteins that interfere with any of the steps of viral pathogenesis. This method is popularly known as pathogen-derived resistance (PDR). It can be either protein or nucleic acid-derived resistance. When designing a PDR-mediated antiviral strategy, it should always be considered that the viral protein selected for PDR should not interfere with the essential host function. Coat protein, movement protein, and replicase protein of viruses have been utilized for PDR to develop virus-resistant plants. RNAi-mediated resistance against the viral genome has been also used.

6.2.2.1 COAT PROTEIN-MEDIATED VIRAL RESISTANCE

Coat protein (CP) of most of the viruses functions as a structural protein that is involved in the assembly and transmission of the virus. It is also reported that transgenic expression of CP can interfere with viral movement protein hence it negatively affects cellular movement of the virus (Bendahmane et al., 2002). Additionally, it has been also found that, when expressed transgenically, CP can also interfere with virus assembly (Asurmendi et al., 2004). In some cases of CP-mediated resistance, CP subunits re-coat the nascent disassembled viral genome thus decreasing the availability of viral RNA for translation (Lu et al., 1998; Beachy, 1999). The gene encoding the coat protein of TMV was used to make transgenic tobacco plants with improving resistance to TMV and its closely related viruses (Powell et al., 1986; Beachy, 1999). Transgenically expressed CP of TMV showed delayed appearance of symptoms and reduced viral titers. In another case, two copies of the CP gene of *Potato virus Y* (PVY) were introduced into the potato and the resultant transgenic potato plant showed resistance to infections by PVY in two successive field tests (Malnoe et al., 1994). The CP gene of *Papaya ringspot virus* (PRSV) was ectopically expressed in papaya plants. They showed resistance against the severe PRSV Hawaii isolate (Tennant et al., 1994). The resistance was manifested as a delay in the symptom development and symptom attenuation with PRSV Hawaii isolates. Likewise, several other virus-resistant crops have been engineered based on CP (Table 6.1).

6.2.2.2 REPLICASE PROTEIN-MEDIATED VIRAL RESISTANCE

Plant viruses encode replicase protein that is involved in viral replication. The *Rep* gene encoding this protein has been utilized to develop virus-resistant plants. It is postulated that the transgenic expression of the *Rep* gene interferes with the virus replication probably due to the binding of replicase protein to host factors or other viral proteins that are indispensable for viral replication (Beachy, 1997). Plant biologists have used full-length, truncated or mutated *Rep* genes from various viruses. *Nicotiana tabacum* plants expressing nucleotides from 3472 to 4916 of TMV strain U1 encoding an important protein of the replicase complex, showed resistance to infection with RNA and virions of TMV U1 (Table 6.1) (Golemboski et al.,1990). Similarly, the transgenic *Nicotiana tabacum* plants expressing a truncated replicase protein of CMV exhibited improved resistance to 12 CMV strains (Zaitlin et al., 1994). Similarly, transgenic *Nicotiana tabacum* 'Turkish Samsun NN' plants containing *nuclear inclusion b (NIb)* gene sequence of O strain of *Potato virus Y* (PVYO) showed resistance to PVYO infection (Audy et al., 1994).

6.2.2.3 MOVEMENT PROTEIN-BASED VIRAL RESISTANCE

For the establishment of successful infection, viruses have to move from the area of infection to other parts of the host. To achieve the goal of systemic infection, plant viruses encode movement protein (MP) that facilitate the inter-cellular and systemic movement of the viruses (Table 6.1). Movement proteins regulate the size of the plasmodesmata by modulating the size exclusion limit and enable the trafficking of the viral nucleoprotein complex. A mutated version of movement protein reduces the virus movement and, thus, the infectivity. This strategy was used to engineer several virus-resistance crops. Plants expressing a mutated 30 KDa protein of TMV showed enhanced resistant to TMV and several other viruses like *Peanut chlorotic virus, Alfalfa mosaic virus*, CMV, *Tobacco*

TABLE 6.1 The application of approaches based on RNAi mechanism and targeting to viral proteins for plant protection against viruses.

Protein targeted/ Method	Plant	Source virus	Genome	Family	Virus resistance	References
CP	Tobacco	TMV	ssRNA	*Virgaviridae*	TMV	Clark et al., 1990
	Tobacco	PVX	ssRNA	*Alphaflexiviridae*	PVX	Hemenway et al., 1988
	Nicotiana benthamiana	TMV	ssRNA	*Virgaviridae*	TMV	Bendahmane et al., 1997
	Peanut	*Tobacco streak virus*	*ssRNA*	*Bromoviridae*	*Tobacco streak virus*	Mehta et al., 2013
Rep	Tobacco	TMV	ssRNA	*Virgaviridae*	TMV	Golemboski et al.,1990
	Tobacco	CMV	ssRNA	*Bromoviridae*	CMV	Zaitlin et al., 1994
MP	Tobacco	TMV	ssRNA	*Virgaviridae*	TMV	Lapidot et al., 1993
	Potato	PVX	ssRNA	*Alphaflexiviridae*	PVX, *Potato virus M, Potato virus S*	Seppanen et al., 1997
	Tobacco	TMV	ssRNA	*Virgaviridae*	*Tobacco rattle tobravirus, Tobacco ringspot nepovirus, Alfalfa mosaic alfamovie, Peanut chlorotic streak caulimovirus, Cucumber mosaic cucumovirus*	Cooper et al., 1995
	Nicotiana benthamiana	TMV	ssRNA	*Virgaviridae*	TMV	Kotlizky et al., 2001

TABLE 6.1 *(Continued)*

Protein targeted/ Method	Plant	Source virus	Genome	Family	Virus resistance	References
RNAi	Cucumber	CMV	ssRNA	*Bromoviridae*	CMV	Wang et al., 2010
	Tobacco	TMV	ssRNA	*Virgaviridae*	TMV	Powell et al., 1989
	Potato	PVX	ssRNA	*Alphaflexiviridae*	PVX	Hamilton and Baulcombe 1999
	Turnip	TCV	ssRNA	*Tombusviridae*	TCV	Deleris et al., 2006
	Nicotiana benthamiana	*Tomato yellow ring virus*	ssRNA	*Bunyaviridae*	*Tomato yellow ring virus*	Hassani-Mehraban et al., 2009
	Cassava	ACMV	ssDNA	*Geminiviridae*	ACMV	Chellappan et al., 2004
	Cauliflower	*Cauliflower mosaic virus*	dsDNA	*Caulimoviridae*	*Cauliflower mosaic* virus	Blevins et al., 2006
	Tomato	*Tomato golden mosaic virus*	ssDNA	*Geminiviridae*	*Tomato golden mosaic* virus	Hagen et al., 2008
	N. benthamiana	ACMV	ssDNA	*Geminiviridae*	ACMV	Patil et al., 2017
	Tobacco	*Soybean mosaic virus, Bean yellow mosaic virus*	ssRNA	*Potyviridae*	*Soybean mosaic virus, Bean yellow mosaic virus*	Thu et al., 2016
	Solanum tuberosum	PVX, PVY, *and Potato virus S*	ssRNA	*Alphaflexiviridae, Potyviridae, Betaflexiviridae*	PVX, PVY, *and Potato virus S*	Hameed et al., 2016

rattle virus, and *Tobacco ringspot virus* (Cooper et al., 1995; Duan et al., 1997; Hou et al., 2000). In another case, *Nicotiana occidentalis* transgenic plants harboring and expressing the gene for MP of *Apple chlorotic virus* and its truncated derivatives exhibited increased resistance to *Grapevine berry inner necrosis virus* (Yoshikawa et al., 2006).

6.2.2.4 RNAI-MEDIATED CROP PROTECTION

Plants employ RNAi defense mechanisms against invading viruses. Researchers have taken advantage of the RNAi mechanism to engineer virus-resistant crops using viral genome sequences. In this case resistance against viruses is achieved through RNAi-mediated degradation or suppression of the viral genome (Table 6.1). This strategy has been applied against both RNA and DNA viruses. In one of the initial studies, transgenic Arabidopsis plants expressing artificial miRNA (amiRNA) were raised (Niu et al., 2006). Two viral genes encoding suppressors of gene-silencing were targeted by the amiRNA method. Transgenic plants showed enhanced resistance against the *Turnip mosaic virus* (TuMV) and *Turnip yellow mosaic virus* (TYMV). In another study, 2b ORF of CMV was targeted by amiRNA based on Arabidopsis miR171. The transgenic *N. tabaccum* plant exhibited improved resistance to CMV (Qu et al., 2007). Transgenic rice plants resistant to *Rice stripe virus* (RSV) were developed using an inverted repeat RNAi construct against the gene encoding nucleocapsid protein (NCP) of RSV (Li et al., 2016). *Cotton leaf curl virus* (CLCuV) causes heavy damage to cotton. The RNAi-based methodology was used to engineer CLCuV resistant plants. Transgenic tobacco expressing sense and antisense RNA targeting viral DNA replicase gene, showed improved resistance against CLCuV (Asad et al., 2003). In another study, transgenic tomato plants containing an ectopic *Rep* gene showed resistance against the *Tomato leaf curl virus* (Yang et al., 2004). Although RNAi-mediated resistance has been found effective against RNA viruses, the degree of resistance varies. This could be because of the secondary structure in viral RNA that probably inhibits the access of the RISC complex. The RNAi methodology does not provide complete resistance against DNA viruses. DNA viruses like geminiviruses remain active during pathogenesis as RNAi cannot degrade viral DNA.

6.3 CRISPR/CAS9-MEDIATED VIRUS RESISTANCE

The type II clustered regularly interspaced short palindrome repeat (CRISPR)/Cas9 (CRISPR-associated) system of *Streptococcus pyogenes* has emerged as one of the most versatile and efficient genome editing tool (Jinek et al., 2012). The CRISPR/Cas9 system cleaves the foreign DNA (Figure 6.2). The mechanism involves the integration of small DNA fragments derived from foreign DNA at a CRISPR locus. These foreign DNA fragments are

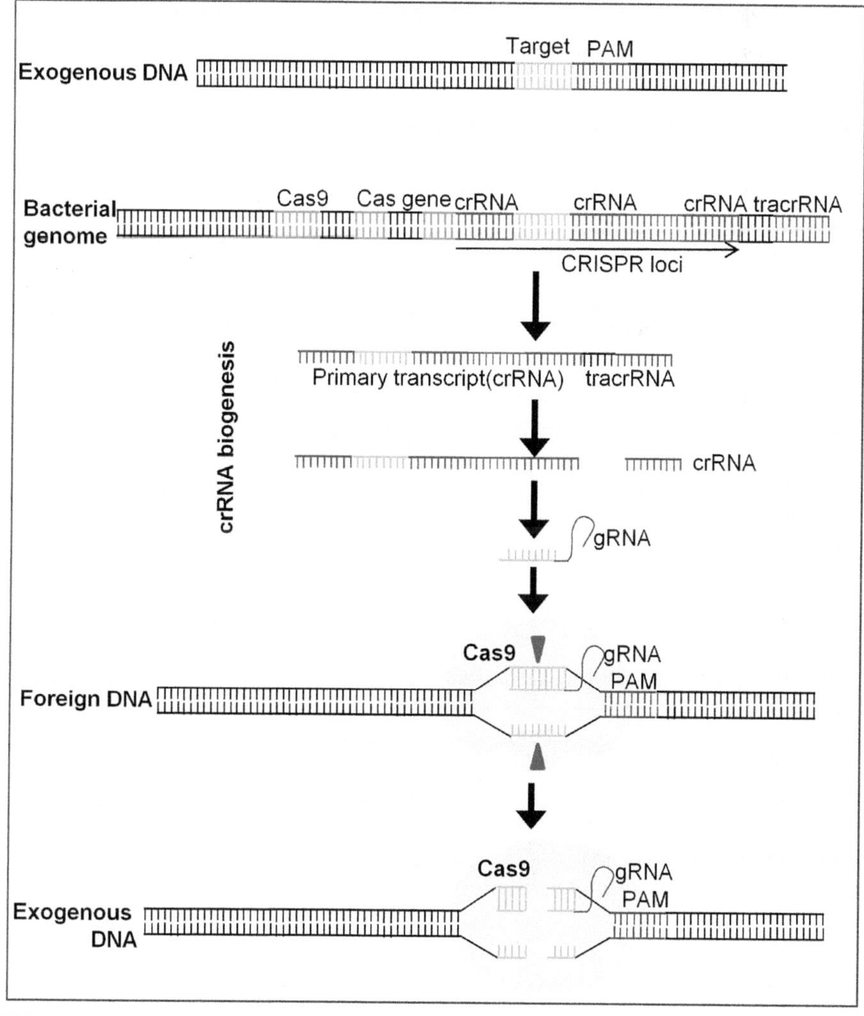

FIGURE 6.2 Mechanism of CRISPR/Cas9-mediated defense response in bacteria against foreign DNA.

called spacers. The transcription of CRISPR produces primary transcripts that are subsequently processed. It leads to the formation of CRISPR RNAs (crRNA). They are approximately 40 nucleotides long. The crRNA combines with trans-activating CRISPR RNA (tracrRNA) and thereby activates the Cas9 nuclease. The activated complex cuts dsDNA sequence, protospacer, in the invading foreign DNA (Barrangou et al., 2007). The specific cleavage activity requires the presence of protospacer-adjacent motif (PAM) having the sequence 5'-NGG-3' or 5'-NAG-3', which is located at the downstream of the target DNA (Gasiunas et al., 2012; Jinek et al., 2012; Ran et al., 2013).

CRISPR/Cas9 tool has successfully been applied in plants like *Arabidopsis thaliana, N. benthamiana*, and rice (Feng et al., 2013; Nekrasov et al., 2013; Shan et al., 2013; Xie and Yang, 2013). This tool has also been used to develop virus-resistant plants (Baltes et al., 2015; Ji et al., 2015). The efficiency and sensitivity of the CRISPR/Cas9 based tool have successfully been demonstrated against DNA as well as RNA viruses (Table 6.2).

6.3.1 CRISPR/CAS9-MEDIATED RESISTANCE AGAINST DNA VIRUSES

The earlier reports of the use of the CRISPR/Cas9 technique were against geminiviruses which contain single-stranded circular genomic DNA. The ssDNA of geminiviruses is converted into dsDNA replicative forms that are targeted by the CRISPR/Cas9. This tool was used against a monopartite geminivirus, *Beet severe curly top virus* (BSCTV). The transgenic tobacco and Arabidopsis plants showed resistance against BSCTV (Ji et al., 2015). In another study, this tool was used to target the *Bean yellow dwarf virus* (BeYDV). Furthermore, the transgenic *N. benthamiana* plants exhibited reduced BeYDV symptoms (Baltes et al., 2015). The application of the CRISPR/Cas9 tool to improve plant immunity against TYLCV has been demonstrated by targeting protein-coding and noncoding sequences of TYLCV. It was observed that targeting the "origin of the replication' region in the intergenic region of TYLCV was most effective via the CRISPR/Cas9 tool. Thc immunity against TYLCV was improved as significantly attenuated symptoms of infection developed (Ali et al., 2015).

TABLE 6.2 The application of CRISPR/Cas9 tool based crop protection against plant viruses.

viruses	Genome	Family	Plants	Target region	References
TYLC *China virus and Tobacco curly shoot virus*	ssDNA	*Geminiviridae*	*Nicotiana benthamiana*	Rep	Chen et al., 2014
TYLCV	ssDNA	*Geminiviridae*	*Arabidopsis thaliana*	Rep, IR	IR Mori et al., 2013
BSCTV	ssDNA	*Geminiviridae*	*A. thaliana*	Rep, IR	Sera, 2005
BSCTV	ssDNA	*Geminiviridae*	*N. benthamiana* and *A. thaliana*	CP, IR, and Rep	Ji et al., 2015
BYDV	ssDNA	*Geminiviridae*	*N. benthamiana*	LIR and Rep/RepA	Baltes et al., 2015
TYLCV, *Beet curly top virus, Merremia mosaic virus*	ssDNA	*Geminiviridae*	*N. benthamiana*	CP, IR, and Rep	Ali et al., 2015
Cotton leaf curl Kokhran virus, TYLC *Sardinian virus, Beet curly top virus, Merremia mosaic virus*	ssDNA	*Geminiviridae*	*N. benthamiana*	CP, IR, and Rep	Ali et al., 2016
TMV	ssRNA	*Potyviridae*	*A. thaliana*	Host factor eIF(iso)4E	Pyott et al., 2016
CVYV, *Zucchini yellow mosaic virus,* PRMV	ssRNA	*Potyviridae*	*Cucumis sativus*	Host factor eIF4E	Chandrasekaran et al.,2016
Rice tungro spherical virus	ssRNA	*Sequiviridae, Virgaviridae*	Rice	eIF4G	Macovei et al., 2018

Intergenic region (IR), coat protein (CP), replication-associated protein (Rep), long intergenic region (LIR)

6.3.2 CRISPR/CAS9 MEDIATED RESISTANCE AGAINST RNA VIRUSES

The CRISPR/Cas9 system has not yet directly been used against RNA viruses instead some host factors have been targeted to enhance the immunity against RNA viruses. These viruses like potyviruses need host proteins like eIF4E (eukaryotic translation initiation factor 4E) and paralogue eIF(iso)4E for multiplication. These factors have previously been characterized as recessive resistance alleles against potyviruses. The CRISPR/Cas9 technology has been applied to mutate eIF4E in cucumber plants that subsequently enhanced resistance against *Cucumber vein yellowing virus* (CVYV), *Zucchini yellow mosaic virus* and *Papaya ringspot mosaic virus*-W (PRSV-W) (Chandrasekaran et al., 2016). In another study, eIF(iso)4E was mutated in Arabidopsis through the CRISPR/Cas9 tool. The resultant plant exhibited resistance to TuMV. The mutation in eIF(iso)4E did not affect the plant vigor and enhanced resistance against TuMV (Pyott et al., 2016).

6.4 CONCLUSION AND FUTURE PERSPECTIVE

Plant viruses pose a major problem to crop cultivation worldwide. In the past few years, several new viruses have been reported which has raised a major concern amongst plant virologists. On the other side, the methodology to engineer virus-resistant plants has not been evolved with the speed of the emerging new viruses. Initially, many efforts were made to develop virus-resistant plants based on pathogen-derived proteins. Later on, the RNAi-based method was developed which was efficient against RNA viruses. In the recent period, the CRISPR/Cas9 tool has been discovered for editing genome. This technology is successfully being utilized for improving resistance against DNA viruses. In field conditions, crops face simultaneous attacks by multiple viruses of different types. It is important to design an effective strategy that can confer resistance to multiple viruses.

ACKNOWLEDGMENTS

PS would like to thank financial and Infrastructure support received from the University Grants Commission, India for Start-up Grant, and the University of Rajasthan, Jaipur respectively. HR would like to acknowledge the DST-INSPIRE Faculty grant (DST/INSPIRE/04/2016/001118) from the Department of Science and Technology, Govt. of India.

KEYWORDS

- **Virus**
- **Resistance**
- **PTGS**
- **RNAi**
- **CRISPR/Cas9**

REFERENCES

Albar, L., Bangratz-Reyser, M., Hebrard, E., Ndjiondjop, M. N., Jones, M., & Ghesquiere, A., (2006). Mutations in the eIF(iso)4G translation initiation factor confer high resistance of rice to rice yellow mottle virus. *Plant Journal, 47*, 417–426.

Ali, Z., Abulfaraj, A., Idris, A., Ali, S., Tashkandi, M., & Mahfouz, M. M., (2015). CRISPR/Cas9-mediated viral interference in plants. *Genome Biology, 16*, 238.

Ali, Z., Ali, S., Tashkandi, M., Zaidi, S. S. E. A., & Mahfouz, M. M., (2016). CRISPR/Cas9-mediated immunity to Gemini viruses: Differential interference and evasion. *Scientific Reports, 6*, 26912.

Asad, S., Haris, W. A. A., Bashir, A., Zafar, Y., Malik, K. A., Malik, N. N., & Lichtenstein, C. P., (2003). Transgenic tobacco expressing Gemini viral RNAs are resistant to the serious viral pathogen causing cotton leaf curl disease. *Archives of Virology, 148*(12), 2341–2352.

Asurmendi, S., Berg, R. H., Koo, J. C., & Beachy, R. N., (2004). Coat protein regulates formation of replication complexes during tobacco mosaic virus infection. *Proceedings of the National Academy of Sciences USA, 101*, 1415–1420.

Audy, P., Palukaitis, P., Slack, S. A., & Zaitlin, M., (1994). Replicase-mediated resistance to potato virus Y in transgenic tobacco plants. *Molecular Plant-Microbe Interaction, 7*, 15–22.

Baltes, N. J., Hummel, A. W., Konecna, E., Cegan, R., Bruns, A. N., Bisaro, D. M., & Voytas, D. F., (2015). Conferring resistance to Gemini viruses with the CRISPR-Cas prokaryotic immune system. *Nature Plants, 1*, 15145.

Barrangou, R., Fremaux, C., Deveau, H., Richards, M., Boyaval, P., Moineau, S., Romero, D. A., & Horvath, P., (2007). CRISPR provides acquired resistance against viruses in prokaryotes. *Science, 315*, 1709–1712.

Baulcombe, D., (2004). RNA silencing in plants. *Nature, 431*, 356–363.

Beachy, R. N., (1997). Mechanisms and applications of pathogen-derived resistance in transgenic plants. *Current Opinion in Biotechnology, 8*, 215–220.

Beachy, R. N., (1999). Coat-protein-mediated resistance to tobacco mosaic virus: Discovery mechanisms and exploitation. *Philosophical Transactions of the Royal Society B Biological Sciences, 354*, 659–664.

Bendahmane, M., Chen, I., Asurmendi, S., Bazzini, A. A., Szecsi, J., & Beachy, R. N., (2007). Coat protein-mediated resistance to TMV infection of *Nicotiana tabacum* involves multiple modes of interference by coat protein. *Virology, 366*, 107–116.

Bendahmane, M., Szecsi, J., Chen, I., Berg, R. H., & Beachy, R. N., (2002). Characterization of mutant tobacco mosaic virus coat protein that interferes with virus cell-to-cell movement. *Proceedings of the National Academy of Sciences USA, 99*, 3645–3650.

Bhattacharjee, S., Zamora, A., Azhar, M. T., Sacco, M. A., Lambert, L. H., & Moffett, P., (2009). Virus resistance induced by NB-LRR *proteins inv*olves Argonaute4-dependent translational control. *Plant Journal, 58*, 940–951.

Bisaro, D. M., & Voytas, D. F., (2015). Conferring resistance to Gemini viruses with the CRISPR–Cas prokaryotic immune system. *Nature Plants, 1*, 15145.

Blevins, T., Rajeswaran, R., Shivaprasad, P. V., Beknazariants, D., Si-Ammour, A., Park, H. S., Park, H. S., et al., (2006). Four plant Dicers mediate viral small RNA biogenesis and DNA virus induced silencing. *Nucleic Acids Research, 34*, 6233–6246.

Borsani, O., Zhu, J., Verslues, P. E., Sunkar, R., & Zhu, J. K., (2005). Endogenous siRNAs derived from a pair of natural CIS-antisense transcripts regulate salt tolerance in *Arabidopsis*. *Cell, 123*, 1279–1291.

Brodersen, P., Sakvarelidze-Achard, L., Bruun-Rasmussen, M., Dunoyer, P., Yamamoto, Y. Y., Sieburth, L., & Voinnet, O., (2008). Widespread translational inhibition by plant miRNAs and siRNAs. *Science, 320*, 1185–1190.

Carr, J. P., & Loebenstein, G., (2010). Natural and engineered resistance to plant viruses, part II. preface. *Advances in Virus Research, 76*, p. vii.

Chandrasekaran, J., Brumin, M., Wolf, D., Leibman, D., Klap, C., Pearlsman, M., Sherman, A., et al., (2016). Development of broad virus resistance in non-transgenic cucumber using CRISPR/Cas9 technology. *Molecular Plant Pathology, 17*, 1140–1153.

Chellappan, P., Vanitharani, R., Pita, J., & Fauquet, C. M., (2004). Short interfering RNA accumulation correlates with host recovery in DNA virus-infected hosts, and gene silencing targets specific viral sequences. *Journal of Virology, 78*, 7465–7477.

Chen, W., Qian, Y., Wu, X., Sun, Y., Wu, X., & Cheng, X., (2014). Inhibiting replication of *Begomoviruses* using artificial zinc finger nucleases that target viral-conserved nucleotide motif. *Virus Genes, 48*, 494–501

Clark, W. G., Register, J. C., Nejidat, A., Eichholtz, D. A., Sanders, P. R., Fraley, R. T., & Beach, R. N., (1990). Tissue-specific expression of the TMV coat protein in transgenic tobacco plants affects the level of coat protein-mediated virus protection. *Virology, 179*, 640–647.

Collier, S. M., Hamel, L. P., & Moffett, P., (2011). Cell death mediated by the N-terminal domains of a unique and highly conserved class of NB-LRR protein. *Molecular Plant Microbe Interaction, 24*, 918–931.

Cooper, B., Lapidot, M., Heick, J. A., Dodds, J. A., & Beachy, R. N., (1995). A defective movement protein of TMV in transgenic plants confers resistance to multiple viruses whereas the functional analog increases susceptibility. *Virology, 206*, 307–313.

Culver, J. N., & Padmanabhan, M. S., (2007). Virus-induced disease: Altering host physiology one interaction at a time. *Annual Reviews of Phytopathology, 45*, 221–243.

Deleris, A., Gallego-Bartolome, J., Bao, J., Kasschau, K., Carrington, J. C., & Voinnet, O., (2006). Hierarchical action and inhibition of plant dicer-like proteins in antiviral defense. *Science, 313*, 68–71.

Dodds, P. N., & Rathjen, J. P., (2010). Plant immunity: Towards an integrated view of plant-pathogen interactions. *Nature Reviews Genetics, 11*, 539–548.

Duan, Y. P., Powell, C. A., Purcifull, D. E., Broglio, P., & Hiebert, E., (1997). Phenotypic variation in transgenic tobacco expressing mutated Gemini virus movement/pathogenicity (BC1) proteins. *Molecular Plant Microbe Interaction, 10*, 1065–1074.

Eamens, A., Wang, M. B., Smith, N. A., & Waterhouse, P. M., (2008). RNA silencing in plants: Yesterday, today, and tomorrow. *Plant Physiology, 147*, 456–468.

Eybishtz, A., Peretz, Y., Sade, D., Gorovits, R., & Czosnek, H., (2010). Tomato yellow leaf curl virus infection of a resistant tomato line with a silenced sucrose transporter gene LeHT1 results in inhibition of growth, enhanced virus spread, and necrosis. *Planta, 231*, 537–548.

Fabian, M. R., Sonenberg, N., & Filipowicz, W., (2010). Regulation of mRNA translation and stability by microRNAs. *Annual Reviews Biochemistry, 79*, 351–379.

Feng, Z., Zhang, B., Ding, W., Liu, X., Yang, D. L., Wei, P., Cao, F., Zhu, S., Zhang, F., Mao, Y., & Zhu, J. K., (2013). Efficient genome editing in plants using a CRISPR/Cas system. *Cell Research, 23*, 1229–1232.

Gasiunas, G., Barrangou, R., Horvath, P., & Siksnys, V., (2012). Cas9-crRNA ribonucleoprotein complex mediates specific DNA cleavage for adaptive immunity in bacteria. *Proceedings of the National Academy of Sciences USA, 109*, E2579–2586.

Golemboski, D. B., Lomonossoff, G. P., & Zaitlin, M., (1990). Plants transformed with a tobacco mosaic virus nonstructural gene sequence are resistant to the virus. *Proceedings of the National Academy of Sciences USA, 87*, 6311–6315.

Haag, J. R., & Pikaard, C. S., (2011). Multi-subunit, RNA polymerases IV and V: Purveyors of non-coding RNA for plant gene silencing. *Nature Reviews Molecular Cell Biology, 12*, 483–492.

Hagen, C., Rojas, M. R., Kon, T., & Gilbertson, R. L., (2008). Recovery from cucurbit leaf crumple virus (family *Geminiviridae*, genus *Begomovirus*) infection is an adaptive antiviral response associated with changes in viral small RNAs. *Phytopathology, 98*, 1029–1037.

Hameed, A., Tahir, M. N., Asad, S., Bilal, R., Eck, J. V., Jander, G., & Mansoor, S., (2016). RNAi-mediated simultaneous resistance against three RNA viruses in potato. *Molecular Biotechnology, 59*, 73–83.

Hamilton, A. J., & Baulcombe, D. C., (1999). A species of small antisense RNA in posttranscriptional gene silencing in plants. *Science, 286*, 950–952.

Hassani-Mehraban, A., Brenkman, A. B., Van, D. B. N. J., Goldbach, R., & Kormelink, R., (2009). RNAi-mediated transgenic *Tospovirus* resistance broken by intraspecies silencing suppressor protein complementation. *Molecular Plant Microbe Interaction, 22*, 1250–1257.

Hemenway, C., Fang, R. X., Kaniewski, W. K., Chua, N. H., & Tumer, N. E., (1988). Analysis of the mechanism of protection in transgenic plants expressing the potato virus X coat protein or its antisense RNA. *EMBO Journal, 7, 1273–1280.*

Herr, A. J., Jensen, M. B., Dalmay, T., & Baulcombe, D. C., (2005). RNA polymerase IV directs silencing of endogenous DNA. *Science, 308*, 118–120.

Holmes, F. O., (1937). Genes affecting response of *Nicotiana tabacum* hybrids to tobacco-mosaic virus. *Science, 85*, 104–105.

Holmes, F. O., (1954). Inheritance of resistance to viral diseases in plants. *Advances in Virus Research, 2*, 1–30.

Hou, Y. M., Sanders, R., Ursin, V. M., & Gilbertson, R. L., (2000). Transgenic plants expressing Gemini virus movement proteins: Abnormal phenotypes and delayed infection by Tomato mottle virus in transgenic tomatoes expressing the Bean dwarf mosaic virus BV1 or BC1 proteins. *Molecular Plant-Microbe Interaction, 13*, 297–308.

Hutvagner, G., McLachlan, J., Pasquinelli, A. E., Balint, E., Tuschl, T., & Zamore, P. D., (2001). A cellular function for the RNA-interference enzyme Dicer in the maturation of the let-7 small temporal RNA. *Science, 293*, 834–838.

Ji, X., Zhang, H., Zhang, Y., Wang, Y., & Gao, C., (2015). Establishing a CRISPR-Cas-like immune system conferring DNA virus resistance in plants. *Nature Plants, 1*, 15144.

Jinek, M., Chylinski, K., Fonfara, I., Hauer, M., Doudna, J. A., & Charpentier, E., (2012). A programmable dual-RNA-guided DNA endonuclease in adaptive bacterial immunity. *Science, 337*, 816–821.

Jones, J. D., & Dangl, J. L., (2006). The plant immune system. *Nature, 444*, 323–329.

Kang, B. C., Yeam, I., & Jahn, M. M., (2005). Genetics of plant virus resistance. *Annual Review of Phytopathology, 43*, 581–621.

Kanno, T., Huettel, B., Mette, M. F., Aufsatz, W., Jaligot, E., Daxinger, L., Kreil, D. P., et al., (2005). Atypical RNA polymerase subunits required for RNA-directed DNA methylation. *Nature Genetics, 37*, 761–765.

Kanyuka, K., Druka, A., Caldwell, D. G., Tymon, A., McCallum, N., Waugh, R., & Adams, M. J., (2005). Evidence that the recessive bymovirus resistance locus rym4 in barley corresponds to the eukaryotic translation initiation factor 4E gene. *Molecular Plant Pathology, 6*, 449–458.

Kotlizky, G., Katz, A., Van, D. L. J., Boyko, V., Lapidot, M., Beachy, R. N., Heinlein, M., & Epel, B. L., (2001). A dysfunctional movement protein of tobacco mosaic virus interferes with targeting of wild-type movement protein to microtubules. *Molecular Plant Microbe Interact, 14*, 895–904.

Li, L., Cheng, G., Biao, W., Tong, Z., Yang, L., Yu-hua, D., Wen, H., Chun, L., & Xi-feng, W., (2016). RNAi-mediated transgenic rice resistance to Rice stripe virus. *Journal of Integrative Agriculture, 15*(11), 2539–2549.

Liu, X., Yu, C. W., Duan, J., Luo, M., Wang, K., Tian, G., Cui, Y., & Wu, K., (2012). HDA6 directly interacts with DNA methyltransferase MET1 and maintains transposable element silencing in *Arabidopsis*. *Plant Physiology, 158*, 119–129.

Lu, B., Stubbs, G., & Culver, J. N., (1998). Coat protein interactions involved in tobacco mosaic tobamovirus cross-protection. *Virology, 248*, 188–198.

Macovei, A., Sevilla, N. R., Cantos, C., Jonson, G. B., Slamet-Loedin, I., Čermák, T., Voytas, D. F., et al., (2018). *Novel Alleles of Rice eIF4G Generated by CRISPR/Cas9-Targeted Mutagenesis Confer Resistance to Rice Tungro Spherical Virus*.

Malnoe, P., Farinelli, L., Collet, G. F., & Reust, W., (1994). Small-scale field test with transgenic potato, cv. bintje, to test resistance to primary and second infections with *Potato virus Y. Plant Molecular Biology, 25*, 963–975.

Matzke, M., Kanno, T., Daxinger, L., Huettel, B., & Matzke, A. J., (2009). RNA-mediated chromatin-based silencing in plants. *Current Opinion in Cell Biology, 21*, 367–376.

Mehta, R., Radhakrishnan, T., Kumar, A., Yadav, R., Dobaria, J. R., Thirumalaisamy, P. P., Jain, R. K., & Chigurupati, P., (2013). Coat protein-mediated transgenic resistance of peanut (*Arachis hypogaea* L.) to peanut stem necrosis disease through agro bacterium-mediated genetic transformation. *Indian Journal of Virology, 24*, 205–213.

Meister, G., & Tuschl, T., (2004). Mechanisms of gene silencing by double-stranded RNA. *Nature, 431*, 343–349.

Mori, T., Takenaka, K., Domoto, F., Aoyama, Y., & Sera, T., (2013). Inhibition of binding of tomato yellow leaf curl virus rep to its replication origin by artificial zinc-finger protein. *Molecular Biotechnology, 54*, 198–203.

Nekrasov, V., Staskawicz, B., Weigel, D., Jones, J. D., & Kamoun, S., (2013). Targeted mutagenesis in the model plant *Nicotiana benthamiana* using Cas9 RNA-guided endonuclease. *Nature Biotechnology, 31*, 691–693.

Nieto, C., Morales, M., Orjeda, G., Clepet, C., Monfort, A., Sturbois, B., et al., (2006). An eIF4E allele confers resistance to an uncapped and non-polyadenylated RNA virus in melon. *Plant Journal, 48*, 452–462.

Niu, Q. W., Lin, S. S., Reyes, J. L., Chen, K. C., Wu, H. W., Yeh, S. D., & Chua, N. H., (2006). Expression of artificial microRNAs in transgenic *Arabidopsis thaliana* confers virus resistance. *Nat Biotechnol., 24*, 1420–1428.

Onodera, Y., Haag, J. R., Ream, T., Costa, N. P., Pontes, O., & Pikaard, C. S., (2005). Plant nuclear RNA polymerase IV mediates siRNA and DNA methylation-dependent heterochromatin formation. *Cell, 120*, 613–622.

Pallas, V., & Garcia, J. A., (2011). How do plant viruses induce disease? Interactions and interference with host components. *Journal of General Virology, 92*, 2691–2705.

Patil, B. L., Bagewadi, B., Yadav, J. S., & Fauquet, C. M., (2017). Mapping and identification of cassava mosaic Gemini virus DNA-A and DNA-B genome sequences for efficient siRNA expression and RNAi based virus resistance by transient agro-infiltration studies. *Virus Research, 213*, 109–115.

Pellio, B., Streng, S., Bauer, E., Stein, N., Perovic, D., Schiemann, A., Friedt, W., Ordon, F., & Graner, A., (2005). High-resolution mapping of the Rym4/Rym5 locus conferring resistance to the barley yellow mosaic virus complex (BaMMV, BaYMV, BaYMV-2) in barley (*Hordeum vulgare ssp. Vulgare* L.). *Theoretical and Applied Genetics, 110*, 283–293.

Pikaard, C. S., Haag, J. R., Ream, T., & Wierzbicki, A. T., (2008). Roles of RNA polymerase IV in gene silencing. *Trends in Plant Science, 13*, 390–397.

Powell, A. P., Nelson, R. S., De, B., Holfmann, N., Rogers, S. G., Fraley, R. T., & Beachy, R. N., (1986). Delay of disease development in transgenic plants that express the *Tobacco mosaic virus* coat protein gene. *Science, 232*, 738–743.

Powell, P. A., Stark, D. M., Sanders, P. R., & Beachy, R. N., (1989). Protection against tobacco mosaic virus in transgenic plants that express tobacco mosaic virus antisense RNA. *Proceedings of the National Academy of Sciences USA, 86*, 6949–6952.

Pyott, D. E., Sheehan, E., & Molnar, A., (2016). Engineering of CRISPR/Cas9-mediated potyvirus resistance in transgene-free *Arabidopsis* plants. *Molecular Plant Pathology, 17*, 1276–1288.

Qu, J., Ye, J., & Fang, R., (2007). Artificial microRNA-mediated virus resistance in plants. *Journal of Virology, 81*, 6690–6699.

Raffaele, S., Bayer, E., Lafarge, D., Cluzet, S., German, R. S., Boubekeur, T., Leborgne-Castel, N., et al., (2009). Remorin, a solanaceae protein resident in membrane rafts and plasmodesmata, impairs potato virus X movement. *Plant Cell, 21*, 1541–1555.

Raja, P., Sanville, B. C., Buchmann, R. C., & Bisaro, D. M., (2008). Viral genome methylation as an epigenetic defense against Gemini viruses. *Journal of Virology, 82*, 8997–9007.

Ran, F. A., Hsu, P. D., Wright, J., Agarwala, V., Scott, D. A., & Zhang, F., (2013). Genome engineering using the CRISPR-Cas9 system. *Nature Protocol., 8*, 2281–2308.

Ream, T. S., Haag, J. R., Wierzbicki, A. T., Nicora, C. D., Norbeck, A. D., Zhu, J. K., Hagen, G., et al., (2009). Subunit compositions of the RNA-silencing enzymes Pol IV and Pol V reveal their origins as specialized forms of RNA polymerase II. *Molecular Cell, 33*, 192–203.

Reyes, M. I., Nash, T. E., Dallas, M. M., Ascencio-Ibanez, J. T., & Hanley-Bowdoin, L., (2013). Peptide aptamers that bind to Gemini virus replication proteins confer a resistance phenotype to tomato yellow leaf curl virus and tomato mottle virus infection in tomato. *Journal of Virology, 87*, 9691–9706.

Schwessinger, B., & Ronald, P. C., (2012). Plant innate immunity: Perception of conserved microbial signatures. *Annual Review of Plant Biology, 63*, 451–482.

Sera, T., (2005). Inhibition of virus DNA replication by artificial zinc finger proteins. *Journal of Virology, 79*, 2614–2619.

Shan, Q., Wang, Y., Li, J., Zhang, Y., Chen, K., Liang, Z., Zhang, K., et al., (2013). Targeted genome modification of crop plants using a CRISPR-Cas system. *Nature Biotechnology, 31*, 686–688.

Sontheimer, E. J., & Carthew, R. W., (2005). Silence from within: Endogenous siRNAs and miRNAs. *Cell, 122*, 9–12.

Tennant, P. F., Gonsaives, C., Ling, K. S., Fitch, M., Manshardt, R., Slightom, J. L., & Gonsalves, D., (1994). Differential protection against papaya ring spot virus isolates in coat protein gene transgenic papaya and classically cross-protected papaya. *Phytopathology, 84*, 1359–1366.

Thu, L. T. M., Thuy, V. T. X., Duc, L. H., Son, L. V., , Ha, C. H., & Mau, C. H., (2016). RNAi-mediated resistance to SMV and BYMV in transgenic tobacco. *Crop Breeding and Applied Biotechnology, 16*, 213–218.

Truniger, V., & Aranda, M. A., (2009). Recessive resistance to plant viruses. *Adv. Virus Res., 75*, 119–159.

Wang, X. B., Wu, Q., Ito, T., Cillo, F., Li, W. X., Chen, X., Yu, J. L., & Ding, S. W., (2010). RNAi-mediated viral immunity requires amplification of virus-derived siRNAs in *Arabidopsis thaliana. Proceedings of the National Academy of Sciences U S A, 107*, 484–489.

Whitham, S., Dinesh-Kumar, S. P., Choi, D., Hehl, R., Corr, C., & Baker, B., (1994). The product of the tobacco mosaic virus resistance gene N: Similarity to toll and the interleukin-1 *receptor. Cell, 78*, 1101–1115.

Whitham, S., McCormick, S., & Baker, B., (1996). The N gene of tobacco confers resistance to tobacco mosaic virus in transgenic tomato. *Proceedings of the National Academy of Sciences USA, 93*, 8776–8781.

Xie, K., & Yang, Y., (2013). RNA-guided genome editing in plants using a CRISPR-Cas system. *Mol Plant, 6*, 1975–1983.

Yang, Y., Sherwood, T. A., Patte, C. P., Hiebert, E., & Polston, J. E., (2004). Use of Tomato yellow leaf curl virus (TYLCV) Rep Gene Sequences to Engineer TYLCV Resistance in Tomato. *Phytopathology, 94*, 490–496.

Yoshii, K., Konno, A., Goto, A., Nio, J., Obara, M., Ueki, T., Hayasaka, D., Mizutani, T., Kariwa, H., & Takashima, I., (2004). Single point mutation in tick-borne encephalitis virus prM protein induces a reduction of virus particle secretion. *Journal of General Virology, 85*, 3049–3058.

Yoshii, M., Yoshioka, N., Ishikawa, M., & Naito, S., (1998). Isolation of an *Arabidopsis thaliana* mutant in which the multiplication of both cucumber mosaic virus and turnip crinkle virus is affected. *Journal of Virol., 72*, 8731–8737.

Yoshikawa, N., Saitou, Y., Kitajima, A., Chida, T., Sasaki, N., & Isogai, M., (2006). Interference of long-distance movement of grapevine berry inner necrosis virus in transgenic plants expressing a defective movement protein of apple chlorotic leaf spot virus. *Phytopathology, 96*, 378–385.

Zaitlin, M., Anderson, J. M., Perry, K. L., Zhang, L., & Palukaitis, P., (1994). Specificity of replicase-mediated resistance to cucumber mosaic virus. *Virology, 201*, 200–205.

Zhang, P., & Gruissem, W., (2003). Efficient replication of cloned African cassava mosaic virus in cassava leaf disks. *Virus Research, 9*, 47–54.

CHAPTER 7

Plant Responses to Weeds, Pests, Pathogens, and Agrichemical Stress Conditions

RITUPARNA SAHA[1], DEBALINA BHATTACHARYA[2], and MAINAK MUKHOPADHYAY[3]

[1]*Department of Biochemistry, University of Kolkata, Kolkata – 700019, West Bengal, India*

[2]*Department of Microbiology, Maulana Azad College, Kolkata – 700013, West Bengal, India*

[3]*Department of Biotechnology, JIS University, Kolkata – 700109, West Bengal, India, E-mail: m.mukhopadhyay85@gmail.com*

ABSTRACT

An unconditional increase in globalization, industrialization, and urbanization has led to a dramatic loss of agricultural land and an increase in environmental stress. Food has become scarce and plants have become increasingly sensitive to various kinds of stress, both biotic and abiotic. Biotic stress includes disturbances that occur as a result of other living organisms like fungi, insects, weeds, pests, whereas abiotic stress is naturally occurring or sometimes man-made. Although, plants already have developed responses to abiotic stress; it's the biotic stress that is life-threatening to them. When encountered with any form of stress, plants response to it through a combination of cellular and molecular pathways. A detailed study of which has led to the development of genetically modified (GM) plants but to no avail, as overtime the stress induced by both living and non-living has become resistant to the modified response.

This chapter will include the various responses of the plants to stress induced by both the living and the non-living. The different cellular and molecular pathways involved before and after gene modification will be discussed in detail.

7.1 INTRODUCTION

Globalization, as a whole, has a universal effect across all aspects of the economy. This also includes, if not directly, the whole ecosystem. The ongoing after-effects of globalization have resulted in a gamut of different biodiversity loss, as in agriculture with the loss of crop varieties, or loss of wild species, with enrichment of exotic ones; and the subsequent pollution of air, soil, and water, resulting in global warming. This has ultimately led to social and economic disruption especially in the third world countries, without limiting itself to the economically vulnerable and rural communities (Ehrenfeld, 2003).

The increase in population and commercialization of life's basic needs has increased demands for space and food supply. The combined effect of deforestation and pollution has led to the high activity of both natural and man-made stress to plants, making farmers to rely heavily on pesticides and herbicides. Over time, this had significant effects on both the plants and the type of stress. The weeds, pests, and pathogens have grown tolerant and adapted to the chemicals, whereas the plants have lost its yield capacity. Research into a biophysical and molecular modification of plants for tolerating stress and increased yield has revived hope, but the unavailability of the modified seeds to farmers in the developing countries has set an upsetting pretext, to begin with. This misbalance, in turn, has set a domino effect on the global economy (Mosa et al., 2017).

In the natural environment, plants have to undergo a range of different and complex interactions, which they have achieved through evolution. Along with this, the plants have also acquired several mechanisms to counteract the various stress conditions in nature. Stress induces mainly metabolic distress within the plant system whose effect can be witnessed through the physiological changes and productivity (Rejeb et al., 2014). These responses are complex and intense and include a vast array of different cellular and molecular adaptations. Plants have developed individual responses to different kinds of stress, which under adverse conditions can cross-talk and eventually lead to a cascade of signaling pathways. Throughout the years, stress has been broadly classified into biotic and abiotic stress, which wreaks havoc in the metabolism of plants when exposed to it, leading to physiological disruption (Mosa et al., 2017).

Abiotic stress is most important as it causes more harm leading to >50% losses in the field, whereas biotic stress is a challenge as it leads to more damage through pathogen and herbivore attack. Over the years, research has shown the development of various defense mechanisms in response to stress in a quick and efficient manner (Rejeb et al., 2014). After recognizing the stress, plants introduce a series of molecular mechanisms including the opening of ion channels, activation of signaling cascades, production of reactive oxygen species (ROS) and phytohormones, which ultimately leads to a change in the genetic programming, thus increasing the tolerance of the plant, meanwhile decreasing biological and physiological damage caused by the stress (Kissoudis et al., 2014).

Interestingly, research has shown that plants witness multiple stress when in the field or in natural conditions. This leads to multiple responses, which may be additive, synergistic, or antagonistic. Under continuous stress, plants grow resistant to various stresses and produce a response for single stress, proving the existence of cross-tolerance and the ability of plants to possess a powerful regulatory response against the changing environment (Atkinson and Urwin, 2012).

7.2 BIOTIC AND ABIOTIC STRESS

In the environment, plants encounter a variety of stresses which either limit or hinders their productivity. These stresses have been broadly classified into two types-biotic stress and abiotic stress (Verma et al., 2013). Based on the number of interacting factors, these stresses have also been grouped into different categories-single, multiple, individual, and combined stresses. Single stress is defined by only one stress factor affecting plant development and growth, whereas multiple stress gets interpreted as the impact of two or more stresses affecting the plant physiology over a period of time which can sometimes overlap (combined stress) and sometimes not (individual stress) (Pandey et al., 2017). Research has stated that stress combinations can have both positive and negative impact on the plant. But, identification of the physio-morphological traits which are affected by the individual and combined stresses are required for the development of plant tolerance to both biotic stresses and abiotic stresses (Mittler, 2006).

Abiotic stress is defined by the non-living a factor which when increases beyond their normal levels brings about stress in plants. These factors when remaining within their normal limit causes a positive impact on plants as well. Abiotic stress mainly includes factors like heavy metals, drought, flood, salinity, extremes in temperature, radiation, winds, etc., (Sha Valli Khan et al., 2014). This type of stress has increasingly become a threat to global food security due to drastic climate changes and environmental deterioration caused by human activity. Plants cope with abiotic stress by initiating a response through various cellular, molecular, and physiological changes (Huang et al., 2013; Verma et al., 2013). A generic signal transduction pathway mainly involves with the signal perception by specific receptors, which when activated mediate the signal to downstream effector molecules which either modulates the intracellular Ca^{2+} level or initiates a protein phosphorylation cascade (Tuteja et al., 2011). And finally, the stress-responsive genes are activated through transcription control which produces cellular responses to the specific stress, like maintaining the osmotic equilibrium during salinity or decreasing water loss in response to drought. During a stress-related injury, the plant undergoes through a cellular detoxification process by synthesizing molecular chaperones, enzymes for scavenging ROS, and various other detoxification proteins, ultimately leading to the repair of the stress-induced injury or damage producing tolerance (Zhu, 2016). There are certain key regulators, which bring about interactions and crosstalk between many molecular pathways. One such interaction happens between ROS and hormones like abscisic acid (ABA) and ethylene, which make up the important regulators of plant responses to abiotic stress. ABA signaling is very fast and sometimes happens without any transcriptional regulation, and is a central regulator of many plant responses to different abiotic stresses. Whereas ethylene is also involved in stress responses like drought, flooding, ozone, heat, cold, and wounding. In many of the stress responses ethylene has also been known to interact with ABA to coordinate a very complex response to stress (Cramer et al., 2011).

Biotic stress occurs as a result of the attack on plants by living organisms such as bacteria, fungi, viruses, parasites, weeds, insects, nematodes, and herbivores. Like abiotic stress, biotic stress is also a leading factor causing loss of food crops worldwide (Singla and Krattinger, 2016). The different types of biotic stress faced by an organism vary with the region and climate that the plant is in. Generating tolerance against biotic stress is equally important for the plant so as to complete its life cycle (Sergenat and Renaut, 2010). For this, plants have developed complex mechanisms to help in their survival. A crucial component in this regard is protein phosphatase 2a (PP2A), which controls pathogenic responses in many plant species. Research has established the connections that these multi-functional enzymes have with the signaling pathways that control cell death and immunity, and most importantly primary and secondary metabolism. Over the years, information regarding plant responses to biotic stresses have been obtained from model plants as well as many cultivated species (Durian et al., 2016; Jeandroz and Lamotte, 2017).

As mentioned earlier, biotic, and abiotic stress has a tendency to overlap. During these cases, phytohormones play a crucial role. They through their signaling pathways have the ability to crosstalk and produce a complex response to the attack (Verma et al., 2016). Other molecules involved in this kind of crosstalk are transcription factors (TFs), kinases, and even some genes have also been identified which are involved for producing tolerance to more than a single kind of stress (Fujita et al., 2006; Gupta et al., 2016). Still, the complicated web of crosstalk is not fully understood. Additional research into genomic and transcriptomic approaches are definitely required for further understanding the role of these crosstalk in plant development.

7.3 RESPONSES OF PLANTS TO WEEDS

7.3.1 BIOLOGY OF WEEDS

The increase in food demand has led to an expansion of the area for farming to get high crop yields. But several weed species becomes an effective limiting factor, thus hindering the process. Over the years, farmers have taken fruitful measures for weed control through chemical management. This, in turn, harmed the crop species' as well. The physiology and ecology of crop and weed species were thoroughly researched for developing methods for weed control (Concenco et al., 2012). Studies were done into the competition between crops and weeds, about how crop rotation could be used for suppression of weed species. There is a lingering gap between turning these results into everyday crop management in the field. And researchers have not been able to fully develop parameters applicable in this scenario other than those which when applied harm the crop plant too (Weiner, 1990).

Defining weed species has always been difficult to elaborate. They are normal plants which under certain conditions are considered as weed due to its presence interfering the crop plant currently in terms of competition for food and space. Some classic general features of weed species include-development and fast growth initially, high production of seeds, ability to maintain seed viability even under favorable conditions, germinating seeds, and emerging from the various depth of soil layers (Preston, 2014). Agricultural weeds have survived the longest without detection because over time they have developed something, which is known as crop mimicry. This is defined by the weed species which resemble the crop species at some growth stage, thus avoiding eradication. This type of weed species have survived overtime due to their ability to even produce and germinate their seeds, at the same time just as the crops they are competing with (Murphy et al., 1996). An example is *Echinochloa crus-galli* var. *Oryzicola* and California rice-both of them produce seed at the same time, which is why the rice seeds got contaminated and the weed got spread throughout the rice-producing fields of California (Barrett, 1983).

A special feature that made weed all the more powerful is the ability of their seeds to enter dormancy when conditions are unsuitable for growth. The seeds can pass onto different soil layers or some wait for special inducers like water for their germination (Rathore et al., 2014). Another ability is their seed dispersal process. They can disperse seeds through winds or through clinging the surface of livestock. Agricultural means provide various different ways to disperse and germinate seeds for different weed species (Preston, 2014). For their reproduction purposes, weed species have mostly relied upon self-pollination, thus restricting genetic diversity. They also often use wind or bees as pollinators for dispersing pollens through large distances and allows for mixing of alleles to increase genetic diversity and increasing the survival rate of the weed species (Horvath et al., 2018). These advantages are beneficial to weed species growing in agricultural lands. Thus, weeds form one of the major biotic stresses on agricultural plants and can have a major impact on agricultural production and yield (Barrett, 1983).

7.3.2 COMPETITION AND INTERACTION BETWEEN PLANTS AND WEED SPECIES

Weeds have the most impact on agricultural practices especially in terms of crop yields. It is due to the competition between weeds and crop species for nutrient, light, and space. The competition whether competitive or not depends upon the crop species and its ability to change their physiological and genetic behavior in accordance with the requirement. This leads to differential use of environmental resources like water that affects the CO_2 availability to leaves and ultimately results in a reduction of photosynthetic efficiency (Jornsgard et al., 1996; Concenco et al., 2012). The more similar the competing individuals, the more stress it puts on the resources, as both of them have similar requirements. It is rightfully known as intraspecific competition due to its high occurrence between morphologically similar individuals. Thus, understanding the developments of weed species and crop plants can help to plan and ultimately subsidize the development of weed species (Blackman and Templeman, 1938).

Competition is one of the most complex phenomena as it is dependent on a variety of factors like biological, environmental, and proximity between the weed and plant. Proximity factors are best described in terms of the spatial arrangement of the plants, plant density, and the proportion of species. For example, the competition between Sugar Beet and weeds like Barnyard grass (*Echinochloa crus-galli* (L.) Beauv.) and lambsquarters (*Chenopodium album* L.) were one of the most troublesome, harming the production of sugar beet in the Pacific Northwest (Dawson, 1964). The competition also depends on whether the crop species were established before or after. If before, then the crop species did not allow the weed species to propagate later, as the crop plant claims for more light during the initial stages (Radosevich, 1987). During these cases, the crop species get established first and the tone of competition decreases, even sometimes the weed species gets fully eradicated. Competition is of various types-when both, i.e., crop and weed species belong to the same species which leads to having the same nutritional requirements it is known as intraspecific competition, and when it happens between two distinct individuals, it is known as interspecific competition (Nieto et al., 1968; Van Heemst, 1985).Thus, the various features of completion as established trough research are as follows: the competition is the highest during the initial stages for both the crop species and the weed species, the infestation of weed species whether moderate or high depends upon when the crop species and the weed species are established, it is not only light, space, CO_2 and major nutrients they compete for-some weed species have also been known to go after the acute minerals, which form the micronutrients, present in the soil, to aid in their germination and thus, inhibit or decrease the growth of other plant species (Dawson, 1965; Satorre and Snaydon, 1992).

Weeds can disrupt pastureland harming livestock and preventing other useful plant species from growing. Weeds can also affect crops by reducing the quality of the end product. For example, the low erucic acid and glucosinolate content often found in canola (oilseed rape) are due to the contaminating weed like that from the *Brassica* species – *Raphanus raphanistrum* which takes up the erucic acid and glucosinolate for its own seed harvest (Blackshaw et al., 2002). Uncontrolled growth of weed species affects the final product of livestock too, in the sense that livestock producing milk ca graze weeds, which may produce some unwanted flavor due to some undesirable compounds consumed by the livestock through weeds. Also, important to know is that weeds can also harbor pests and pathogens, which can have an altogether disastrous effect on the environment (Preston, 2014).

7.3.3 WEED MANAGEMENT PRACTICES

Weed management is greatly required to reduce the incidence of weed species, if not completely eradicating them. The main motives behind weed management practices are: to reduce weed density to tolerable levels, reduction to the damage inflicted by a weed species on the crop plant and shifting the growth and composition of weed species towards more manageable ones. Management practices like chemical, non-chemical, and even manual practices are mostly adopted, with preference to non-chemical and manual practices due to the effect chemical herbicides have on the crop plants beyond certain levels.

Studies have shown weeds become more competitive with crops when the soil has a high nutrient level. Weeds take up higher concentrations of nitrogen, potassium, phosphorus, magnesium, and calcium more than crops, thus depriving the soil of its nutrients and affecting the crop yields. This has led to the development of fertilization strategies, like the deep band application of fertilizers, applying fertilizers containing nitrification inhibitors or that contains ammonium or urea, which increases the concentration of nitrogen in the soil and in turn inhibits the growth of weeds which are sensitive to ammonia or urea (Di Tomaso, 1995). For a long time, farmers were extensively using herbicides for weeds due to lack of other options. Like the field trial that took place in the year 1999 and 2000 in the northern Guinea savanna of Nigeria for reduction of early, weed competition in maize. A combination of metolachlor and atrazine herbicide in different concentrations was used with medium maize density, which helped to reduce weed density immensely (Chikoye et al., 2004). An integrated weed management system is extensively required to increase the profitability of crop production along with maximizing weed control while diminishing the use of herbicide. A planned approach was established using the herbicide alachlor and metribuzin combined with one or two between-row cultivation, which allowed for a 50–75% reduction in herbicide use without reduction of weed control or the crop yield (Buhler et al., 1991). Studies into gene flow with molecular biology have helped produce herbicide-resistant seeds which offer new options for weed control. Like the herbicide-resistant traits obtained by cultivated rice through genetic, modification helps in control of the weed red rice, which is to wreak havoc during rice cultivation, effecting its yield and productivity (Gealy et al., 2003). Another weed management practice evolved through time is the use of crop rotation and preventing the use of herbicide containing the chemical the weed is resistant to. This proved to be quite advantageous in the case of controlling the glyphosate-resistant horseweed (*Conyza canadensis*) (Davis et al., 2009). The success of crop rotation is due to the fact that weed suppression seems to be based on the usage of crop sequences that create differing patterns of resource competition, soil interference, and an altogether unstable and frequently changing environment which proves to be harmful to the weed species (Liebman and Dyck, 1993).

A rational alternative which has gained increased attention in a short span of time is the use of allelopathy because of rising concerns in the use of the pesticide. Allelopathy provides a rational weed control through the production and release of allelochemicals from leaves, flowers, seeds, stems, and roots of living or dead plant materials. The idea is that cover crops or mulches synthesize enough allelochemicals, which suppresses weed species and they often exhibit selectivity like any synthetic herbicide (Weston, 1996). Allelochemics are released as secondary metabolites and often show a great range of diversity in its chemical nature but is mostly characterized into phenolics and terpenoids. These natural plant products have a wide range of advantages over chemical herbicides and possess a wide range of weed-suppressing ability (Singh et al., 2003). Several groups of chemicals defined as allelochemicals have already been identified and commercialized for its use as natural herbicide (Bhowmik and Inderjit, 2003). A more extensive study is required into allelopathy as a weed management tool for crop production.

The widespread use of herbicides for controlling weed populations has raised concerns for farmers taking up organic farming. But with the increase in demand for organic produce, several weed control options have also come up and has since developed, with new implements which can help in the future (Bond and Grundy, 2001). These sustainable management systems create challenges as well, like the increase in weed severity due to the use of herbicide-resistant crops, which facilitates more use of herbicides and ultimately, leads to the degradation of environmental quality (Mortensen et al., 2012). A more detailed study is required to increase the understanding of weed dynamics for herbicide-resistant research, which can further help in agricultural weed management (Neve et al., 2009).

7.4 RESPONSES OF PLANTS TO PESTS

7.4.1 BIOLOGY AND DISTRIBUTION OF PESTS

Pests are insects or animals which have detrimental effects on crops, plants, and livestock, and have since become a global threat to food security. An extensive portion of agricultural productions is lost to these organisms, with post-harvest incurring more losses (Bebber et al., 2013). The distribution of pests has always been found associated with the distribution of human activities. This introduction of pests to new places threatens the native species and thus, plays a role in alteration of the ecosystem (Liebhold et al., 1995). Crops are engineered in ways that farmers could get an increased yield, with not providing resistance to diseases and pests. Pests which have grown habituated to the plants' wild counterparts harm the engineered version as well (Lovett et al., 2006). Farmers have been obliged to shift their agricultural land with harmful losses to the biodiversity of the previous area. Pests have successfully caught with them as well providing no respite as well. It has become important to study the distribution of pests and extrapolate their probable route, and take measures as necessary (Gibbs and Wainhouse, 1986). Scientists have found the distribution of pests involves a number of factors like biophysical and socio-economic. More information is required to correlate the biogeographical distribution of pests with the factors which it depends on (Bebber et al., 2014).

Today it has become more important to study the distribution and impacts of pests as a factor of climate change. This change has brought a typical change to the sensitivities of plants and crops, making them more vulnerable to pests and the modification in their interaction (Bebber et al., 2013; Kriticos et al., 2013). Reports about the rising CO_2 levels and temperature in certain regions have started affecting the genetics of pests. Pests are now showing increased capacity for generation, recombination, and variations in pathogenicity and fitness. The change and diversity in the interaction between plants and pests are becoming less understood. A robust understanding and prediction in this field will help to reduce food security issues (Gregory et al., 2009).

7.4.2 DYNAMIC RESPONSES OF PLANTS AGAINST PESTS

Plants and pests have co-existed for millions of years, and have developed relationships which affect both parties at various levels. Their relationship changes from time to time and is a region as well as climate specific. Insects have always served as pollinators by dispersing pollens, and have mutual benefits whereas the other aspect involves the insect predation of plants, against which the plants have developed resistance to (Gatehouse, 2002). The plants have a diverse range of strategies to either evade or resist the insect herbivores, starting from biochemical defenses to physical properties to habitat choice. Even there are differences between one member of species to another, adopting subtly different strategies for co-existing or resisting the pests (Bergman and Tingey, 1979). Plant defenses can be both constitutive and induced, where the former is species-specific, and the latter can be found in all plants. And all the forms of the immune system works through intricate and complex signaling system, which includes a range of molecules likejas-monate, systemin, hydrogen peroxide, and even volatiles (Jermy, 1984).

The intensity of the plant responses varies with the densities of attack as well, with some insect infection bringing about host mortality. For example, when the relationship between the physiology of *Pinuscontorta* var. *latifolia* and the behavior of the bark beetle, *Dendroctonus ponderosae* was studied, it was found that when the attack density was low, the trees responded by secreting inhibitory or toxic compounds producing necrotic lesions and confining the infection. And when the attack is to surpass a limit, it exhausts the defenses of the tree and ultimately brings about mortality (Raffa and Berryman, 1983). The most studied host plant interaction is between rice (*Oryza sativa* L.) and the brown planthopper (*Nilaparvata lugens*) (BPH). Rice is one of the ancient and the most genetically diverse food crop. BPH does damage by feeding and consecutively, transmitting stunt virus diseases. So, when BPH resistant rice seeds were produced, the pests modified its genotype as well (Sogawa, 1982).

Another response system developed by plants is through genetics. The plant immune system recognizes some specific molecules called elicitors, during a pest's infection. These molecules then initiate several signaling cascades which ultimately induce specific genes for producing a cohesive and complex response either to resist or evade the pests' infection. Research has found several disease resistance genes and has studied their interaction with pathogen a virulence genes (Keen, 1990). This gene-for-gene coevolution concept was a major model for research, but recent studies challenge this concept with the idea that natural selection within the plant population forms the major driving force, including the effects of genetic drift and gene flow (Thompson and Burdon, 1992). Though pests have changed their genotypes in order to break host plant resistance (HPR), still a number of genes are found like the aphid resistant genes and alleles, out of which two dominant genes have been known to encode a protein that helps in specifically recognizing aphids (Dogimont et al., 2010).

Plants also produce resistance in the form of chemical responses, which includes secondary metabolites and phytohormones. This includes allelochemicals, which are synthesized and produced to manipulate predator-prey interaction in order to bring about pest control. The plants which are prone to pests' infection, produces volatile allelochemicals that attract predators for which the pests are prey, thus preventing infection (Dicke et al., 1990). Hormones like ABA have been studied for its ability to play a role in jasmonate-induced defenses against herbivores, but a more detailed study is required into how plants coordinate the production of these chemicals with pests' infection (Thaler and Bostock, 2004).

HPR is an economic, advantageous, and effective method of pest control. But considerable research is required for the identification and transfer of resistant genes to crops, which can adjust to different ecosystems and produces a high yield (Sharma and Ortiz, 2002).

7.4.3 ARTIFICIAL MODIFICATIONS FOR TOLERANCE AGAINST PESTS

For artificial resistance, farmers for a long time have always relied on synthetic insecticides, which protect crops against insects. But this had many negative effects-especially on the environment; while the insect's overtime has overcome the resistance as well. And certain components in the insecticides above a certain concentration prove to be quite toxic for the plants. So, modified pest management is readily required which includes the use of chemical insecticides in combination with other biological-based techniques (Brattsten et al., 1986).

Molecular biology has provided some respite in the sense, that scientists have now developed gene modification or gene transfer techniques, which enables a plant to gain an artificial tolerance against insect infection. Since then, genetic modification has brought about a diverse range of resistance to plants. Secondary plant metabolites deter insects and also cause interference in the development of plant diseases. But the amount in which they are synthesized is so small, that the plant is unable to resist itself from the disease for long. Gene cloning from various sources like *Bacillus thuringiensis*, which is known to produce a toxin-a trypsin inhibitor and wild bean that produces an insecticidal lectin-like protein; have produced effective protection to plants. By cloning these specific genes, special signaling pathways are introduced to produce an amplified response to insect infection (Dawson et al., 1989). Truncated forms of insect control genes from *Bacillus thuringiensis* var. *kurstaki* HD-1 *(cryIA(b)* and HD-73 *(cryIA(c)* when expressed in cotton plants provided effective control from lepidopteran insect pests (Perlak et al., 1990). The insertion of *cryIIIA* gene which also encodes an insect control protein, from *Bacillus thuringiensis* var. *tenebrionis* into Russet Burbank potato produced resistance and protected from damage by Colorado potato beetles (*Leptinotarsa decemlineata* (Say)) (Perlak et al., 1993).

Other developments of transgenic plants include expressing a truncated-endotoxin gene, *cryIA(b)* of *Bacillus thuringiensis*(Bt) which is known to have specific activity against lepidopteran insects into japonica rice. The modified plants were more resistant to pests like the striped stemborer (*Chilo suppressalis*) and leaf folder (*Cnaphalocrosis medinalis*) in comparison to the untransformed rice plants (Fujimoto et al., 1993). Genetically modified (GM) crops exhibiting insect resistance due to the presence of Bt endotoxins is a most common trait in many experimental crops. Extensive testing has proved that there are no adverse effects related to the wide-spread and long-term use of Bt plants against non-target plant-feeding insects. But further research is required to study the effects of GM crops on soil biota, soil processes, and others (O'Callaghan et al., 2005).

7.5 RESPONSES OF PLANTS TO PATHOGENS

7.5.1 BIOLOGY OF PATHOGENS

Plant pathogens are infectious organisms, which bring about diseases and cause the degradation of the plant's conditions. New analytical approaches including high throughput DNA sequencing have helped scientists to track how pathogens have emerged and evolved, especially in agro-ecosystems. The development of new agricultural practices and crop species has led the pathogens present in their wild counterparts to pass onto the domesticated crops and evolve rapidly through horizontal gene transfer and hybridization (Stukenbrock and McDonald, 2008). Pathogens have both direct and indirect effects on the plant community. A direct effect is explained by the cases where the pathogens either prevent the recruitment of seeds or causes reduction in the adult or seedling plants; whereas an indirect effect occurs when there is an outbreak parasite from herbivores leads to a reduction of grazing pressure (Dobson and Crawley, 1994).

Plant pathogens have the ability to evolve rapidly through interspecific hybridization as can be observed when exotic pathogens are introduced, which leads to a disease epidemic. The difference being that when the pathogen was in its endemic location, it mostly undergoes routine selection, favoring the maintenance of either stable or sometimes fluctuating population structure over time. But in its new location, it gets subjected to new selection pressure with new host population, new vectors, and a different climate, thus leading to a rapid evolution (Brasier, 2001). RNA silencing has played a central role in host defense in plants against pathogens. In turn, pathogens have now developed RNA silencing suppressors, suggesting that disruption of host silencing is a part of the virulence strategy for many plant pathogens. Supporting evidence has illustrated that plants have also developed specific defense mechanism against RNA-silencing suppression by pathogens, proving a never-ending molecular race between the pathogen and the host (Pumplin and Voinnet, 2013).

Aerial dispersal of pathogens on the global and continental level has a huge impact on plant diseases. The dispersion of fungal spores in long distances by the wind can spread plant diseases across continents and can even establish in areas where the host plant is not even present. Thus, more study and research into the epidemiology and population genetics is required for taking up preventive measures (Brown, 2002).

7.5.2 PHYSIOLOGICAL AND GENETIC RESPONSES DEVELOPED BY PLANTS AGAINST PATHOGENS

Microorganisms, for a large part, have played a role in decomposing dead material, where only a few of them have evolved from saprophytic organisms to biotrophic pathogens. Now, these pathogens have the ability to colonize the living plants whereas the plants produce a natural immune response to resist the attack (Kombrink and Somssich, 1995). Plants responses to infection through synthesis and production of a variety of compounds which plays a central role in the different modes of disease resistance. These compounds include proteases, chitinases, phospholipases, lignin, antifungal proteins, hydroxyproline, and glycine-rich glycoproteins, oligosaccharides, hydrogen peroxide, and ROS (Kuc, 1997).

The plant system possesses an efficient and strong innate immune system which detects and wards off potentially dangerous microbes. This is achieved through the perception of pathogen- or microbe-associated molecular patterns (PAMPS or MAMPS) that is present at the plant's cell surface and the responses are known as PAMP-triggered immunity (PTI) (Boller and He, 2009). These elicitors are very important in mediating the effects of plant resistance although pathogen virulence factors have now become much evolved to suppress these elicitors and ultimately their response.

Artificial plant breeding and molecular cloning of disease resistance genes (R genes) which enable the plants to avert a diverse range of pathogens, has revealed that the proteins encoded by these genes have several common features. This proves that plants overtime has evolved a common signaling pathway for producing resistance to a diverse range of pathogens. Thus, characterization and research into these molecular signals can help to develop novel strategies for controlling plant disease (Staskawicz et al., 1995; Alvarado and Scholthof, 2009).

The ability of the plant to evade or get rid of the pathogens depends upon its ability to recognize the pathogen as early as possible. Many a plant responds to the early onset of attack by a hypersensitive reaction (HR) wherein the plants localize the infection through tissue necrosis or release of toxic molecules. The rapidity of the HR process allows the activation of preformed molecules rather than inducing disease resistance genes. One of these quick processes includes the fast release of active oxygen species (AOS), which is also known as the oxidative burst. AOS is mainly toxic intermediates which are a result of the successive one-electron reduction of molecular O_2 and include the superoxide anion (O_2^-), hydroxyl radical (OH) and hydrogen peroxide (H_2O_2). The AOS response forms a first defense response with HR and has been observed in a variety of plant species (Mehdy, 1994). Another molecule, which was recently identified was nitric oxide (NO) whose functions include modulation of hormones, wounding, and defense responses, including the regulation of cell death. Evidence also suggests that NO signaling also interacts with salicylic acid (SA) and jasmonic acid (JA) signaling pathways in coordinating a cumulative attack (Wendehenne et al., 2004). Sugar signaling is another mode of plant response to pathogen attack, in that it acts as signaling molecules which interact with the signaling network of hormones, which in turn regulates the plant immune system. Sugars also enhance the oxidative burst and increase the lignification of cell walls, stimulate the synthesis of flavonoids and induce certain pathogenesis-related proteins, thus bringing about a higher level of initial and fast response to the infection (Morkunas and Ratajczak, 2014).

Plants also respond to pathogen attacks with protease inhibitors, which gets both regulated and induced during infection (Ryan, 1990). Even plant DNA-binding TFs like the WRKY are also involved in developing a plant response to the pathogen. Research has shown that *Arabidopsis WRKY3* and *WRKY* when expressed in transgenic plants play a positive role in developing plant resistance to necrotrophic pathogens, which is further helped by the fact that these factors are nuclear-localized and have a sequence-specific DNA-binding activity (Lai et al., 2008). Ubiquitination is the most popular regulatory mechanism which has recently been reported to play a role in plant resistance to pathogens as well. Since then, several ubiquitin ligases have been identified as defense regulators. Although no specific targets have been identified for these ubiquitin ligases, still this unique protein modification system may play an immense role in regulating plant defense mechanisms (Devoto et al., 2003).

Plant-pathogen infection and response is a diverse and complex mechanism. The molecules and signaling pathways involved are still not completely identified. A comprehensive and detailed study is required for further identification and detailed mechanism of infection and resistance that develops as a result. Artificial modifications and gene cloning techniques can help to move forward and gather information in this regard.

7.6 RESPONSES OF PLANTS TO AGRICHEMICALS

7.6.1 *CHEMISTRY AND ADVERSE EFFECTS OF AGRICHEMICALS*

Agrichemicals agricultural chemicals are chemicals which are either used for promoting certain features of the plant or for protecting the plant against microorganisms or viruses. Agricultural chemicals like fertilizers, pesticides, and herbicides are chemically synthesized. These contain chemicals or minerals which are useful for plants or which can harm the insect, pest or pathogen by killing it. In the early days, farmers began to use it whenever there was an unwanted effect on the plant due to some stress or the other. But over time, overuse, and diversity of insects, pests, and pathogens have started to adversely affect the crops and leads to detrimental effects on the total environment.

Different derivatives of urea and carbamic acid are used as agrichemicals in pesticides and herbicides, because of their broad effectiveness and low toxicity to higher animals. This dimethyl phenyl urea, N-methyl-1-naphthyl-carbamate, and other related compounds have been tested on and considered to be safe, at low levels to humans, if they consume it through the food crop it was used on (Elespuru and Lijinsky, 1973). Though chemical herbicides, pesticides, and fungicides are much needed, there have been reports of their potential mutagenic and carcinogenic effects, especially to the farmers who use it seasonally if not daily. So, studies are conducted for the cytogenetic effects of these agrichemicals in workers exposed to it (Sobti et al., 1982).

Agrichemicals like pesticides and fertilizers have also contaminated groundwater throughout the world, in, and around farming areas. This brings about millions of people in danger and the cost of monitoring these chemicals in groundwater is going to cost billions. To avoid this, farmers can take up low-input methods, which can if not stop but reduce the environmental damage (Fleming, 1987). The source, transport, and fate of agrichemicals are mostly controlled by a combination of agricultural activities, environmental conditions, biological features, and chemical properties. Extensive research should be made on the field scale to understand the magnitude of transport of these chemicals to the water systems (Capel et al., 2008).

7.6.2 PHYSIOLOGICAL AND GENETIC RESPONSES OBSERVED IN PLANTS AGAINST AGRICHEMICALS

When using agrichemicals, it is always better to understand the response and study the physiology of crop plants as well. For example, chili (*Capsicum annuum* L.) is an important vegetable crop in India. But due to being an upland crop, it faces tremendous competition from weeds. And chemical control of weed has taken over manual weeding because of various reasons. The use of chemical herbicides has many unwanted consequences as well, like the reduction in photosynthetic rate in the crop plant whereas an increase in resistance of the weeds which it was intending to kill. Repeated use of herbicides has caused the genetic change, thus leading to instability. And these changes are altering the integrity of the plant at the gene or nucleotide level, thus producing a cyto-genetic response and reduction in crop yield (Reddy and Rao, 1982). There are many such effects that have been reported over the years, but research has still not been able to produce an effective result to reduce this form of response from plants. GM crops are an option to reduce the use of agrichemicals, but the evolved and more resisted form of insects, weeds, and pathogens infect the modified crops as well, making the farmers use agrichemicals to evade and protect the crop as a result. A more extensive study is required to develop and reduce the use of agrichemicals in agro-ecosystems.

7.7 CONCLUSION AND FUTURE PERSPECTIVE

The change in climate and as a result in the environment has brought a drastic change in the modes, types, and frequency of stress in the plant systems. And agroecosystems have been the most affected, as global food demands has raised. Research has shown the possible effects of climate change on crop and livestock yields and the economical consequences as a result of it. The rising temperature was the most dominant abiotic factor identified to have a direct effect on the development and survival of plants. Temperature also has brought changes in the biology of insects, pests, and pathogens as well, bringing about a genotypic and phenotypic variation, which is unknown to farmers. This has changed the strategies taken up by the pest and pathogens to exploit and infect plants. There are still many challenges which the biologists are facing in order to combat the effect of stress on plants. Future research needs to focus on more about the changing patterns of the stress, the response of the plants globally, and to increase awareness about the changing patterns for long-term effects.

KEYWORDS

- **abiotic stress**
- **agrichemical stress**
- **biotic stress**
- **pathogens**
- **pests**
- **weed species**

REFERENCES

Alvarado, V., & Scholthof, H. B., (2009). Plant response against invasive nucleic acids: RNA silencing and its suppression by plant viral pathogens. *Seminars in Cell and Developmental Biology, 20*(9), 1032–1040. doi: https://doi.org/10.1016/j.semcdb.2009.06.001.

Atkinson, N. J., & Urwin, P. E., (2012). The interaction of plant biotic and abiotic stresses: From genes to the field. *Journal of Experimental Botany, 63*(10), 3523–3543. doi: https://doi.org/10.1093/jxb/ers100.

Barrett, S. H., (1983). Crop mimicry in weeds. *Economic Botany, 37*(3), 255–282. doi: https://doi.org/10.1007/BF02858881.

Bebber, D. P., Holmes, T., & Gurr, S. J., (2014). The global spread of crop pests and pathogens. *Global Ecology and Biogeography, 23*(12), 1398–1407. doi: https://doi.org/10.1111/geb.12214.

Bebber, D. P., Holmes, T., Smith, D., & Gurr, S. J., (2014). Economic and physical determinants of the global distributions of crop pests and pathogens. *New Phytologist, 202*(3), 901–910. doi: https://doi.org/10.1111/nph.12722.

Bebber, D. P., Ramotowski, M. A. T., & Gurr, S. J., (2013). Crop pests and pathogens move pole wards in a warming world. *Nature Climate Change,3*, 985–988. doi: https://doi.org/10.1038/nclimate1990.

Bergman, J. M., & Tingey, W. M., (1979). Aspects of interaction between plant genotypes and biological control. *Bulletin of the Entomological Society of America,25*(4), 275–279. doi: https://doi.org/10.1093/besa/25.4.275.

Bhowmik, P. C., & Inderjit, (2003). Challenges and opportunities in implementing allelopathy for natural weed management. *Crop Protection, 22*(4), 661–671. doi: https://doi.org/10.1016/s0261-2194(02)00242-9.

Blackman, G. E., & Templeman, W. G., (1938). The nature of the competition between cereal crops and annual weeds. *The Journal of Agricultural Science,28*(2), 247. doi: https://doi.org/10.1017/s0021859600050656.

Blackshaw, R. E., Lemerle, D., Mailer, R., & Young, K. R., (2002). Influence of wild radish on yield and quality of canola. *Weed Science, 50*(3), 344–349. doi: https://doi.org/10.1614/0043–1745(2002)050[0344:iowroy]2.0co;2.

Boller, T., & He, S. Y., (2009). Innate immunity in plants: An arms race between pattern recognition receptors in plants and effectors in microbial pathogens. *Science, 324*(5928), 742–744. doi: https://doi.org/10.1126/science.1171647.

Bond, W., & Grundy, A. C., (2001). Non-chemical weed management in organic farming systems. *Weed Research, 41*(5), 383–405. doi: https://doi.org/10.1046/j.1365-3180.2001.00246.x.

Brasier, C. M., (2001). Rapid evolution of introduced plant pathogens via interspecific hybridization. *BioScience, 51*(2), 123. doi: https://doi.org/10.1641/0006-3568(2001)051[0123:reoipp]2.0.co;2.

Brattsten, L. B., Holyoke, C. W., Leeper, J. R., & Raffa, K. F., (1986). Insecticide resistance: Challenge to pest management and basic research. *Science, 231*(4743), 1255–1260. doi: https://doi.org/10.1126/science.231.4743.1255.

Brown, J. K. M., (2002). Aerial dispersal of pathogens on the global and continental scales and its impact on plant disease. *Science, 297*(5581), 537–541. doi: https://doi.org/10.1126/science.1072678.

Buhler, D. D., Gunsolus, J. L., & Ralston, D. F., (1991). Integrated weed management techniques to reduce herbicide inputs in soybean. *Agronomy Journal, 84*(6), 973–978. doi: https://doi.org/10.2134/agronj1992.00021962008400060013x.

Capel, P. D., McCarthy, K. A., & Barbash, J. E., (2008). National, holistic, watershed-scale approach to understand the sources, transport, and fate of agricultural chemicals. *Journal of Environmental Quality, 37*(3), 983. doi: https://doi.org/10.2134/jeq2007.0226.

Chikoye, D., Schulz, S., & Ekeleme, F., (2004). Evaluation of integrated weed management practices for maize in the northern Guinea savanna of Nigeria. *Crop Protection, 23*(10), 895–900. doi: https://doi.org/10.1016/j.cropro.2004.01.013.

Concenco, G., Aspiazu, I., Ferreira, E. A., Galon, L., & Da Silva, A. F., (2012). Physiology of crops and weeds under biotic and abiotic stresses. *Applied Photosynthesis,* 257–280. doi: https://doi.org/10.5772/30691.

Cramer, G. R., Urano, K., Delrot, S., Pezzotti, M., & Shinozaki, K., (2011). Effects of abiotic stress on plants: A systems biology perspective. *BMC Plant Biol., 11*, 163. doi: https://doi.org/10.1186/1471-2229-11-163.

Davis, V. M., Gibson, K. D., Bauman, T. T., Weller, S. C., & Johnson, W. G., (2009). Influence of weed management practices and crop rotation on Glyphosate-resistant horseweed (*Conyzacanadensis*) population dynamics and crop yield-years III and IV. *Weed Science, 57*(4), 417–426. doi: https://doi.org/10.1614/ws-09-006.1.

Dawson, G. W., Hallahan, D. L., Mudd, A., Patel, M. M., Pickett, J. A., Wadhams, L. J., & Wallsgrove, R. M., (1989). Secondary plant metabolites as targets for genetic modification of crop plants for pest resistance. *Pesticide Science, 27*(2), 191–201. doi: https://doi.org/10.1002/ps.2780270209.

Dawson, J. H., (1964). Competition between irrigated field beans and annual weeds. *Weeds, 12*(3), 206. doi: https://doi.org/10.2307/4040730.

Dawson, J. H., (1965). Competition between irrigated sugar beets and annual weeds. *Weeds, 13*(3), 245. doi: https://doi.org/10.2307/4041038.

Devote, A., Muskett, P. R., & Shirasu, K., (2003). Role of ubiquitination in the regulation of plant defense against pathogens. *Current Opinion in Plant Biology, 6*(4), 307–311. doi: https://doi.org/10.1016/s1369-5266(03)00060-8.

Di Tomaso, J. M., (1995). Approaches for improving crop competitiveness through the manipulation of fertilization strategies. *Weed Science,43*, 491–497. doi: https://doi.org/10.1017/S0043174500081522.

Dicke, M., Sabelis, M. W., Takabayashi, J., Bruin, J., & Posthumus, M. A., (1990). Plant strategies of manipulating predator-prey interactions through allelochemicals: Prospects for application in pest control. *Journal of Chemical Ecology, 16*(11), 3091–3118. doi: https://doi.org/10.1007/bf00979614.

Dobson, A., & Crawley, M., (1994). Pathogens and the structure of plant communities. *Trends in Ecology and Evolution, 9*(10), 393–398. doi: https://doi.org/10.1016/0169-5347(94)90062-0.

Dogimont, C., Bendahmane, A., Chovelon, V., & Boissot, N., (2010). Host plant resistance to aphids in cultivated crops: Genetic and molecular bases, and interactions with aphid populations. *ComptesRendusBiologies,333*(6/7), 566–573. doi: https://doi.org/10.1016/j.crvi.2010.04.003.

Durian, G., Rahikainen, M., Alegre, S., Brosche, M., & Kangasjarvi, S., (2016). Protein phosphatase 2A in the regulatory network underlying biotic stress resistance in plants. *Front Plant Sci., 7*, 812. doi: https://doi.org/10.3389/fpls.2016.00812.

Ehrenfeld, D., (2003). Globalization: Effects on biodiversity, environment, and society. *Conservation and Society, 1*(1), 99–111.

Elespuru, R. K., & Lijinsky, W., (1973). The formation of carcinogenic nitroso compounds from nitrite and some types of agricultural chemicals. *Food and Cosmetics Toxicology, 11*(4), 807–817. doi: https://doi.org/10.1016/s0015-6264(73)80364-5.

Fleming, M. H., (1987). Agricultural chemicals in groundwater: Preventing contamination by removing barriers against low-input farm management. *American Journal of Alternative Agriculture, 2*(03), 124. doi: https://doi.org/10.1017/s0889189300001776.

Fujimoto, H., Itoh, K., Yamamoto, M., Kyozuka, J., & Shimamoto, K., (1993). Insect-resistant rice generated by the introduction of a modified-endotoxin gene of *Bacillus thuringiensis*. *Nature Biotechnology, 11*(10), 1151–1155. doi: https://doi.org/10.1038/nbt1093-1151.

Fujita, M., Fujita, Y., Noutoshi, Y., Takahashi, F., Narusaka, Y., Yamaguchi-Shinozaki, K., & Shinozaki, K., (2006). Crosstalk between abiotic and biotic stress responses: A current view from the points of convergence in the stress signaling networks. *Current Opinion in Plant Biology, 9*(4), 436–442. doi: https://doi.org/10.1016/j.pbi.2006.05.014.

Gatehouse, J. A., (2002). Plant resistance towards insect herbivores: A dynamic interaction. *New Phytologist, 156*(2), 145–169. doi: https://doi.org/10.1046/j.1469-8137.2002.00519.x.

Gealy, D. R., Mitten, D. H., & Rutger, J. N., (2003). Gene flow between red rice (*Oryza sativa*) and herbicide-resistant rice (*O. sativa*): Implications for weed management. *Weed Technology 17*(3), 627–645. doi: https://doi.org/10.1614/wt02-100.

Gibbs, J. N., & Wainhouse, D., (1986). Spread of forest pests and pathogens in the Northern Hemisphere. *Forestry: An International Journal of Forest Research, 59*(2), 141–153. doi: https://doi.org/10.1093/forestry/59.2.141.

Gregory, P. J., Johnson, S. N., Newton, A. C., & Ingram, J. S. I., (2009). Integrating pests and pathogens into the climate change/food security debate. *Journal of Experimental Botany 60*(10), 2827–2838. doi: https://doi.org/10.1093/jxb/erp080.

Gupta, P., Sharma, R., Sharma, M. K., Sharma, M. P., Satpute, G. K., Garg, S., Singla-Pareek, S. L., & Pareek, A., (2016). Signaling cross talk between biotic and abiotic stress responses in soybean. *Abiotic and Biotic Stresses in Soybean Production, 1*, 27–52. doi: https://doi.org/10.1016/B978-0-12-801536-0.00002-5.

Horvath, D. P., Bruggeman, S. A., Miller, J. M., Anderson, J. V., Dogramaci, M., Scheffler, B. E., Hernandez, A. G., Foley, M. E., & Clay, S., (2018). Weed presence altered biotic stress and light signaling in maize when weeds were removed early in the critical weed-free period. *Plant Direct, 2*, 1–15. doi: https://doi.org/10.1002/pld3.57.

Huang, J., Levine, A., & Wang, Z., (2013). Plant abiotic stress. *The Scientific World Journal.* doi: https://dx.doi.org/10.1155/2013/432836.

Jeandroz, S., & Lamotte, O., (2017). Plant responses to biotic and abiotic stresses: Lessons from cell signaling. *Front Plant Sci.,8*, 1772. doi: https://doi.org/10.3389/fpls.2017.01772.

Jermy, T., (1984). Evolution of insect/host plant relationships. *The American Naturalist, 124*(5), 609–630. doi: https://doi.org/10.1086/284302.

Jornsgard, B., Rasmussen, K., Hill, J., & Christiansen, J. L., (1996). Influence of nitrogen on competition between cereals and their natural weed populations. *Weed Research, 36*(6), 461–470. doi: https://doi.org/10.1111/j.1365–3180–1996.tb01675.x.

Keen, N. T., (1990). Gene-for-gene complementarity in plant-pathogen interactions. *Annual Review ofGenetics, 24*(1), 447–463. doi: https://doi.org/10.1146/annurev.ge.24.120190.002311.

Kissoudis, C., Van, D. W. C., Visser, R. G. F., & Van, D. L. G., (2014). Enhancing crop resilience to combined abiotic and biotic stress through the dissection of physiological and molecular crosstalk. *Front. Plant Sci., 5*, 207. doi: https://doi.org/10.3389/fpls.2014.00207.

Kombrink, E., & Somssich, I. E., (1995). Defense responses of plants to pathogens. *Advances in Botanical Research*, 1–34. doi: https://doi.org/10.1016/s0065-2296(08)60007-5.

Kriticos, D. J., Leriche, A., Palmer, D. J., Cook, D. C., Brockerhoff, E. G., Stephens, A. E. A., & Watt, M. S., (2013). Linking climate suitability, spread rates, and host-impact when estimating the potential costa of invasive pests. *Plos One*. doi: https://doi.org/10.1371/journal.pone.0054861.

Kuc, J., (1997). Molecular aspects of plant responses to pathogens. *Acta Physiologiae Plantarum, 19*(4), 551–559. doi: https://doi.org/10.1007/s11738-997-0053-2.

Lai, Z., Vinod, K. M., Zheng, Z., Fan, B., & Chen, Z., (2008). Roles of *Arabidopsis* WRKY3 and WRKY4 transcription factors in plant responses to pathogens. *BMC Plant Biology, 8*(1), 68. doi: https://doi.org/10.1186/1471-2229-8-68.

Liebhold, A. M., Macdonald, W. L., Bergdahl, D., & Mastro, V. C., (1995). Invasion by exotic forest pests: A threat to forest ecosystems. *Forest Science, 41*(1), a0001–z0001. doi: https://doi.org/10.1093/forestscience/41.s1.a0001.

Liebman, M., & Dyck, E., (1993). Crop rotation and intercropping strategies for weed management. *Ecological Applications, 3*(1), 92–122. doi: https://doi.org/10.2307/1941795.

Lovett, G. M., Canham, C. D., Arthur, M. A., Weathers, K. C., & Fitzhugh, R. D., (2006). Forest ecosystem responses to exotic pests and pathogens in eastern North America. *Bioscience, 56*(5), 395–405. doi: https://doi.org/10.1641/0006-3568(2006)056[0395:FERTEP]2.0.CO;2 (accessed on 13 June 2020).

Mehdy, M. C., (1994). Active oxygen species in plant defense against pathogens. *Plant Physiology, 105*, 467–472.

Mittler, R., (2006). Abiotic stress, the field environment and stress combination. *Trends Plant Sci., 11*(1), 15–19. doi: https://doi.org/10.1016/j.tplants.2005.11.002.

Morkunas, I., & Ratajczak, L., (2014). The role of sugar signaling in plant defense responses against fungal pathogens. *Acta Physiologiae Plantarum, 36*(7), 1607–1619. doi: https://doi.org/10.1007/s11738–014–1559-z.

Mortensen, D. A., Egan, J. F., Maxwell, B. D., Ryan, M. R., & Smith, R. G., (2012). Navigating a critical juncture for sustainable weed management. *Bioscience, 62*(1), 75–84. doi: https://doi.org/10.1525/bio.2012.62.1.12.

Mosa, K. A., Ismail, A., & Helmy, M., (2017). Introduction to plant stresses. *Plant Stress Tolerance*. Springer, Cham. doi: https://doi.org/10.1007/978-3-319-59379-1_1.

Murphy, S. D., Yakubu, Y., Weise, S. F., & Swanton, C. J., (1996). Effect of planting patterns and Inter-row cultivation on competition between corn (*Zea mays*) and late emerging weeds. *Weed Science, 44*(4), 865–870. doi: https://www.jstor.org/stable/4045752.

Neve,P.,Vila-Aiub,M.,&Roux,F.,(2009).Evolutionary-thinkinginagriculturalweedmanagement. *Phytologist, 184*(4), 783–793. doi: https://doi.org/10.1111/j.1469-8137.2009.03034.x.

Nieto, H. J., Brondo, M. A., & Gonsalez, J. T., (1968). Critical periods of the crop growth cycle for competition from weeds. *International Journal of Pest Management: Part C, 14*(2), 159–166. doi: https://doi.org/10.1080/05331856809432576.

O'Callaghan, Glare, T. R., Burgess, E. P. J., & Malone, L. A., (2005). Effects of plants genetically modified for insect resistance on nontarget organisms. *Annual Review of Entomology, 50*(1), 271–292. doi: https://doi.org/10.1146/annurev.ento.50.071803.1303.

Pandey, P., Irulappan, V., Bagavathiannan, M. V., & Senthil-Kumar, M., (2017). Impact of combined abiotic and biotic stresses on plant growth and avenues for crop improvement by exploiting physio-morphological traits. *Front. Plant Sci.,8*, 537. doi: https://doi.org/10.3389/fpls.2017.00537.

Perlak, F. J., Deaton, R. W., Armstrong, T. A., Fuchs, R. L., Sims, S. R., Greeplate, J. T., & Fischhoff, D. A., (1990). Insect-resistant cotton plants. *Nature Biotechnolog., 8*(10), 939–943. doi: https://doi.org/10.1038/nbt1090–939.

Perlak, F. J., Stone, T. B., Muskopf, Y. M., Petersen, L. J., Parker, G. B., McPherson, S. A., Wyman, J., Love, S., et al., (1993). Genetically improved potatoes: Protection from damage by Colorado potato beetles. *Plant Molecular Biology, 22*(2), 313–321. doi: https://doi.org/10.1007/bf00014938.

Preston, C., (2014). Plant biotic stress: Weeds. *Encyclopedia of Agriculture and Food Systems, 4*, 343–348. doi: https://doi.org/10.1016/B978-0-444-52512-3.00169-8.

Pumplin, N., & Voinnet, O., (2013). RNA silencing suppression by plant pathogens: Defense, counter-defense and counter-counter-defense. *Nature Reviews Microbiology, 11*(11), 745–760. doi: https://doi.org/10.1038/nrmicro3120.

Radosevich, S. R., (1987). Methods to study interactions among crops and weeds. *Weed Technology, 1*(3), 190–198. doi: https://doi.org/10.1017/s0890037x00029523.

Raffa, K. F., & Berryman, A. A., (1983). The role of host plant resistance in the colonization behavior and ecology of bark beetles (Coleoptera: Scolytidae). *Ecological Monographs, 53*(1), 27–49. doi: https://doi.org/10.2307/1942586.

Rathore, M., Singh, R., Choudhury, P. P., & Kumar, B., (2014). Weed stress in plants. *Abiotic Stresses in Major Pulses: Current Status and Strategies*. doi: https://doi.org/10.1007/978-81-322-1620-9_14.

Reddy, S. S., & Rao, G. M., (1982). Cytogenetic effects of agricultural chemicals. II. Effects of herbicides lasso and basagran on chromosomal mechanism in relation to yield and yield components in chili (*Capsicum annuum* L.). *Cytologiam, 47*(2), 257–267. doi: https://doi.org/10.1508/cytologia.47.257.

Rejeb, I. B., Pastor, V., & Mauch-Mani, B., (2014). Plant responses to simultaneous biotic and abiotic stress: Molecular mechanisms. *Plants (Basel), 3*(4), 458–475. doi: https://doi.org/10.3390/plants3040458.

Ryan, C. A., (1990). Protease inhibitors in plants: Genes for improving defenses against insects and pathogens. *Annual Review of Phytopathology, 28*(1), 425–449. doi: https://doi.org/10.1146/annurev.py.28.090190.002233.

Satorre, E. H., & Snaydon, R. W., (1992). A comparison of root and shoot competition between spring cereals and *Avenafatua* L. *Weed Research, 32*(1), 45–55.doi: https://doi.org/10.1111/j.1365-3180.1992.tb01861.x.

Sergeant, K., & Renaut, J., (2010). Plant biotic stress and proteomics. *Current Proteomics 7*(4), 275–297. doi: https://doi.org/10.2174/157016410793611765.

Sha, V. K. P. S., Nagamallaiah, G. V., Dhanunjay, R. M., Sergeant, K., & Hausman, J. F., (2014). Abiotic stress tolerance in plants. *Emerging Technologies and Management of Crop Stress Tolerance*, 23–68. doi: https://doi.org/10.1016/b978-0-12-8008875-1.00002-8.

Sharma, H. C., & Ortiz, R., (2002). Host plant resistance to insects: An eco-friendly approach for pest management and environmental conservation. *Journal of Environmental Biology, 23*(2), 111–135.

Singh, H. P., Batish, D. R., & Kohli, R., (2003). Allelopathic interactions and allelochemicals: New possibilities for sustainable weed management. *Critical Reviews in Plant Sciences,22*(3/4), 239–311. doi: https://doi.org/10.1080/713610858.

Singla, J., & Krattinger, S. G., (2016). Biotic stress resistance genes in wheat. *Reference Module in Food Science*. doi: https://doi.org/10.1016/B978-0-08-100596-5.00229-8.

Sobti, R. C., Krishan, A., & Pfaffenberger, C. D., (1982). Cytokinetic and cytogenetic effects of some agricultural chemicals on human lymphoid cells *in-vitro*: Organophosphates. *Mutation Research, 102*(1), 89–102. doi: https://doi.org/10.1016/0165-1218(82)90149-5.

Sogawa, K., (1982). The rice brown planthopper: Feeding physiology and host plant interactions. *Annual Review of Entomology 27*(1), 49–73. doi: https://doi.org/10.1146/annurev.en.27.010182.000405.

Staskawicz, B. J., Ausubel, F. M., Baker, B. J., Ellis, J. G., & Jones, J. D. G., (1995). Molecular genetics of plant disease resistance. *Science, 268*(5211), 661–667. doi: https://doi.org/10.1126/science.7732374.

Stukenbrock, E. H., & McDonald, B. A., (2008). The origins of plant pathogens in agro-ecosystems. *Annual Review of Phytopathology, 46*(1), 75–100. doi: https://doi.org/10.1146/annurev.phyto.010708.154114.

Thaler, J. S., & Bostock, R. M., (2004). Interactions between abscisic-acid-mediated responses and plant resistance to pathogens and insects. *Ecology, 85*(1), 48–58. doi: https://doi.org/10.1890/02–0710.

Thompson, J. N., & Burdon, J. J., (1992). Gene-for-gene co-evolution between plants and parasites. *Nature, 360*(6400), 121–125. doi: https://doi.org/10.1038/360121a0.

Tuteja, N., Singh, G. S., & Tuteja, R., (2011). Plant Responses to abiotic stresses: Shedding light on salt, drought, cold, and heavy metal stress. *Omics and Plant Abiotic Stress Tolerance*, 39–64. doi: https://doi.org/10.2174/978160805058111110101.

Van, H. H. D. J., (1985). The influence of weed competition on crop yield. *Agricultural Systems, 18*(2), 81–93. doi: https://doi.org/10.1016/0308-521x(85)90047-2.

Verma, S., Nizam, S., & Verma, P. K., (2013). Biotic and abiotic stress signaling in plants. *Stress Signaling in Plants: Genomics and Proteomics Perspective, 1*, 25–49. doi: https://doi.org/10.1007/978-1-4614-6372-6_2.

Verma, V., Ravindran, P., & Kumar, P. P., (2016). Plant hormone-mediated regulation of stress responses. *BMC Plant Biol., 16*, 86. doi: https://doi.org/10.1186/s12870-016-0771-y.

Weiner, J., (1990). Asymmetric competition in plant populations. *Trends in Ecology and Evolution, 5*(11), 360–364. doi: https://doi.org/10.1016/0169-5347(90)90095-U.

Wendehenne, D., Durner, J., & Klessig, D. F., (2004). Nitric oxide: A new player in plant signaling and defense responses. *Current Opinion in Plant Biology, 7*(4), 449–455. doi: https://doi.org/10.1016/j.pbi.2004.04.002.

Weston, L. A., (1996). Utilization of allelopathy for weed management in agroecosystems. *Agronomy Journal, 88*(6), 860–866. doi: https://doi.org/10.2134/agronj1996.00021962003600060004x.

Zhu, J., (2016). Abiotic stress signaling and responses in plants. *Cell, 167*(2), 313–324. doi: https://doi.org/10.1016/j.cell.2016.08.029.

CHAPTER 8

Stomatal Adaptive Response in Plants Under Drought Stress

HASTHI RAM[1], AMANDEEP KAUR[1], NISHU GANDASS[1], MEGHA KATOCH[1], SHWETA ROY[2], NIRBHAY KUMAR KUSHWAHA[3], and PRAVEEN SONI[4]

[1]*National Agri-Food Biotechnology Institute (NABI), Sector-81, SAS Nagar, Mohali, Punjab – 140306, India, E-mail: hrmundel84@gmail.com (H. Ram)*

[2]*National Institute of Plant Genome Research, New Delhi – 110067, India*

[3]*School of Life Sciences, Jawaharlal Nehru University, New Delhi – 110067, India*

[4]*Department of Botany, University of Rajasthan, Jaipur – 302004, Rajasthan, India, E-mail: praveen.soni15@gmail.com*

ABSTRACT

In most of the plants opening of the keep their stomata open during the day-time to uptake CO_2 for the photosynthesis, and side effect of keeping their stomata open during the day is that they lose 97% of the water they uptake to transpiration via stomata Plant hormone abscisic acid (ABA) has been considered as very crucial in drought-induced stomatal closing. The underlying mechanism of ABA triggered stomata closing is quite well understood. In succinct, during drought, plants induce accumulation and signaling of ABA in guard cells through various ways. Increased ABA signaling leads to increase of cytosolic Ca+2 levels, which consequently activates various ion channels at plasma membrane and tonoplast, which leads to efflux of K+ and various anions from the cell, leading to loss of turgor in guard cells and closing of the stomata. Although C3 plants have mechanisms to close their stomata during day-time when they are faced with water-deficit conditions, (but in prolonged drought conditions, continuous closing of stomata reduces plant's capacity to: 1) regulate its temperature, (2) nutrient absorption by roots, and (3) photosynthesis, which leads to their death. Plants with crassulacean acid metabolism (CAM) have a mechanism for night-time carbon fixation, so they keep their stomata closed during day-time to avoid this high rate of day time transpiration.

8.1 INTRODUCTION

"Stoma" is originally derived from the ancient word "στόμα," which means "inlet." Small pores are found in the stomata in the epidermis of leaf and stem, through which water vapor and gases are also passed. Stomata mainly present on the lower part of the leaf and new stomata are generated continually during leaf development (Zhao et al., 2015). The stomata term refers to the paired guard cells and the pore itself. Stomata are present in all land plants groups except liverworts; however, their number, size, and distribution vary considerably. The epidermal cells are polygonal in shape and their walls are waxy anticlinal (Karthikeyan et al., 2012). Pores are bordered by a couple of specific cells called guard cells, and specialized additional cells surround the two guard cells. Stomata are differentiated on the basis of their size, shape, and arrangement of the additional cells. Sometime stomata are present as stomatal crypts, which are at a lower surface of the leaf epidermis which forms a chamber-like shape that contains more than one stomata. They are an irregular celled type, missing distinct additional cells. The guard cells are responsible for governing the opening size. Dicots and non-grass monocots have kidney liked shaped stomata. Stomata permit the absorption of carbon dioxide necessary for photosynthesis from the air. The way to this mechanism is to suppose to be that the air inside the leaf is soaked with water vapor, with the immersion vapor pressure at that point computed by an exponential association with leaf temperature.

Drought is most significant abiotic stresses affecting the output yield of most of the crops. Drought is a period of in which below-average precipitation occurs in a given region, which results in prolonged shortages in the water supply. Various plants adopt various mechanisms to adapt to drought conditions, which broadly can be categorized into two mechanisms: drought avoidance and drought tolerance. Drought avoidance introduces to mechanisms in which plants complete their life cycle without facing drought, whereas drought tolerance mechanisms involve facing the drought with changes in their physiological activities. Drought leads to the defeat of plant growth or death decrease productivity. Cell growth is mostly affected (inhibition of cell wall and protein synthesis). So there is a need to focus on stomata handling drought stress. Plant growth in water stress shows their evolution for endurance to drought by drought escape by completing its life cycle before drought by modulating their vegetative and reproductive growth based on water availability by two mechanisms, i.e., rapid phonological and developmental plasticity. The rapid phonological strategy involves producing minimal seeds before water depletion and having no special adaptation. Plant with developmental plasticity produce less flowers under water stress but large number of seeds during wet seasons. Water saver plants work by reducing transpiration and transpiration area by involving maintenance of cell through turgor pressure, osmotic adjustment, and cellular elasticity. Improving crop yield is essential for growing population; there exist the variety of different mechanisms for drought escape, avoidance, or tolerance in many plant populations. Extreme drought resistance is found in resurrection plants, which possess strong drought escape abilities. These plants can survive for years in drought conditions for optimizing growth for survival. Crop plants are grown for high yield and during their exposure to random short-term drought stress, they respond quickly to limit damage and continue to grow and give yield, therefore bringing adaptive conditions. Therefore, bringing drought adaptive mechanisms from plant adapted to grow in extreme conditions is not a feasible option as it may result in a yield penalty under normal as well as drought conditions. Identification of mechanisms and traits and genes regulating yield under drought stress that are from normal conditions is an important task. For example, transgenic plants expressing OsNAC5, OsNAC9, OsNAC10 transcription factors (TFs) showed an increase of 5–26% yield under normal conditions are overlooked because currently the emphasis is given to yield under drought stress. Adaptation through drought resistance involves morpho-physiological alterations in adaptations processes, which are controlled by molecular mechanisms regulating gene expression. Genotypes differing in drought adaptive conditions serve as great resources to drought adaptation in crop plants which in turn can be exploited to improve drought resistance.

8.2 MECHANISM OF STOMATAL DEVELOPMENT

In plant development, Stomata is identified by a section of cell divisions. Entrance towards stomatal is originate through the dissection of an comparable epidermal post-embryonic cell to allow two irregular dimensions of daughter cells and the compact cell resulting to appearance in equivalve selection, this is called a meristemoid, short-lived cell of stem has similar properties. This undergoes various 'enlarge' asymmetric divisions in a meristemoid and which further reproduce to form a meristemoid and then sister cells are producing in extra numbers, previous discriminating towards the guard mother cell (GMC) and these guard cells are oval in shaped. It undergoes a single symmetric division because of their single symmetric division. They have the capability to synthesize the pair of guard cells, and which give rise to stomatal pore within the cells and these sister's cells mature within the epidermal cells, but it again split unsymmetrical to construct extra satellite meristemoids. Transcription element is proteins which bind to DNA-regulatory sequences, it generally restricted in the 5'-upstream zone to target genes, and which can regulate the rate of gene transcription. In this case, it may give rise to increased or decreased gene transcription, subsequent altered cellular function, and protein synthesis, TFs are also play crucial role such as (BHLH) transcription strands (dividing 90% amino acid comparability similar and 40% similarity total in their BHLH), speechless (SPCH), FAMA, and MUTE are sequentially mandatory for the change among the crucial cell types in the stomatal lineage (Pillitteri et al., 2007; MacAlister et al., 2007; Ohashi-Ito and Bergmann, 2006). SPCH obtain MMC regulates the unsymmetrical entrance division of these cells, also the consequent asymmetric increasing and spacing divisions that enlarge the lineage (Robinson et al., 2011; MacAlister et al., 2007, 2011; Pillitteri et al., 2007). Eliminate MUTE stem cell performance by encouraging the discrimination of meristemoids into GMCs (MacAlister et al., 2007), and FAMA assist the terminal characterization and cell division of GMCs into GCs (Ohashi-Ito and Bergmann, 2006). Mutant phenotype and expression patterns compare to these roles within the stomatal lineage cells, SPCH reveals only in MMCs. Meristemoids, and in *such* mutant, stomatal lineage cells are not open so the epidermis involves only the pavement cells. The expression of the bHLH TFs also provides beneficial intuition towards the nature of their movements. Overexpression of MUTE produces epidermis which is properly full with stomata which involving some cells that are transdifferentiated from pavement cells (Pillitteri et al., 2007). MUTE cannot make GCs due to lack of FAMA, but it can decay the need for SPCH, demonstrating the recent unsymmetrical divisions which are not absolutely needed for construct GC fates (Pillitteri et al., 2007). FAMA can also produce ectopic GCs, but they are not accurately paired, expressing a second role of FAMA in regulating cell division (Ohashi-Ito and Bergmann, 2006).

MYB TF family involved to regulating various processes like responding to abiotic and biotic stresses, enlargement, differentiation, defense, metabolism, etc., application of the MYB TFs FOUR LIPS and MYB88 approve the required for compact cell-cycle constitution throughout the GMC-to-GC transition (Xie et al., 2010). Overexpression of SPCH, in comparison to the output with FAMA and MUTE, does not form additional stomata (MacAlister et al., 2007). This phenotype disclosed two additional levels of regulation: first and second. First, is the post-translational phosphoregulation of SPCH protein itself and second is those precursor cell fates can be diverted midway through the pathway (Lampard et al., 2008). The technique accountable for these additional levels of control is explained later in this chapter. To understand the molecular control of stomatal development different types of model organisms are used. One of the major factors in stomata is EPFs, a family of excrete peptide signals called epidermal patterning factors and which have an idea to contain the receptor-like protein, TMM (too many mouths) and a putative leucine-rich repeat (LRR) receptor-like protein kinase (Rychel et al., 2010; Bergmann et al., 2007). These EPFs and TMM well known reported receptors that binds and activates an intracellular mitogen-activated protein kinase cascade, and phosphorylates and destabilizes a basic helix-loop-helix transcription factor required, early in leaf development, for cells to enter the stomatal lineage.

Previous studies have been described as three-act of EPF on the stomatal development pathway. Two pathways act as inhibitors of stomatal development, and the third one acts as an activator in the stomatal development process. EPF1 and EPF2 act to inhibit the formation of stomatal precursors by performing distinct but overlapping functions (Lampard et al., 2008; Hara et al., 2009). The main function EPF1 is in orienting cell divisions and prevents stomata from forming in clusters or pairs. The double mutant lacking both EPF1 and EPF2 has an additive phenotype. The *epf1epf2* plants exhibit greatly increased stomatal densities, and also has stomatal pairing and additional arrested cells. Plants manipulated to constitutively over-express EPF1 or EPF2 (*EPF1*OE and *EPF2*OE) have very few stomata. The third peptide, EPFL9/STOMAGEN, has essentially the opposite function to EPF1 and EPF2, the promotion of stomatal development (Hara et al., 2007; Kondo et al., 2010). This activator peptide is secreted from mesophyll cells and is proposed to prevent inhibitory peptides such as EPF1 or EPF2 from binding their receptor and inhibiting stomatal development.

8.3 EFFECTS OF DROUGHT ON STOMATAL MORPHOLOGY

Stomatal behave like drought which can vary widely both within and across species. Few plant species, like cowpea, and cassava are normally accessible to decreasing water capability, and stomatal transmission and transpiration decrease so more that the leaf water potential remains totally constant at the same time as drought. Stomatal delivery is normally closely related to soil water status and only parts of the plants that can precisely influence by the soil water status are the root system. Actually, stomatal closure can also be caused by the dehydrating root system part. Signals other than ABA, such as pH and inorganic ion reconstruction, in between the roots and shoots play a role in long-distance signaling. Short duration drought period, appear after a few weeks of water insufficiency, plants change their some physiological procedures, for example, they can change the size of their stomata. Yadollahi et al. (2011), has shown that drought significantly affect length and width and also thickness of the stomata in almonds (Prunusdulcis). Zhao et al. (2015), has demonstrated that decreased soil water content essentially rejuvenated age of maize in stomatal, lead about a vast stomatal increment thickness however a diminishing in stomatal size and gap. Zhang et al. (2004) in Platanusa Cerifolia and Liu et al. (2006) in jujube found an in the favorable connection between the length of the stomata under various drought conditions and stomatal thickness conditions. Yadollahi et al. (2011) found in the deficiency of relation between drought stress and stomatal density. They recommend that plant potency use other methods for drought resistance like stomatal delivery. They terminated stomata density are not proper evidence for drought resistance screening of almond seedlings; however, the size of stomata could be a suitable index for drought resistance. On the other hand, studies of the drought stress results to increasing stomatal density (Zhang et al., 2006; Xu and Zhou, 2008) in other plants. However, in other studies, it was shown stomatal size decreased with drought stress which designated the changes that occur in stomatal morphology could enlarge the plant adaptation to drought stress orders (Cutler et al., 1977; Spence et al., 1986). However, Spence et al. (1986) submitted that stoma caused by the small size of guard cells and stoma remain open under even drought conditions which convey the balance gain by carbon in photosynthesis and the inhibition of unconstrained water deficiency via transpiration adaptive response under dry conditions is prime adaption. Stomatal complex features such as density, size, and circulation may affect gas exchange and their interactions with key ecological factors such as water conditions. A developmental respond of drought is the production of a thicker cuticle which decreased water loss from the epidermis. Waxes can be retainer to both on the surface in the cuticle layer and the inner cuticle layer in the wax may be more important to controlling the rate of water loss in various ways that are more complicated than just increasing the amount of wax present. A thicker cuticle also helps to decrease CO_2 permeability; however, it will not change leaf photosynthesis because the epidermal cells underneath the cuticle are nonphotosynthetic. Cuticular transpiration reported for only 5 to 10% of the complete leaf transpiration, when stress is extremely severe or if the cuticle has been damaged. Because of their significant (e.g., by wind-driven sand). Shortfall of water also active energy diversion from leaves. The evaporative heat loss lowers leaf temperature, however, during water stress when transpiration is limited, the leaf heats up. It is very crucial for plants to maintain leaflet temperature that is very lower than the temperature of air, and that involving evaporation of more large quantities of water. That's why plants have adaptations that help to cool leaves by process other then evaporation, for example, leaf size is change, and leaf orientation or conserving water is very effective. For example, many arid-zone plants having very small leaves, which help reduce the resistance of the boundary layer to the transfer of heat from the leaf to the air. Leaves of some plants have tightly filled hairs, which keeps leaves cooler by reflecting radiation, but it also reflects the visible wavelengths that are active in photosynthesis and thus it decreases carbon assimilation. Some desert plants, for example, white brittlebush have dimorphic leaves to avoid excessive heating.

8.4 MECHANISM OF DROUGHT TRIGGERED STOMATAL CLOSING

Stomata are important for plant fertility and outlive. Every stoma function is to limit the loss of water due to evaporation and gases exchange, mainly CO_2 and O_2, which is necessary to control photosynthesis and regulate water loss by modulating the transpiration level. The factors involved in the process of stomata development were important for plant motion from water to land during evolution. Stomatal growth requires an equilibrium between proliferation and cell specification. The differentiation of stomata needed three different types of precursor cells: the meristemoid mother cell (MMC), meristemoids, and the GMC. In the end, the differentiation of the stoma itself within the structure of the guard cells during stomatal growth takes place. Recent research results have shown the action mode of stomata depends on various environmental factors and intracellular signals such as CO_2 concentration, biotic, and abiotic stresses, and additionally different plant hormones.

The first reaction of most of the plant towards drought is the closing of their stomata to prevent the water loss via transpiration (Mansfield and Atkinson, 1990; Berry et al., 2010; Casson and Hetherington, 2010; Brodribb and McAdam, 2011; Torres Ruiz et al., 2013; Clauw et al., 2015; Nemeskeri et al., 2015). This faster reaction is regulated by various signaling pathways such as abscisic acid (ABA), ethylene, auxins, and cytokinins (CKs) (Nemhauser et al., 2006; Huang et al., 2008). Generally, ABA is positive regulators of stomatal closure, while auxin and CKs are liable for stomatal opening. The way of action of ethylene hinges upon the tissue and conditions (Nemhauser et al., 2006; Huang et al., 2008).

ABA acknowledged ion movement across the cell membrane by exporting K+ and Cl^- ions through ion efflux pathways. This discharge water out of the cell through osmosis (Vahisalu et al., 2008) consequently, decreasing the internal solute concentration. The loss of water from guard cells lowers the volume of the cell, resulting in the loss of turgor pressure and closing of the stoma (Joshi-Saha et al., 2011).

Under drought stress conditions, ABA concentration reaches to high which causes ion efflux and an inhibition of sugar uptake by the guard cells, therefore reducing the stomatal apertures. ABA concentrations can increase up to 30-fold in the response under stress condition (Outlaw, 2003). Stomatal regulation is involved in the maintenance of the photosynthesis process in plants under stress conditions (Mahajan and Tuteja, 2005; Koyro et al., 2012; Perez Martin et al., 2014). Under Drought stress, the supply of carbon dioxide reduced significantly from the environment due to the closure of stomata, and photorespiration is increased, which maintains the carboxylating function of RuBisCO. Excess utilization reduced the equivalents in the chloroplast that causes a significant increase in oxygen-free radicals which produce oxidative damage in chloroplasts (Manzoni et al., 2011; Lisar et al., 2012).

One important signaling molecule present in the guard cell is reactive oxygen species (ROS), which is important in the signal transduction pathway. At high concentration of ROS, cause oxidative damage to the protein and DNA. Increased levels of ABA in guard cells cause H_2O_2 production; these oxidants modulate the activity of various phosphatase and kinases and alter its signaling pathway.

8.5 ABA TRIGGERS CHANGES IN ION HOMEOSTASIS IN THE GUARD CELLS, WHICH LEADS TO STOMATAL CLOSURE UNDER STRESS

During drought, plant hormone ABA triggers stomata closing, and the underlying mechanism is quite well understood (3). ABA signaling includes ABA receptors, phosphatases (PP2Cs) and kinases (SnRK2s) (Ma et al., 2009;

Park et al., 2009; Santiago et al., 2009; Nishimura et al., 2010). Phosphorylation and downstream activation are regulated by the inactivation of PP2Cs such as ABI1 and ABI2 by the complex ABA-receptor and of target phosphatases-SnRK2, such as SnRK2.2/D, SnRK2.3/E, and SnRK2.6/OST1/E, which are accountable for the regulation of ABA signaling and abiotic stress response (Fujii and Zhu, 2009; Fujita et al., 2009; Umezawa et al., 2009). Ion channels and the proton pump are controlled by the Kinases. ABA stops the functions of a proton pump such as H^+-ATPase. In succinct (Figure 8.1), during drought, plants induce accumulation and signaling of ABA in guard cells through various ways, which leads to increase of cytosolic Ca+2 levels, through the delivery of second messenger inositol-1,4,5-triphosphate (IP_3), which consequently trigger various ion channels at the plasma membrane and tonoplast leading to efflux of K+ and various anions from the cell, leading to reduction of turgor in guard cells and closing of the stomata.

During drought stress, Ca^{2+}-dependent protein kinases (CDPKs) get activates and helps to control stomatal closing in an ABA-dependent manner. The action of PP2Cs such as ABI1 is inhibited after receiving ABA signals by the receptors. ABI1 recognizes as a negative regulator of CPK21 (Ca^{2+} dependent protein kinase 21) which phosphorylates SLAC1 (slow anion channel-associated 1) leads to the generation of anion and the efflux of K^+ (Geiger et al., 2010). An increase in Ca^{2+} level in cytosolic face activates the Ca^{2+}-dependent pathways that turn to inhibit K^+ import and trigger membrane depolarization. Calcium-dependent kinases-CPK3 (calcium-dependent protein kinase 3) and CPK6 (calcium-dependent protein kinase 6) were identified by Mori et al. (2006) during water stress as a positive regulator of ABA signaling in the guard cells. Inactivation of these genes leads to a reduction in the action of S-type channels by ABA and Ca^{2+}, the deterioration of the ABA regulation of Ca^{2+} permeable channels and it decreased sensitivity of stomata in ABA. Disruption of the regulatory subunit *RCN1* (*roots curl in NP)* for encoding the gene PP2A (protein phosphatase 2A) led to the activation of ABA activation of anion channels and stomatal to ABA sensitivity decreases (Kwak et al., 2002).

ABA level increased the response to drought stress activates a signaling pathway that requires a sequence of events like an increase of the cytosolic Ca^{2+} level, which rotates to activate the anion channels (S-type and R-type), results of the depolarization occur in the plasma membrane. Activation of various kinases was responsible for the efflux of K^+ from the stomatal guard cells. Also with K^+ efflux, an efflux of water is also observed. These events together dropdown the turgor pressure of the guard cells and cause stomatal closure under drought conditions.

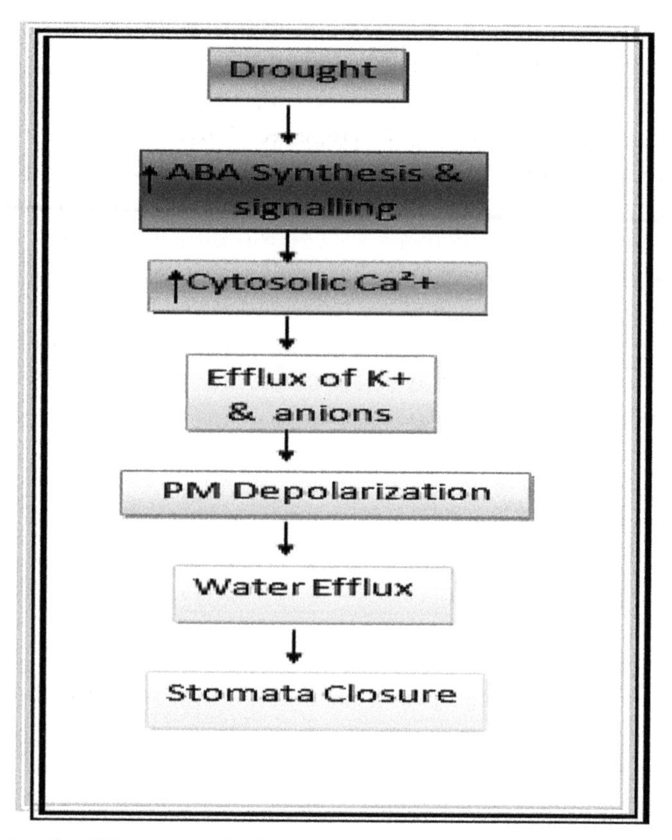

FIGURE 8.1 Simplified schematic of drought-induced stomatal closing.

8.6 STOMATAL OPENING CLOSING MECHANISM IN CAM PLANTS

Importance of CAM plants over C3 plants during drought conditions: Water availability is such an important issue for C3 plants, for, e.g., most of the crops; however, it is not such important for plants with crassulacean acid metabolism (CAM), for, e.g., many arid plants. These plants are called CAM plants. The reason behind C3 plants is that for the intake of CO_2 for the process of photosynthesis they need to keep their stomata open in course of day-time, and side effect of keeping their stomata open during the day is that they lose 97% of the water they uptake to transpiration via stomata (Goss et al., 1973). Unlike C3 plants, CAM plants have a mechanism for night-time carbon fixation, so they keep their stomata closed during day-time to avoid this high rate of day time transpiration. Although C3 plants have mechanisms to close their stomata during day-time when they are faced with water-deficit conditions, but during prolonged drought, continuous closing of stomata leads to their death because of their inability to 1) regulate temperature, 2) nutrient absorption by roots, and 3) photosynthesis.

CAM mechanism during CAM photosynthesis, CAM plants uptake CO_2 from the atmosphere during the night time and with help of phosphoenol-pyruvate carboxylase (PEPC) fix it into oxaloacetate (OAA) in the cytosol. OAA turns down to malate with help of NAD(P)-malate dehydrogenase (MDH) and stored in vacuoles as malic acid. During the day time, stored malic acid is transported out of vacuole as malate and gets changed to OAA, and then these C4 acids undergo decarboxylation by either NAD(P)-malic enzyme or PEP carboxykinase depending on the plant species (Costa et al., 2015). The released CO_2 is concentrated around the RUBISCO and from there it enters into the C3 carbon fixation cycle or Calvin-cycle (Figure 8.2).

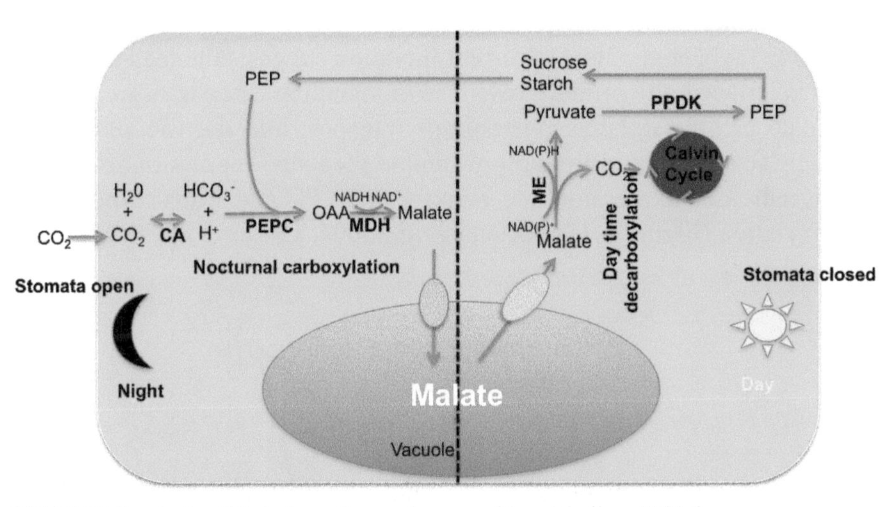

FIGURE 8.2 A simplified view of crassulacean acid metabolism (CAM).

Since the CAM pathway is an adaption to arid plants, its application to crops to confer drought tolerance is widely recognized, but so far, there have been very little effort to transfer CAM components into C3 plants (Winteretal et al., 1985; Lipka et al., 1999). More recently, people started to work towards a common aim to get the whole CAM pathway transferred into C3 plants (Kebeish et al., 2012). Although nocturnal carboxylation and day-time decarboxylation processes are quite well understood in CAM plants, but underlying process regarding their inverse stomata cycling is not clear. Hence, without understanding the molecular components of inverse stomata cycling, the idea of transfer of the CAM pathway is not expected to be fruitful.

8.7 DIURNAL REGULATION OF CAM PATHWAY

Current literature suggests that the progression of the CAM pathway needs advancement in transporters as well as the regulatory capacity of key enzymes for temporal separation of carboxylation and decarboxylation processes. Typically circadian control of carbon flux through PEPC is observed as a major component supporting the day and night separation of carboxylation processes that define CAM (Borland et al., 2014). Circadian control of PEPC activity is mainly mediated by its reversible phosphorylation by PEP carboxylase kinase (PPCK) (Nimmo et al., 2000). PPCK show rhythmic expression over the diel-cycle only in CAM performing tissues/species, not in C3 performing tissues/species (Carter et al., 1991; Borland et al., 2004; Taybi et al., 2004). Current literature suggests the idea that when CAM is induced, the primary response of signaling cascade is an increase in malate or other related metabolite's transport into the vacuole during the night, causing de-repression of malate mediated repression of PPCK, which increases expression and activity of PEPC and leads to nocturnal carboxylation (Figure 8.3) (Borland et al., 2004; Carter et al., 1991; Taybi

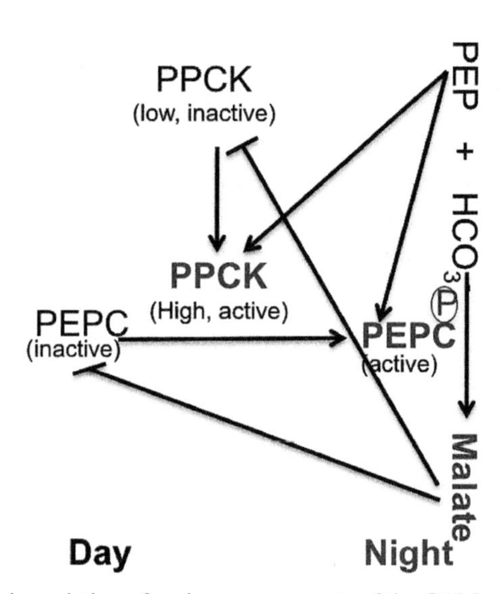

FIGURE 8.3 Diurnal regulation of various components of the CAM pathway.

et al., 2004). However, still direct evidence of the effect of malate or other metabolites on PPCK expression or CAM is lacking. Furthermore, in most of the previous studies, malate/metabolite levels were determined in the whole-leaf extracts that doesn't differentiate between cytosolic and vacuolar malate levels. Since, CAM is, in summary, a temporal separation of metabolites across the tonoplast; it seems very likely that the metabolites have a prominent regulatory role regarding their temporal separation across the tonoplast.

8.8 CONSEQUENCES AND FUTURE POSSIBILITIES

Closing of Stomata is the first reaction of plants against drought and stress. Soil moisture content is mostly related to it as compared to leaf water status. ABA which is produced in dehydrating roots is mainly used to control stomatal responses to drought stress. A direct interaction between the shrinking of stomata and the xylem ABA content is well established. Other aspects that have an important role in the regulation of stomata in plants include development in the leaf to air vapor pressure deficit, the status of plant nutritional, xylem sap pH, plant hydraulic conductance in addition farnesyl transferase activity, and reduction in the content of water. Plants species are also responsible for the effects of drought stress on stomatal morphology. Hence, it is important to do analysis on reactions of different species or their genotypes to water stress, regarding stomatal status. The contribution of ABA is popular in stomatal response to water stress; however, the role of other hormones, including brassinosteroids (BRs), salicylic acid (SA) and cytokinin, need to be explored. Interactions of other environmental factors, such as heat, light, etc., in managing stomatal reaction to water stress also need to be investigated in great detail. As compared to our clear knowledge of drought and light-mediated stress responses in stomatal guard cells, the molecular mechanisms behind stomatal signaling and CO_2 sensing are largely unknown. CAM photosynthesis is a major variant of the photosynthesis and has been repeatedly detected as providing high potential for sustainable output under climate change (Borland et al., 2011, 2014, 2015; Owen and Griffiths, 2014; Yang et al., 2015). However, there is still much more to be found about the proper working of CAM stomata. Application of the CAM pathway for improvement of C3 crops is being currently explored, which could provide fruitful results if it become successful.

KEYWORDS

- **guard mother cell**
- **malate dehydrogenase**
- **meristemoid mother cell**
- **reactive oxygen species**
- **too many mouths**
- **transcription factors**

REFERENCES

Berry, J. A., Beerling, D. J., & Franks, P. J., (2010). Stomata: Key players in the earth system, past and present. *Curr. Opinion Plant Biol., 13*, 233–240. doi: 10.1016/j.pbi.2010.04.013.

Borland, A. M., & Taybi, T., (2004). Synchronization of metabolic processes in plants with crassulacean acid metabolism. *Journal of Experimental Botany, 55*(400), 1255–1265. doi.org/10.1093/jxb/erh105.

Borland, A. M., Hartwell, J., Weston, D. J., Schlauch, K. A., Tschaplinski, T. J., Tuskan, G. A., Yang, X., & Cushman, J. C., (2014). Engineering crassulacean acid metabolism to improve water-use efficiency. *Trends Plant Sci., 19*(5), 327–338. doi: 10.1016/j.tplants.2014.01.006.

Brodribb, T. J., & McAdam, S. A., (2011). Passive origins of stomatal control in vascular plants. *Science, 331*, 582–585. doi: 10.1126/science.1197985.

Carter, P. J., Hugh, G. N., Charles, A. F., & Malcon, B. W., (1991). Circadian rhythms in the activity of a plant protein kinase. *EMBO J., 10*, 2063–2068.

Casson, S. A., & Hetherington, A. M., (2010). Environmental regulation of stomatal development. *Curr. Opinion Plant Biol., 13*, 90–95. doi: 10.1016/j.pbi.2009.08.005.

Clauw, P., Coppens, F., De Beuf, K., Dhondt, S., Van, D. T., Maleux, K., & Inze, D., (2015). Leaf responses to mild drought stress in natural variants of *Arabidopsis thaliana. Plant Physiol., 167*(3), 800–816. doi: 10.1104/pp.114.

Costal, J. M., Monnet, F., Jannaud, D., Leonhardt, N., Ksas, B., Ilja, R. M., Pantin, F., & Gentyet, B., (2015). Open all night long: The dark side of stomatal control. *Plant Physiology, 167*, 289–294. doi: org/10.1104/pp.114.253369.

Cushman, J., Tillett, R. L., Wood, J. A., Branco, J. M., & Schlauch, K. A., (2008). Large-scale mRNA expression profiling in the common ice plant, *Mesembryanthemum crystallinum*, performing C3 photosynthesis and crassulacean acid metabolism (CAM). *J. Exp. Bot., 59*, 1875–1894. doi: 10.1093/jxb/ern008.

Cutler, J. M., Rains, D. W., & Loomis, R. S., (1977). The importance of cell size in the water relations of plants. *Physiol. Plant, 40*, 225–260. doi.org/10.1111/j.1399-3054.1977.tb04068.x.

Fujii, H., & Zhu, J. K., (2009). *Arabidopsis* mutant deficient in 3 abscisic acid-activated protein kinases reveals critical roles in growth, reproduction and stress. *Proc. Natl. Acad. Sci. U.S.A., 106*, 8380–8385. doi: 10.1073/pnas.0905855106.

Fujita, Y., Nakashima, K., Yoshida, T., Katagiri, T., Kidokoro, S., & Kanamori, N., (2009). Three SnRK2 sprotein kinases are the main positive regulators of abscisic acid signaling in response to water stress in *Arabidopsis*. *Plant Cell Physiol., 50*, 2123–2132. doi: 10.1093/pcp/pcp147.

Gaastra, P., (1959). Photosynthesis of crop plants as influenced by light, carbon dioxide, temperature, and stomatal diffusion resistance. *Meded. Landbouwhogeschool, Wageningen., 59*, 1–68.

Geiger, D., Scherzer, S., Mumm, P., Marten, I., Ache, P., & Matschi, S., (2010). Guard cell anion channel SLAC1 is regulated by CDPK protein kinases with distinct Ca^{2+} affinities. *Proc. Natl. Acad. Sci. U.S.A., 107*, 8023–8028. doi: 10.1073/pnas.0912030107.

Goss, J. A., (1973). New York, Pergamon Press.

Joshi-Saha, A., Valon, C., & Leung, J., (2011). Abscisic acid signal off the starting block. *Mol. Plant, 4*, 562–580. doi: 10.1093/mp/ssr055.

Karthikeyan, R., Venkatesh, P., & Chandrasekhar, N., (2012). Morpho anatomical studies of leaves of *Abutilon indicum* (Linn.) sweet. *Asian Pacific Journal of Tropical Biomedicine, 2*(2), S464–S469. doi.org/10.1016/S2221-1691(12)60255-X.

Koyro, H. W., Ahmad, A., & Geissler, N., (2012). Abiotic stress responses in plants. In: Ahmad, P., & Prasad, M. N. V., (eds.), *An Overview in Environmental Adaptations and Stress Tolerance of Plants in the Era of Climate Change.* Springer, Berlin.

Kwak, J. M., Maser, P., & Schroeder, J. I., (2008). The clickable guard cell, version II: Interactive model of guard cell signal transduction mechanisms and pathways. *ArabidopsisBook 6*, e0114.

Lampard, G. R., MacAlister, C. A., & Bergmann, D. C., (2008). *Arabidopsis* stomatal initiation is controlled by MAPK-mediated regulation of the bHLH SPEECHLESS. *Science, 322*(5904), 1113–1116. doi: 10.1126/science.1162263.

Lipka, V., Hausler, R. E., & Rademacher, T., (1999). *Solanum tuberosum* double transgenic expressing phosphoenolpyruvate carboxylase and NADP-malic enzyme display reduced electron requirement for CO_2 fixation. *Plant Science, 144*(2), 93–105. doi: 10.1016/S0168–9452(99)00063-1.

Lisar, S. Y. S., Motafakkerazad, R., Hossain, M. M., & Rahman, I. M. M., (2012). Water stress in plants: Causes, effects and responses. In:Rahman, I. M. M., (ed.), *Water Stress* (pp. 1–14). InTech Publication.

Liu, S., Liu, J., Cao, J., Bai, C., & Shi, R., (2006). Stomatal distribution and character analysis of leaf epidermis of jujube under drought stress. *Journal of Anhui Agric. Sci., 34*, 1315–1318.

Ma, Y., Szostkiewicz, I., Korte, A., Moes, D., Yang, Y., & Christmann, A., (2009). Regulators of PP2C phosphatase activity functions as abscisic acid sensors. *Science, 324*, 1064–1068. doi: 10.1126/science.1172408.

MacAlister, C. A., Ohashi-Ito, K., & Bergmann, D. C., (2007). Transcription factor control of asymmetric cell divisions that establish the stomatal lineage. *Nature, 445*(7127), 537. doi: 10.1038/nature05491.

Mahajan, S., & Tuteja, N., (2005). Cold, salinity and drought stresses: An overview. *Arch Biochem. Biophys., 444*, 139–158. doi: 10.1016/j.abb.2005.10.018.

Mansfield, T. J., & Atkinson, C. J., (1990). Stomatal behavior in water stressed plants. In: Alscher, R. G., & Cumming, J. R., (eds.), *Stress Responses in Plants: Adaptation and Acclimation Mechanisms* (pp. 241–264). Wiley-Liss, New York.

Manzoni, S., Vico, G., Katul, G., Fay, P. A., Polley, W., Palmroth, S., & Porporato, A., (2011). Optimizing stomatal conductance for maximum carbon gain under water stress: A meta-analysis across plant functional types and climates. *Funct. Ecol., 25*, 456–467. doi: org/10.1111/j.1365-2435.2010.01822.x.

Mori, I. C., Murata, Y., Yang, Y., Munemasa, S., Wang, Y. F., & Andreoli, S., (2006). CDPKs CPK6 and CPK3 function in ABA regulation of guard cell S-type anion and Ca^{2+}-permeable channels and stomatal closure. *PLoS Biol., 4*, e327. doi: 10.1371/journal.pbio.0040327.

Nemeskeri, E., Molnar, K., Vigh, R., Nagy, J., & Dobos, A., (2015). Relationships between stomatal behaviour, spectral traits and water use and productivity of green peas (*Pisum sativum* L.) in dry seasons. *Acta Physiol. Plant, 37*, 1–16. doi: 10.1007/s11738-015-1776-0.

Nimmo, J., (2000). The regulation of phosphoenolpyruvate carboxylase in CAM plants. *Trends Plant Sci., 5*(2), 75–80. doi: 10.1016/S1360-1385(99)01543-5.

Nishimura, N., Sarkeshik, A., Nito, K., Park, S. Y., Wang, A., & Carvalho, P. C., (2010). PYR/PYL/RACR family members are major *in-vivo* ABI1 protein phosphatase 2C interacting proteins in *Arabidopsis*. *Plant J., 61*, 290–299. doi: 10.1111/j.1365-313X.2009.04054.x.

Ohashi-Ito, K., & Bergmann, D. C., (2006). *Arabidopsis* FAMA controls the final proliferation/differentiation switch during stomatal development. *The Plant Cell, 18*(10), 2493–2505. doi: 10.1105/tpc.106.046136.

Outlaw, W. H. Jr., (2003). Integration of cellular and physiological functions of the guard cells. *CRC Crit. Rev. Plant Sci., 22*, 503–529. doi: 10.1080/713608316.

Park, S. Y., Fung, P., Nishimura, N., Jensen, D. R., Fujii, H., & Zhao, Y., (2009). Abscisic acid inhibits type 2C protein phosphatases via the PYR/PYL family of START proteins. *Science, 324*, 1068–1071. doi: 10.1126/science.1173041.

Perez-Martin, A., Michelazzo, C., Torres-Ruiz, J. M., Flexas, J., Fernandez, J. E., Sebastiani, L., & Diaz-Espejo, A., (2014). Regulation of photosynthesis and stomatal and mesophyll conductance under water stress and recovery in olive trees: Correlation with gene expression of carbonic anhydrase and aquaporins. *J Exp Bot., 65*(12), 3143–3156. doi: 10.1093/jxb/eru160.

Pillitteri, L. J., Sloan, D. B., Bogenschutz, N. L., & Torii, K. U., (2007). Termination of asymmetric cell division and differentiation of stomata. *Nature, 445*(7127), 501. doi: 10.1038/nature05467.

Robinson, S., De Reuille, P. B., Chan, J., Bergmann, D., Prusinkiewicz, P., & Coen, E., (2011). Generation of spatial patterns through cell polarity switching. *Science, 333*(6048), 1436–1440. doi: 10.1126/science.1202185.

Santiago, J., Dupeux, F., Round, A., Antoni, R., Park, S. Y., & Jamin, M., (2009). The abscisic acid receptor PYR1 in complex with abscisic acid. *Nature, 462*, 665–668. doi: 10.1038/nature08591.

Spence, R. D., Wu, H., Sharpe, P. J. H., & Clark, K., (1986). Water stress effects on guard cell anatomy and the mechanical advantage of the epidermal cells. *Plant Cell Environ., 9*, 197–202. doi.org/10.1111/1365-3040.ep11611639.

Taybi, T., Nimmo, H. G., & Borland, A. M., (2004). Expression of phosphoenolpyruvate carboxylase and phosphoenol pyruvate carboxylase kinase genes. Implications for genotypic capacity and phenotypic plasticity in the expression of crassulacean acid metabolism. *Plant Physiol., 135*, 587–598. doi: org/10.1104/pp.103.036962.

Taybi, T., Patil, S., Chollet, R., & Cushman, J. C., (2000). A minimal serine/threonine protein kinase circadianly regulates phosphoenol pyruvate carboxylase activity in crassulacean acid metabolism-induced leaves of the common ice plant. *Plant Physiology, 123*, 1471–1481. doi.org/10.1104/pp.123.4.1471.

Thompson, A. J., Jackson, A. C., Symonds, R. C., Mulholland, B. J., Dadswell, A. R., Blake, P. S., Burbidge, A., & Taylor, I. B., (2000). Ectopic expression of a tomato

9-cis-epoxycarotenoid dioxygenase gene causes over-production of abscisic acid. *Plant Journal, 23*(3), 363–374. doi.org/10.1046/j.1365-313x.2000.00789.x.

Torres-Ruiz, J. M., Diaz-Espejo, A., Morales-Sillero, A., Martín-Palomo, M. J., Mayr, S., Beikircher, B., & Fernandez, J. E., (2013). Shoot hydraulic characteristics, plant water status, and stomatal response in olive trees under different soil water conditions. *Plant Soil, 373*, 77–87. doi: 10.1007/s11104-013-1774-1.

Umezawa, T., Sugiyama, N., Mizoguchi, M., Hayashi, S., Myouga, F., & Yamaguchi-Shinozaki, K., (2009). Type 2C protein phosphatases directly regulate abscisic acid-activated protein kinases in *Arabidopsis. Proc. Natl. Acad. Sci. U.S.A., 106*, 17588–17593. dio:10.1073/pnas.0907095106.

Vahisalu, T., Kollist, H., Wang, Y. F., Nishimura, N., Chan, W. Y., & Valerio, G., (2008). SLAC1 is required for plant guard cell S-type anion channel function in stomatal signaling. *Nature, 452*, 487–491. doi: 10.1038/nature06608.

Winter, K., (1985). *Photosynthetic Mechanisms and the Environment* (pp. 329–387).

Xu, Z., & Zhou, G., (2008). Responses of leaf stomatal density to water status and its relationship with photosynthesis in a grass. *J. Exp. Bot., 59*, 3317–3325. doi: 10.1093/jxb/ern185.

Yadollahi, A., Arzani, K., Ebadi, A., Wirthensohn, M., & Karimi, S., (2011). The response of different almond genotypes to moderate and severe water stress in order to screen for drought tolerance. *Scientia Horticulturae, 129*, 403–413. doi: 10.1016/j.scienta.2011.04.007.

Zhang, Y. P., Wang, Z. M., Wu, Y. C., & Zhang, X., (2006). Stomatal characteristics of different green organs in wheat under different irrigation regimes. *Acta Agron. Sin., 32*, 70–75.

Zhao, W., Sun, Y., Kjelgren, R., & Liu, X., (2015). Response of stomatal density and bound gas exchange in leaves of maize to soil water deficit. *Acta Physiologiae Plantarum, 37*(1), 1–9, 1704. doi: 10.1007/s11738-014-1704-8.

CHAPTER 9

Understanding the Epigenetic Regulation of Plants Under Stress and Potential Application in Agriculture

SHWETA ROY[1], NIRBHAY KUMAR KUSHWAHA[2], HASTHI RAM[3], and PRAVEEN SONI[4*]

[1]*Department of Cell and Molecular Biology, Uppsala University, Uppsala-75237, Sweden*

[2]*Department of Plant Biology, Swedish University of Agricultural Sciences, Uppsala-75007, Sweden*

[3]*National Agri-Food Biotechnology Institute, Mohali – 140306, India*

[4]*Department of Botany, University of Rajasthan, Jaipur – 302004, India*

[*]*Corresponding author. E-mail: praveen.soni15@gmail.com*

ABSTRACT

Being sessile organisms, plants cannot evade from various biotic and abiotic stresses so they have developed various adaptations and intricate stress response mechanisms to protect themselves. Plant genome undergoes a transcriptional reprogramming to provide resistance against various abiotic and biotic stresses. Epigenetic regulation involves histone modifications, DNA methylation, and small RNAs. It plays an essential role in controlling gene expression through chromatin modification. Till now several epigenetic regulators have been identified as stress response regulators. Recent studies show that chromatin modifications can also help in creating stress memory for survival against recurrent stresses. It has also been proved to be a transgenerational stress memory which is passed to the progenies. Therefore, epigenetic regulators can be key elements for engineering stress-resistant plants. At the same time, these regulators broadly alter the stress response transcriptome so they can be useful in not only a particular stress but combination of different biotic and abiotic stresses.

9.1 INTRODUCTION

All organisms face various environmental challenges in their lifetime. Unlike animals, plants cannot evade those challenges due to their sessile nature. Therefore, in due course of evolution, they have acquired various adaptations and developed defense mechanisms to overcome biotic and abiotic challenges. Whole plant genome undergoes a transcriptional reprogramming to provide resistance or tolerance against the adverse condition. Eukaryotic genomic DNAs carrying genetic information in association with histone proteins are organized in a regulatory structure of chromatin (Luger et al., 1997). All genetic information of DNA never gets expressed at any given time, some of the genes are expressed while other remains silenced. This specific gene expression mechanism is a result of fine-tuning between DNA and histone modifications without changing the primary gene sequences, the phenomenon known as epigenetics. Epigenetic modification can play a key role in that adverse circumstances as it can affect a large number of genes without changing the underlying genetic compositions as well as due to its reversible nature helps in maintaining the basal level gene expression, once the stress is relieved. While most of the modifications are reversible in nature, some of them are stable and provide long-term stress memory to combat recurrent stresses. Recent evidence suggests that these modifications can be inherited across meiotic cell divisions thus, to the next generation. Therefore, the genes coding for epigenetic regulators can be potential targets for generation of stress-resistant plants.

9.2 TYPES OF EPIGENETIC MODIFICATION

Epigenetic modifications can be largely divided into post-translational modifications of histone proteins, DNA or nucleic acid-based modifications, and noncoding RNA based modifications. Details about these modifications and their role in stress management are discussed below.

9.2.1 HISTONE MODIFICATIONS

Covalent modifications at the N terminal tail of histone proteins alter nucleosome positioning and compaction; therefore, regulate higher-order chromatin structure. Thus, they can play a key role in gene regulation. Acetylation and methylation of histones can be a crucial switch between a permissive and repressive state of chromatin. Other post-translational modifications (PTMs) of histones like phosphorylation, ubiquitination, sumoylation, glycosylation, and ADP-Ribosylation affect DNA-histone interaction within and between nucleosomes which affects DNA accessibility and transcription factor recruitment. These modifications regulate various cellular processes like transcription, replication, cell cycle, and DNA repair (Kouzarides et al., 2007). Types of histone modifications and their role in stress responses are discussed below.

9.2.1.1 HISTONE ACETYLATION

All four core histone proteins can be acetylated and deacetylated at different positions (Lusser et al., 2001). Acetylation occurs at certain lysine residues of core histone proteins H2A, H2B, H3, and H4. Acetylation of lysine at the N terminal tail neutralizes its positive charge and results in alteration of DNA histone interaction. Generally, it is considered to form a more open structure of chromatin and making it more accessible for transcription machinery proteins (Shahbazian et al., 2007). Histone acetylation is catalyzed by histone acetyltransferases (HATs) which transfer acetyl moiety of acetyl-coenzyme to the target lysine residue. Acetylation mark can be removed by the enzymatic action of histone deacetylases (HDACs). HATs and HDACs control cellular activities and their interplay regulates overall histone acetylation level in the cell (Bhaumik et al., 2007).

9.2.1.2 HISTONE METHYLATION

Two members of the core histone complex, H3 and H4 can be methylated at specific lysine or arginine residues and in some cases simultaneously at both residues (Kouzarides et al., 2007). Lysine residues can be methylated as mono, di, and trimethylated forms at 4, 9, 27, 36, and 79 positions of H3 and K20 of H4 by the enzymatic activity of histone methyltransferases (HMTs). Whereas arginine can be mono or dimethylated by arginine methyltransferase (PRMT1) enzyme (Shilatifard et al., 2006). Almost all methyltransferases have the SET domain which is required for its enzymatic activity. Unlike HATs, HMTs are more specific for their substrate (Bhaumik et al., 2007).

Similar to acetylation, histone lysine methylation can also be reversed by histone demethylases (HDMs) which can be either amine oxidases, e.g., Lysine-specific histone demethylase1 (LSD1) or hydroxylases such as Jumanji (JmjC) domain proteins (Klose et al., 2006). Histone methylation can have different biological implications depending on the type and position of modified lysine or arginine. In general, H3K4 mono-, di-, or tri-methylation and H3K36 methylation are associated with the active state of chromatin and H3K27 and H3K9 methylation results in transcription repression and heterochromatin formation respectively (Zhou et al., 2009). Similar to histone acetylation status, methylation status is also very dynamic and depends on the activity of HMTs and HDMs (Xiao et al., 2016).

9.2.1.3 HISTONE UBIQUITINATION

Histone ubiquitination is positively correlated with active transcription regulation (Weake et al., 2008). Histone H3, H2A, and H2B are known to be monoubiquitinated at specific lysine residues (Schnell et al., 2003). In *Arabidopsis*, ubiquitin carrier protein 1/2 (UBC1/UBC2) function as E2 conjugating enzyme and histone mono-ubiquitination 1/2 (HUB1/HUB2) are orthologues of an E3 ubiquitin ligase. One of the studies in plants revealed HUB1 or HUB2 in coordination with UBC1 or UBC2 form monoubiquitination machinery and promotes H3K4 trimethylation and regulates the flowering time in *Arabidopsis* (Cao et al., 2008). A recent study has revealed the role of the monoubiquitination of H2B in the rapid modulation of expressions of many genes during the photomorphogenesis (Bourbousse et al., 2012). A list of all major modifications on core histone molecules and their effects on transcription are summarized in Table 9.1.

9.2.2 NUCLEIC ACID/DNA-BASED MODIFICATION

Unlike histones, methylation is the only known covalent modification in DNA till date. In eukaryotes, the fifth carbon of cytosine is subjected to methylation that modulates chromatin architecture and regulates gene expression, DNA recombination, and genetic imprinting (Bird et al., 2002). Cytosine can be methylated in three sequence contexts; 5' CpG 3', 5' CpHpG 3', and 5' CpHpH 3'. Cytosine in CpG and CpHpG are called symmetric cytosines and CpHpH is called asymmetric cytosine (Finnegan et al., 2000). Methyltransferase1 (MET1) maintains CG methylation, Chromomethylase 3 (CMT3)

TABLE 9.1 List of Major Modifications at N-terminal Tail of Core Histone Proteins

Histone protein	Histone modifications			
	Lysine methylation	Arginine methylation	Acetylation	Ubiquitination
H3	K4	R2	K9	
	K9	R8	K14	
	K27	R17	K18	
	K36	R26	K23	
	K79		K27	
			K36	
			K56	
H4	K20	R3	K5	
		R17	K8	
		R19	K12	
			K16	
			K20	
H2A			K5	K120
			K12	
			K15	
			K29	
H2B			K119	K143

*Blue color denotes activation of transcription, red color denotes repression of transcription and black color refers to modifications that can result in either activation or repression.

maintains CHG methylation and Domain Rearranged Methyl Transferase 1/2 (DRM1 and DRM2) mediates RNA directed *de novo* DNA methylation (Finnegan et al., 2000). Presence of methylated cytosine is considered as a component of transcriptionally silent chromatin and acts as a mechanism to halt the expression of transposable elements (Martienssen et al., 2001). DNA methylation can be inhibited by chemicals like 5-azacytidine and zebularine which bind and inhibit the active site of methyltransferases. These chemicals can induce genome-wide hypomethylation which can result in altered growth and development.

9.2.3 NONCODING RNA-BASED MODIFICATION

Noncoding RNA (ncRNA) plays important role in the regulation of gene expression at both transcriptional and post-transcriptional levels without translating into proteins (Buhler et al., 2008; Kloc et al., 2008). According to sizes, they can be divided into two categories: (i) short ncRNA (<30 nucleotides) which includes microRNA (miRNA), small interfering RNA (siRNA), Piwi-interacting RNAs (piRNA); and (ii) long ncRNA which is generally more than 200 nucleotides.

miRNA binds to their complementary mRNA target to induce cleavage and degradation of the transcript. siRNA mediated epigenetic regulation occurs at the post-transcriptional level. They help in degradation of the target mRNA as well as guide the RNA directed DNA methylation to induce epigenetic modifications.

lncRNA recruits chromatin-modifying complexes at the site of regulation. In rice, antisense RNA is synthesized during stressful environmental conditions (Gan et al., 2012). Another example of well-known lncRNA is an antisense RNA present in *Arabidopsis* where flowering inhibitor *FLC* gene is regulated by its own antisense transcript COOLAIR (Swiezewski et al., 2009) and an alternative transcript formed from the first intron of FLC gene known as COLDAIR (Heo et al., 2011). Both lncRNA are transcribed in prolonged winters and recruit PRC2 complex for histone modification via H3k27me3 enrichment at FLC locus to repress the expression and initiate flowering. A schematic model of all three epigenetic mechanisms and their interplay in the regulation of gene expression is shown in Figure 9.1.

9.3 EPIGENETIC REGULATIONS UNDER ABIOTIC STRESSES

Plants perceive environmental changes and respond towards them in terms of molecular, structural, and physiological modifications. Recent studies show a correlation in histone modification, DNA methylation, and other epigenetic changes with the expression pattern of various stress-responsive genes (Kim et al., 2008; Roy et al., 2018; Kushwaha et al., 2017). These reports clearly indicate the role of epigenetic regulation in stress tolerance of plants. Along with that, stresses can create a stress memory which can be retained for a prolonged time and sometimes inherited to the next generations also. At the same time, stresses can sometimes alter a broad spectrum of genes which might be useful in tolerance against different stresses. In our chapter we have discussed three most detrimental abiotic stresses; drought, high temperature, and salinity in the context of epigenetic modifications.

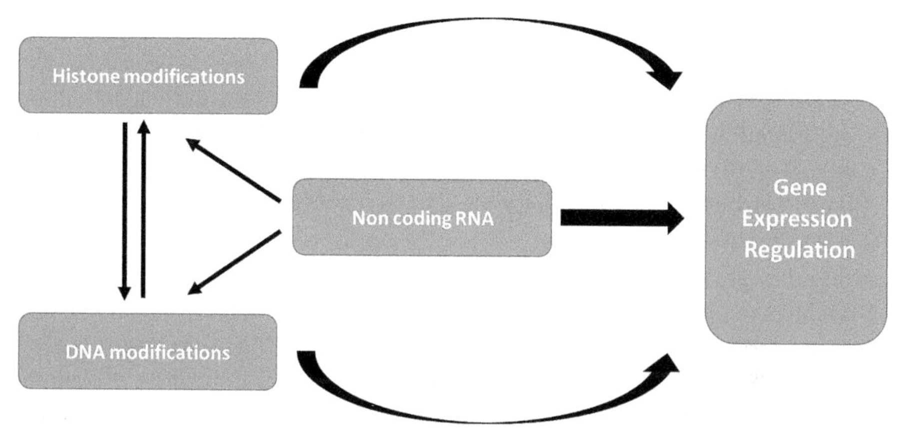

FIGURE 9.1 A schematic model of all three epigenetic mechanisms and their interplay in regulation of gene expression.

9.3.1 DROUGHT STRESS

Over-exploitation of groundwater due to increasing population and utilizable water depletion is a general observation around the world. Plants cannot escape from water-deficient habitat; therefore, structural, physiological, and molecular changes become essential for their growth and survival. Minimizing transpiration rate, accumulation of cell protectant proteins, and osmolytes, as well as transcriptional reprogramming, are some of them. Accumulation of drought-responsive phytohormone ABA is induced during water deficiency. Reports suggest that ABA induces histone modification in chromatin (Kim et al., 2012). In *Arabidopsis* seedlings H3K9, H3K27, H3K23 acetylation which is generally considered as an active state of chromatin, increases in drought-responsive genes/loci after a short duration of stress (Kim et al., 2008). Furthermore, drought stress in mature *Arabidopsis* plants revealed a strong correlation between H3K4 methylation status and the expression pattern of stress-responsive genes (Van Dijk et al., 2010). A similar phenomenon of H3K4me3 enrichment at loci of stress-responsive genes has been observed in rice seedlings (Zong et al., 2013). *Arabidopsis ATX1* which trimethylates H3K4 shows greater sensitivity towards drought in a loss of function mutant. *ATX1* directly regulates *NCED3* transcription which is a key enzyme of ABA biosynthesis. Expression of *NCED3* is significantly compromised in *atx1* suggesting the requirement of *ATX1* mediated H3K4me3 accumulation at *NCED3* locus, ABA accumulation, and consequently drought tolerance (Ding et al., 2011). Similarly, in rice, few histone acetyltransferases (*OsHAC703*, *OSHAG703*, *OSHAf701*, and *OsHAM701*) are notably induced during drought stress. Their induction is correlated with global induction of H3K9, H3K18, H3K27, and H3K5 acetylation (Fang et al., 2014). A similar observation was seen in Barley where, ABA treatment induced three GNAT-MYST family HAT genes (*HvMYST*, *HvELP3*, and *HvGCN5*) (Papaefthimiou et al., 2010). Under strong drought conditions, H3K4me3 and H3K9ac modifications at loci of drought-responsive genes *RD29* and *RD20* take place with high enrichment than moderate drought stress. This suggests that modifications depend on the intensity of stress. To reset the basal level expression during the drought recovery phase, H3K9ac modification decreases at *RD29* and *RD20* genes loci (Kim et al., 2008). Simultaneously, a stress memory is generated via H3K4me3 histone modification to overcome the recurrent drought phases in a rapid and robust manner (Oh et al., 2008; Zhang et al., 2009; Ding et al., 2011, 2012; Kim et al., 2012).

Whole genome analysis of drought-induced *Arabidopsis* seedlings showed altered DNA methylome which was more evident in the coding regions of stress-responsive genes. Differential DNA methylation status may contribute to the altered gene expression profile upon drought challenge (Colaneri et al., 2013). In rice, DNA methylation changes are more significant in drought-tolerant varieties which also indicate the role of DNA methylation in drought tolerance (Wang et al., 2010).

9.3.2 HEAT STRESS

Global warming has increased the annual temperature by more than 1°C in the last few decades which is a serious threat for crop yield. Increase in temperature by 1°C can cause a ~10% loss in total yield of rice (Peng et al., 2004). Understanding of epigenetic and genetic regulation of heat response can help in engineering plants for high-temperature tolerance. Recent findings revealed the role of many heat sensors, heat shock proteins (HSPs) and factors (HSFs), phytohormones, secondary metabolites, in acclimatization against high temperatures (Bokszczanin et al., 2013; Qu et al., 2013). In *Chlamydomonas reinhrdtii*, heat stress induces H3 and H4 acetylation at promoters of active genes (Strenkert et al., 2011). The H3k9me2 level at *OsFIE1* locus changes even during moderate heat stress in developing rice seeds (Folsom et al., 2014).

In cotton anthers, expression of genes encoding histone modifiers like histone methyltransferase, histone monoubiquitination gene (HUB) and histone demethylase jumanji C (jmjC) domain-containing proteins are downregulated (Min et al., 2014). Global methylation status can be affected differently by heat stress in different species. High temperature induces global methylation level in *Arabidopsis*. Genes encoding DNA methyltransferases such as *DRM2*, *NRPD1*, and *NRPE1* are upregulated in response to heat stress. Contrary to *Arabidopsis*, heat-sensitive genotypes of *Brassica napus* show high DNA methylation than heat-tolerant genotypes (Gao et al., 2014). However, high temperature significantly decreases expression of *DRM1* and *DRM3* during anther developing stages in cotton which results in a global decrease of methylation levels (Min et al., 2014). *OsFIE1* expression in developing seed depends upon the duration and intensity of heat stress treatment. It increases during the moderate heat stress while decreases at a severe temperature at later time points. *CMT3* expression varies at different temperatures; it reduces with an increase in temperature. Above reports clearly, indicate towards a direct correlation between stress challenge and epigenome modification by various epigenetic regulators.

9.3.3 SALINITY STRESS

Evidence is available about epigenetic changes during almost every type of stress. Increase in the global population with a decrease in the cultivable land has amplified the burden of food security which can be addressed by increasing crop productivity only. Generation of salt-tolerant plants can be an answer for that as saline lands can be used for agricultural purpose. Analysis of stress-responsive genes and their epigenetic regulation can increase our understanding or knowledge about plants' tolerance mechanism. Mutation in *ADA2b* gene which modulates HAT activity renders *Arabidopsis* plant hypersensitive for salt stress (Kaldis et al., 2011) indicating HAT activity in salt tolerance. *ada2b* mutant has reduced H3 and H4 acetylation at *RAB18*, *RD29b*, and *COR6.6* genes which explains the hypersensitivity to salt stress (Kaldis et al., 2011). In soybean, there is a complex mechanism of salt tolerance where DNA methylation and histone modification decides the final outcome of tolerance (Song et al., 2012). Salt stress-responsive transcription factors are induced by DNA demethylation as well as histone modification (Song et al., 2012). Histone deacetylase *hda6* mutant and RNAi lines are hypersensitive to salt stress in *Arabidopsis* (Chen et al., 2010). Another member of histone deacetylase complex *HDAC1* is responsible for repression of ABA signaling genes *ABA1*, *RAB18*, and *RD29A* under high salinity stress (Perrella et al., 2013). In maize, cell wall-related and salinity responsive genes like *ZmEXPB2* and *ZmXET1* are induced by enrichment of H3K9 acetylation. Simultaneously the expression of two HAT genes *ZmHATB* and *ZmGCN5* is also induced under salt stress (Li et al., 2014). Above reports suggest that fine tuning of HAT and HDAC activity can play a crucial role in overall salinity stress response of plants.

9.4 EPIGENETIC MODIFICATION AND PLANT IMMUNITY

Plant stress response is a sequential event which comprises of stress perception, signaling cascades, transcriptional reprogramming and final response to the stress. The response towards a biotic stress is initiated after the recognition of the pathogen by the plant. Several pathogen-associated molecular patterns or microbe-associated molecular patterns (PAMPs/MAMPs) of pathogens are recognized by plants specific receptors and via a rapid signaling cascade, the stress resistance response known as PAMP-triggered immunity (PTI) is initiated. In order to suppress PTI, successful pathogens inject effectors molecules into the host cell. In response to effectors, host plants have evolved cognate cytosolic receptor molecules to activate Effector-Triggered Immunity (ETI) (Dangl and Jones et al., 2006). Higher plants contain MAPK to transduce the signaling and activation of defense responsive transcription factors. Phytohormones play a major role in regulating defense response. Salicylic acid (SA), jasmonic acid (JA), ethylene (ET), auxin, gibberellic acid (GA), abscisic acid (ABA), brassinosteroid (BR), strigolactones, etc. modulate growth, development, and stress response of plants. Their synergistic and antagonistic interplay largely affect the defense outcome. In general, SA responsive signaling is activated after challenge with biotrophic pathogens and JA/Et signaling regulates necrotrophic pathogen defense response of plants. Recent exhaustive studies on plant immunity revealed that it is a dynamic and complex event which is initiated just after recognition of pathogen. Defense response comes within 6 hours of infection (Lewis et al., 2015). At the same time, plants can differentiate pathogens containing different PAMPs or those ones having both PAMPs and effector secretion. After discrimination they activate specific defense responses against them.

Epigenetic regulation by chromatin modification plays a major role in the establishment or induction of defense response (Ding et al., 2015). Studies on several mutants of histone modification enzymes in *Arabidopsis* and rice showed altered defense response against different pathogens strengthen the role of these regulators in plant immunity.

9.4.1 EPIGENETIC REGULATOR AND PLANT IMMUNITY AGAINST BIOTROPHIC PATHOGENS

In general, histone acetylation leads to an open chromatin state which facilitates the binding of transcription factors and other proteins involved in transcription. Acetylation activity of histone acetyltransferases (HATs) and deacetylation by histone deacetylases (HDACs) are responsible for the status of histone and thereby the expression of the nearby genes. Role of many HATs and HDACs are studied in plants defense. In *Arabidopsis, HDA19* is a negative regulator of defense against biotrophic pathogens. *hda19* mutants show enhanced SA accumulation and higher expression of defense marker genes *PR1* and *PR2*. Further experiments proved that HDA19 directly binds to *PR1* and *PR2* genes and affect the acetylation status of H3 (Choi et al., 2012). A study by another group revealed the interaction of HDA19 with two WRKY family TFs, WRKY38 and WRKY62. Similar to *hda19* defense phenotype, WRKY38 and WRKY62 also show enhanced resistance against *Pseudomonas syringae pv tomato* DC3000 (*Pst* DC3000). Another HDAC, *HDA6* is also a negative regulator of SA mediated defense signaling. *hda6* mutant is resistant for *Pst* DC3000 and displays constitutive upregulation of SA defense signaling marker genes such as *PR1*, *PR2*, *WRKY18*, *WRKY70*, *WRKY38*, *EFR* and *FRK1* (Wang et al., 2017). Chromatin immunoprecipitation revealed increased H3Ac at all these genes loci (Wang et al., 2017).

Similar to *HDA19* and *HDA6*, another histone deacetylase *AtSRT2*, a homolog of yeast *SIR2* gene, is a negative regulator of *Pst* DC3000 immunity. *srt2* mutant has increased expression of *PR1* and SA biosynthesis related genes; *PAD4*, *EDS5*, and *SID2*. Results demonstrated that *SRT2* affect plant immunity by regulating SA biosynthesis in plants which is a key element for defense signaling against bacterial pathogens (Wang et al., 2010). Contrary to the previously described HDACs, a MAPK activated HDAC, *HD2B* is a positive regulator. *hd2b* mutant is susceptible for *Pst*hrc. *HD2B* regulates plant immunity by playing a key role in the reprogramming of gene expression after detection of bacterial PAMP flg22. Similar reports are also present from other crop plants, like rice histone deacetylase *HDT701* affects global H4 acetylation level. *hdt701* RNAi lines have increased global H4 acetylation level and resistant for *Xanthomonas oryzae* (*Xoo*). In coordination with the phenotype, *hdt701*RNAi lines have lower expression of PAMP receptor genes and defense-related genes as well as accelerated ROS production in presence of flg22 (Ding et al., 2012) Thus, *HDT701* is a negative regulator of basal defense.

Similar to HDACs, histone acetyltransferases are also important for regulation of plant immunity. *Arabidopsis* acetyltransferase *HAC1-1* has been shown to be important in priming of defense genes after recurrent abiotic stresses and challenge with a biotrophic pathogen. These recurrent abiotic stresses and biotrophic pathogen induce epigenetic marks accumulation responsible for the permissive state of transcription on PTI responsive genes. Finally, *hac1-1* failed to accumulate those epigenetic marks on the marker genes as well as did not show any change in the defense profile after recurrent environmental challenge (Singh et al., 2014). Thus, indicating the role of *HAC1-1* in plant immunity.

Unlike histone acetylation, which is generally considered as a permissive state of transcription, histone methylation affects transcription according to its position on the histone. Final methylation status depends on the activity of both methyltransferase and demethylases which affects many aspects of plant development and immunity.

Methylation at H3K4 is performed by Trithorax protein. *Arabidopsis* homolog of TRX protein, ATX1 is involved in the maintenance of global H3K4 trimethylation (Alvarez-Venegas et al., 2003). *atx1* mutant has reduced global H3K4me3 as well as altered defense phenotype. ATX1 directly binds to *WRKY70* and promotes its expression. *atx1* mutant has reduced *WRKY70* and *PR1* expression, which is accompanied with reduced resistance for *Pst* DC3000 hrc (Alvarez-Venegas et al., 2007). Another SET domain-containing histone methyltransferase *SDG8* is required for maintaining immunity of *Arabidopsis*. *sdg8* mutant is susceptible for *Psm*, *Pst* DC3000, *Pst AvrRpm1* and *PsmAvrB*. Expression of R gene *RPM1* is also compromised in *sdg8* mutant (Palma et al., 2010). In another study, similar results were obtained where both *SDG8* and *SDG25* were found as positive regulators of ETI and PTI. Both genes play role in regulating expression of *CCR2* and *CER3*, important for carotenoid and cuticle biosynthesis respectively. Recently, a SET domain-containing histone methyltransferase MEDEA is found to be a negative regulator of both hemibiotrophic and necrotrophic pathogen. Further analysis revealed repressive histone modification at *Arabidopsis* R gene RPS2 for disease suppression (Roy et al., 2018). Some histone demethylases are also studied in the context of plant defense response phenotype. Jumanji domain-containing histone demethylases are reported to have a role in immunity. *Arabidopsis JMJ27*, an H3K9me1/2 demethylase is a positive regulator of defense. *jmj27* mutants are susceptible for *Pst* DC3000 and *Psm* by upregulating expression of two negative regulators, *WRKY25* and *WRKY33* (Dutta et al., 2017). Rice *JMJ705* demethylates H3K27me3 on many defense-related genes such as *PR5*, *PR10*, *OPR7*, *LOX2*, *POX5*, *POX8* and *POX22*. Increased expression of these positive defense regulators results in resistance for bacterial pathogen *Xoo*. Rice H3K4 demethylase *JMJ704* is associated with defense against *Xoo*. *jmj704* mutant has increased occupancy of H3K4me3 on negative regulators *WRKY62*, *NRR* and OS-*11N3*, hence, shows enhanced susceptibility against *Xoo* (Hou et al., 2015).

Apart from acetylation and methylation, histone proteins can undergo ubiquitination also. Ubiquitination is generally considered as activating mark for transcription. *Arabidopsis* H2B monoubiquitinating enzyme HUB1 is required for regulating SA and ET/JA signaling. Although *hub1* mutant does not show any altered defense response against *Pst* DC3000 but regulates expression of *SNC1* and *RPP* cluster genes.

9.4.2 EPIGENETIC MODIFICATION IN NECROTROPHIC PATHOGENS

Unlike biotrophs and hemibiotrophs, resistance for necrotrophic pathogens requires activation of ET/JA signaling. Several reports histone modification regulators are present which alters plant defense against fungus. Histone modifying proteins like *Arabidopsis* histone deacetylase *HDA19* and *HDA6* are involved in regulation of ET/JA pathway. Their expression is induced after treatment with JA, ethylene or ethylene precursors. *Arabidopsis HDA19* positively regulate fungal pathogen defense (Zhou et al., 2005). *hda19 mutant* plants are susceptible to necrotrophic pathogens and overexpression plants are resistant (Zhou et al., 2005; Zhou et al., 2010). Similarly another histone deacetylase *HDA6* also positively regulates fungal immunity. Probably both *HDA6* and *HDA19* are redundant in function and works through suppression of SA signaling. HD2 family HDAC from rice, *HDA701* is an H4 deacetylase and a negative regulator of *Magnaporthe* defense (Ding et al., 2012). Chromatin modifier like Elongator complex, which facilitates transcription by modifying chromatin, is crucial for gene expression. Its elongator subunit 2 (*AtELP2*) regulates histone modification as well as a DNA methylation of defense genes involved in *Alternaria brassisicola* and *Botrytis cinerea* immunity (Wang et al., 2015). *atelp2* mutants are compromised in *WRKY33*, *ORA59* and *PDF1.2* expression (Wang et al., 2013; Wang et al., 2015).

Rice histone demethylase *JMJ705* is highly induced in response to external methyl jasmonate (MeJA) treatment. *jmj705* mutant plants show compromised resistance as well as diminished expression of defense genes (Li et al., 2013). Histone methyltransferases like *SDG8* and *SDG25* are involved in resistance against *B. cinerea* and *A. brassisicola. SDG8* and *SDG25* act by regulating the expression of *CCR2* and *CER3*. Accumulation of H3K4me3 and H3K36me3 declines at *SNC1*, *CCR2* and *CER3* genes in *sdg8* and *sdg25* mutant plants (Lee et al., 2016). *ccr2* and *cer3* mutants also display reduced disease resistance as of *sdg8* and *sdg25* which confirm the role of *SDG8* and *SDG25* in plant immunity. *Arabidopsis* and tomato *HUB1* and *HUB2* are required for resistance to necrotrophic pathogens by regulating ET and SA signaling response (Dhawan et al., 2009; Zhang et al., 2015).

9.4.3 EPIGENETIC REGULATION OF PLANT AGAINST VIRAL PATHOGENS

Plant viruses as obligate parasites and rely on host machinery for their multiplication and propagation. Histone and nucleic acid mediated epigenetic regulation plays a crucial role in viral pathogenesis. The minichromosome of *cabbage leaf curl virus* (*CaLCV*) isolated from infected *Arabidopsis* and *Nicotiana benthamiana* plants showed deposition of H3K4me2 and H3acetyl and repressive modifications like H3K9me2 and H3K27me2 (Raja et al., 2008). In another study, *pepper golden mosaic virus* (*PepGMV*) isolated from symptomatic pepper leaves were enriched in H3K4me3 whereas, the minichromosome of *PepGMV* isolated from recovered leaves showed higher H3K9me2 deposition (Ceniceros-Ojeda et al., 2016). In one of the recent studies, *Chilli leaf curl virus* genome isolated from infected *N. benthamiana* plants were detected with monoubiquitinated H2B and enhanced H3K4me3 marks. The study also revealed that the components of monoubiquitination machinery HUB1 homologs in *N. benthamiana* (*NbHUB1*) and Ubiquitin-Conjugating Enzyme 2 (*NbUBC2*) which are important for histone monoubiquitination on viral minichromosomes. It was also demonstrated that silencing of *NbHUB1* and *NbUBC2* reduce the viral gene expression and pathogenesis (Kushwaha et al., 2017).

The infectivity of plant viruses is also regulated by transcriptional and post-transcriptional gene silencing. *CaLCV* and *beat curly top virus* (*BCTV*) genome were found to be methylated in *Arabidopsis* (Raja et al., 2008). *PepGMV* genome isolated from recovered leaves of pepper was more methylated than the genome isolated from symptomatic pepper plants (Ceniceros-Ojeda et al., 2016). Two different strain of *Tomato Leaf Curl Virus* (*Tomato Leaf Curl New Delhi Virus* and *Tomato Leaf Curl Gujarat Virus*) exhibited DNA methylation in tobacco plants (Basu et al., 2018).

9.5 HERITABLE EPIGENETIC VARIATION AND CROP IMPROVEMENT

Epigenetic changes can be stably inherited across several meiotic divisions and can result in phenotypic variations, known as epialleles. This transgenerational nature of epialleles can be useful in agronomics. In rice, various epialleles are used for enhancing agronomic traits. Rice *epi-d1* allele is responsible for the silencing of *DWARF1* (*D1*) gene which is responsible for the dwarf phenotype. Later studies revealed DNA and repressive histone hypermethylation at the promoter of the *D1* gene in *epi-d1* epimutant is responsible for the dwarf phenotype (Fujimoto et al., 2012). In tomato, hypermethylation at Squamosa Binding Protein (SBP) box transcription factor leads to *cnr* (colorless nonripening) epiallele formation and fruit maturation defect (Manning et al., 2006). This phenotype can be reversed by application of DNA methylation inhibitor 5-azacytidine (Martel et al., 2011). Apart from natural epialleles, these changes can be induced to achieve the desired phenotype. Treatment of rice seeds with 5-azacytidine induced hypomethylation at *XA21G* promoter. This epimutation at *XA21G* led to the development of *Xoo* resistant lines. This phenotype was maintained up to nine generations in rice (Akimoto et al., 2007). Furthermore, epigenetic variations can be used as biomarkers for predicting the agronomic traits.

9.6 FUTURE PERSPECTIVE

Large numbers of genes are differentially regulated under stressful environment by modification in their DNA methylation pattern, histone N terminal modifications or variation in the biogenesis of ncRNA. Deep studies of modification sites and pattern can provide an insight about the genes involved in stress tolerance and adaptation. Those genes can be targeted as candidate genes for more efficient approaches to enhance crop productivity. Moreover, the epigenome of the stress-resistant plant can certainly turn out to be a substantial resource of improved crop varieties. Stable inheritance of epigenetic variation can be exploited in agronomics as well as these epialleles can be useful as a breeding material for the development of better varieties. Furthermore, better understanding and manipulation of epigenome can answer the concern of food safety.

ACKNOWLEDGMENTS

PS would like to thank the financial and infrastructure support received from University Grants Commission, India for Start-up Grant and the University of Rajasthan, Jaipur respectively. HR would like to acknowledge DST-INSPIRE Faculty grant from Department of Science and Technology, Govt. of India.

CONFLICTS OF INTEREST

The authors declare that they have no conflicts of interest.

KEYWORDS

- **Biotic and abiotic stresses**
- **DNA methylation**
- **Epialleles**
- **Epigenetics**
- **Histone modification**

REFERENCES

Ai-Li, Q., Yan-Fei, D., Qiong, J., & Cheng, Z., (2013). Molecular mechanisms of the plant heat stress response. *Biochemical and biophysical research communications, 432*(2), 203–207.

Akimoto, K., Hatsue, K., Hyun-Jung, K., Emiko, O., Cecile, M. S., Yuko, W., & Hiroshi, S., (2007). Epigenetic inheritance in rice plants. *Annals of Botany, 100*(2), 205–217.

Alvarez-Venegas, R., Ayed, A. A., Ming, G., James, R. A., & Zoya, A., (2007). Epigenetic control of a transcription factor at the cross section of two antagonistic pathways. *Epigenetics*, *2*(2), 106–113.

Alvarez-Venegas, R., Stephane, P., Monther, S., Xiaohong, W., Ueli, G., & Zoya, A., (2003). ATX-1, an *Arabidopsis* homolog of trithorax, activates flower homeotic genes. *Current Biology*, *13*(8), 627–637.

Basu, S., Nirbhay, K. K., Ashish, K. S., Pranav, P. S., Vinoth, K. R., & Supriya, C., (2018). Dynamics of a Gemini virus-encoded pre-coat protein and host RNA-dependent RNA polymerase 1 in regulating symptom recovery in tobacco. *Journal of Experimental Botany, 69*(8), 2085–2102.

Bhaumik, S. R., Edwin, S., & Ali, S., (2007). Covalent modifications of histones during development and disease pathogenesis. *Nature Structural and Molecular Biology, 14*(11), 1008.

Bird, A., (2002). DNA methylation patterns and epigenetic memory. *Genes and Development, 16*(1), 6–21.

Bokszczanin, K. L., Sotirios, F., Hamed, B., Arnaud, B., Palak, C., Maria, L. C., Nurit, F., et al., (2013). Perspectives on deciphering mechanisms underlying plant heat stress response and thermo tolerance. *Frontiers in Plant Science, 4*, 315.

Bourbousse, C., Ikhlak, A., François, R., Gérald, Z., Eddy, B., Sandrine, B., Vincent, C., et al., (2012). Histone H2B monoubiquitination facilitates the rapid modulation of gene expression during *Arabidopsis* photomorphogenesis. *PLoS Genetics, 8*(7), e1002825.

Bühler, M., (2009). RNA turnover and chromatin-dependent gene silencing. *Chromosoma, 118*(2), 141–151.

Cao, Y., Dai, Y., Cui, S., & Ma, L., (2008). Histone H2B monoubiquitination in the chromatin of FLOWERING LOCUS C regulates flowering time in *Arabidopsis*. *The Plant Cell, 20*(10), 2586–2602.

Ceniceros-Ojeda, E. A., Rodríguez-Negrete, E. A., & Rivera-Bustamante, R. F., (2016). Two populations of viral minichromosomes are present in a Gemini virus-infected plant showing symptom remission (recovery). *Journal of Virology*, JVI-02385.

Colaneri, A. C., & Alan, M. J., (2013). Genome-wide quantitative identification of DNA differentially methylated sites in *Arabidopsis* seedlings growing at different water potential. *PloS One, 8*(4), e59878.

Dao-Xiu, Z., (2009). Regulatory mechanism of histone epigenetic modifications in plants. *Epigenetics, 4*(1), 15–18.

Dhawan, R., Hongli, L., Andrea, M. F., Synan, A. Q., Hai-Ning, D., Scott, D. B., Ortrun, M. S., & Tesfaye, M., (2009). Histone monoubiquitination1 interacts with a subunit of the mediator complex and regulates defense against necrotrophic fungal pathogens in *Arabidopsis*. *The Plant Cell, 21*(3), 1000–1019.

Ding, B., & Guo-Liang, W., (2015). Chromatin versus pathogens: The function of epigenetics in plant immunity. *Frontiers in Plant Science, 6*, 675.

Ding, B., Maria, D. R. B., Yuese, N., Blake, C. M., & Guo-Liang, W., (2012). HDT701, a histone H4 deacetylase, negatively regulates plant innate immunity by modulating histone H4 acetylation of defense-related genes in rice. *The Plant Cell*, tpc-112.

Ding, Y., Michael, F., & Zoya, A., (2012). Multiple exposures to drought 'train' transcriptional responses in *Arabidopsis*. *Nature Communications, 3*, 740.

Ding, Y., Zoya, A., & Michael, F., (2011). The *Arabidopsis* trithorax-like factor ATX1 functions in dehydration stress responses via ABA-dependent and ABA-independent pathways. *The Plant Journal, 66*(5), 735–744.

Dutta, A., Pratibha, C., Julie, C., & Ramesh, R., (2017). JMJ 27, an *Arabidopsis* H3K9 histone demethylase, modulates defense against *Pseudomonas syringae* and flowering time. *The Plant Journal, 91*(6), 1015–1028.

Fang, H., Xia, L., Greg, T., Jun, D., & Lining, T., (2014). Expression analysis of histone acetyltransferases in rice under drought stress. *Biochemical and Biophysical Research Communications, 443*(2), 400–405.

Finnegan, E. J., & Kovac, K. A., (2000). Plant DNA methyltransferases. In: *Plant Gene Silencing* (pp. 69–81). Springer, Dordrecht.

Folsom, J. J., Kevin, B., Xiaojuan, H., Dong, W., & Harkamal, W., (2014). Rice FIE1 regulates seed size under heat stress by controlling early endosperm development. *Plant Physiology*, 113.

Fujimoto, R., Taku, S., Ryo, I., Kenji, O., Takahiro, K., & Elizabeth, S. D., (2012). Molecular mechanisms of epigenetic variation in plants. *International Journal of Molecular Sciences, 13*(8), 9900–9922.

Gan, Q., Dejun, L., Guozhen, L., & Lihuang, Z., (2012). Identification of potential antisense transcripts in rice using conventional microarray. *Molecular Biotechnology, 51*(1), 37–43.

Gao, G., Jun, L., Hao, L., Feng, L., Kun, X., Guixin, Y., Biyun, C., Jiangwei, Q., & Xiaoming, W., (2014). Comparison of the heat stress induced variations in DNA methylation between heat-tolerant and heat-sensitive rapeseed seedlings. *Breeding Science, 64*(2), 125–133.

Heo, J. B., & Sibum, S., (2011). Vernalization-mediated epigenetic silencing by a long intronic noncoding RNA. *Science, 331*(6013), 76–79.

Hou, Y., Liyuan, W., Ling, W., Lianmeng, L., Lu, L., Lei, S., Qiong, R., Jian, Z., & Shiwen, H., (2015). JMJ704 positively regulates rice defense response against *Xanthomonas oryzae pv. oryzae* infection via reducing H3K4me2/3 associated with negative disease resistance regulators. *BMC Plant Biology, 15*(1), 286.

Jones, J. D. G., & Jeffery, L. D., (2006). The plant immune system. *Nature, 444*(7117), 323.

Jong-Myong, K., Taiko, K. T., Junko, I., Akihiro, M., Hiroshi, K., & Motoaki, S., (2012). Transition of chromatin status during the process of recovery from drought stress in *Arabidopsis thaliana. Plant and Cell Physiology, 53*(5), 847–856.

Jong-Myong, K., Taiko, K. T., Junko, I., Taeko, M., Makiko, K., Akihiro, M., Tetsuro, T., Hiroshi, K., Kazuo, S., & Motoaki, S., (2008). Alterations of lysine modifications on the histone H3 N-tail under drought stress conditions in *Arabidopsis thaliana. Plant and Cell Physiology, 49*(10), 1580–1588.

Kaldis, A., Despoina, T., Oznur, T., & Konstantinos, E. V., (2011). *Arabidopsis thaliana* transcriptional co-activators ADA2b and SGF29a are implicated in salt stress responses. *Planta, 233*(4), 749–762.

Kloc, A., & Robert, M., (2008). RNAi, heterochromatin and the cell cycle. *Trends in Genetics, 24*(10), 511–517.

Klose, R. J., & Adrian, P. B., (2006). Genomic DNA methylation: The mark and its mediators. *Trends in Biochemical Sciences, 31*(2), 89–97.

Kouzarides, T., (2007). Chromatin modifications and their function. *Cell, 128*(4), 693–705.

Kushwaha, N. K., Mansi, B., & Supriya, C., (2017). The replication initiator protein of a Gemini virus interacts with host monoubiquitination machinery and stimulates transcription of the viral genome. *PLoS Pathogens,13*(8), e1006587.

Lee, S., Fuyou, F., Siming, X., Sang, Y. L., Dae-Jin, Y., & Tesfaye, M., (2016). Global regulation of plant immunity by histone lysine methyltransferases. *The Plant Cell*, tpc-00012.

Lewis, L. A., Krzysztof, P., Torres-Zabala, M. D., Siddharth, J., Laura, B., Jonathan, M., Christopher, A. P., et al., (2015). Transcriptional dynamics driving MAMP-triggered immunity and pathogen effector-mediated immunosuppression in *Arabidopsis* leaves following infection with *Pseudomonas syringae* pv tomato DC3000. *The Plant Cell*, tpc-15.

Li, H., Shihan, Y., Lin, Z., Junjun, T., Qi, Z., Fei, G., Pu, W., Haoli, H., & Lijia, L., (2014). Histone acetylation associated up-regulation of the cell wall related genes is involved in salt stress induced maize root swelling. *BMC Plant Biology, 14*(1), 105.

Li, T., Xiangsong, C., Xiaochao, Z., Yu, Z., Xiaoyun, L., Shaoli, Z., Saifeng, C., & Dao-Xiu, Z., (2013). Jumonji C protein JMJ705-mediated removal of histone H3 lysine 27 trimethylation is involved in defense-related gene activation in rice. *The Plant Cell*, tpc-113.

Li-Ting, C., Ming, L., Yu-Yuan, W., & Keqiang, W., (2010). Involvement of *Arabidopsis* histone deacetylase HDA6 in ABA and salt stress response. *Journal of Experimental Botany, 61*(12), 3345–3353.

Luger, K., Armin, W. M., Robin, K. R., David, F. S., & Timothy, J. R., (1997). Crystal structure of the nucleosome core particle at 2.8 Å resolution. *Nature, 389*(6648), 251.

Lusser, A., Doris, K., & Peter, L., (2001). Histone acetylation: Lessons from the plant kingdom. *Trends in Plant Science, 6*(2), 59–65.

Manning, K., Mahmut, T., Mervin, P., Yiguo, H., Andrew, J. T., Graham, J. K., James, J. G., & Graham, B. S., (2006). A naturally occurring epigenetic mutation in a gene encoding an SBP-box transcription factor inhibits tomato fruit ripening. *Nature Genetics, 38*(8), 948.

Martel, C., Julia, V., Petra, T., & James, J. G., (2011). The tomato MADS-box transcription factor ripening inhibitor interacts with promoters involved in numerous ripening processes in a colorless nonripening-dependent manner. *Plant physiology, 157*(3), 1568–1579.

Martienssen, R. A., & Vincent, C., (2001). DNA methylation and epigenetic inheritance in plants and filamentous fungi. *Science, 293*(5532), 1070–1074.

Min, L., Yaoyao, L., Qin, H., Longfu, Z., Wenhui, G., Yuanlong, W., Yuanhao, D., Shiming, L., Xiyan, Y., & Xianlong, Z., (2014). Sugar and auxin signaling pathways respond to high temperature stress during anther development as revealed by transcript profiling analysis in cotton. *Plant Physiology*, 113.

Oh, S., Sunchung, P., & Steven, V. N., (2008). Genic and global functions for Paf1C in chromatin modification and gene expression in *Arabidopsis*. *PLoS Genetics, 4*(8), e1000077.

Palma, K., Stephan, T., Frederikke, G. M., Berthe, K. F., Bjørn, N. H., Peter, B., Daniel, H., Morten, P., & John, M., (2010). Autoimmunity in *Arabidopsis* acd11 is mediated by epigenetic regulation of an immune receptor. *PLoS Pathogens, 6*(10), e1001137.

Papaefthimiou, D., Eleni, L., Aliki, K., Konstantinos, B., & Athanasios, T., (2010). Epigenetic chromatin modifiers in barley: III. Isolation and characterization of the barley GNAT-MYST family of histone acetyltransferases and responses to exogenous ABA. *Plant Physiology and Biochemistry 48*(2, 3), 98–107.

Peng, S., Jianliang, H., John, E. S., Rebecca, C. L., Romeo, M. V., Xuhua, Z., Grace, S. C., et al., (2004). Rice yields decline with higher night temperature from global warming. *Proceedings of the National Academy of Sciences, 101*(27), 9971–9975.

Perrella, G., Lopez-Vernaza, M. A., Craig, C., Emanuela, S., Veronique, G., Christoph, V., Fabian, K., et al., (2013). Histone deacetylase complex1 expression level titrates plant growth and abscisic acid sensitivity in *Arabidopsis*. *The Plant Cell*, tpc-113.

Raja, P., Bradley, C. S., Cody, B. R., & David, M. B., (2008). Viral genome methylation as an epigenetic defense against Gemini viruses. *Journal of Virology, 82*(18), 8997–9007.

Roy, S., Priya, G., Mohit, P. R., Ravi, M., & Ashis, K. N., (2018). The polycomb-group repressor MEDEA attenuates pathogen defense. *Plant Physiology*, 01579.

Schnell, J. D., & Linda, H., (2003). Non-traditional functions of ubiquitin and ubiquitin-binding proteins. *Journal of Biological Chemistry*.

Shahbazian, M. D., & Michael, G., (2007). Functions of site-specific histone acetylation and deacetylation. *Annu. Rev. Biochem., 76*, 75–100.

Shilatifard, A., (2006). Chromatin modifications by methylation and ubiquitination: Implications in the regulation of gene expression. *Annu. Rev. Biochem., 75*, 243–269.

Singh, P., Shweta, Y., Po-Wen, C., Chia-Hong, T., Chun-Wei, Y., Keqiang, W., & Laurent, Z., (2014). Environmental history modulates *Arabidopsis* pattern-triggered immunity in a histone acetyltransferase1–dependent manner. *The Plant Cell*, tpc-114.

Song, Y., Dandan, J., Shuo, L., Peng, W., Qiang, L., & Fengning, X., (2012). The dynamic changes of DNA methylation and histone modifications of salt responsive transcription factor genes in soybean. *PloS One, 7*(7), e41274.

Strenkert, D., Stefan, S., Frederik, S., Schulz-Raffelt, M., & Michael, S., (2011). Transcription factor-dependent chromatin remodeling at heat shock and copper-responsive promoters in *Chlamydomonas reinhardtii. The Plant Cell*, tpc-111.

Sun-Mee, C., Hae-Ryong, S., Soon-Ki, H., Muho, H., Chi-Yeol, K., Jaejin, P., Yong-Hwan, L., et al., (2012). HDA19 is required for the repression of salicylic acid biosynthesis and salicylic acid-mediated defense responses in *Arabidopsis. The Plant Journal, 71*(1), 135–146.

Swiezewski, S., Fuquan, L., Andreas, M., & Caroline, D., (2009). Cold-induced silencing by long antisense transcripts of an *Arabidopsis* Polycomb target. *Nature, 462*(7274), 799.

Van, D. K., Yong, D., Sridhar, M., Jean-Jack, M. R., Rong, L., Jingyi, Y., Peter, L., et al., (2010). Dynamic changes in genome-wide histone H3 lysine 4 methylation patterns in response to dehydration stress in *Arabidopsis thaliana. BMC Plant Biology, 10*(1), 238.

Wang, C., Feng, G., Jianguo, W., Jianli, D., Chunhong, W., & Yi, L., (2010). *Arabidopsis* putative deacetylase AtSRT2 regulates basal defense by suppressing PAD4, EDS5 and SID2 expression. *Plant and Cell Physiology, 51*(8), 1291–1299.

Wang, C., Yezhang, D., Jin, Y., Yanping, Z., Yijun, S., James, C., & Zhonglin, M., (2015). *Arabidopsis* Elongator subunit 2 positively contributes to resistance to the necrotrophic fungal pathogens *Botrytis cinerea* and *Alternaria brassicicola. The Plant Journal, 83*(6), 1019–1033.

Wang, Y., Chuanfu, A., Xudong, Z., Jiqiang, Y., Yanping, Z., Yijun, S., Fahong, Y., David, M. A., & Zhonglin, M., (2013). The *Arabidopsis* elongator complex subunit2 epigenetically regulates plant immune responses. *The Plant Cell*, tpc-113.

Wang, Y., Qin, H., Zhenjiang, W., Hui, W., Shiming, H., Ye, J., Jin, Z., et al., (2017). HISTONE DEACETYLASE 6 represses pathogen defense responses in *Arabidopsis thaliana. Plant, cell and Environment, 40*(12), 2972–2986.

Weake, V. M., & Jerry, L. W., (2008). Histone ubiquitination: Triggering gene activity. *Molecular Cell, 29*(6), 653–663.

Wen-Sheng, W., Ya-Jiao, P., Xiu-Qin, Z., Dwivedi, D., Ling-Hua, Z., Ali, J., Bin-Ying, F., & Zhi-Kang, L., (2010). Drought-induced site-specific DNA methylation and its association with drought tolerance in rice (*Oryza sativa* L.). *Journal of Experimental Botany, 62*(6), 1951–1960.

Xiao, J., Un-Sa, L., & Doris, W., (2016). Tug of war: Adding and removing histone lysine methylation in *Arabidopsis. Current Opinion in Plant Biology, 34*, 41–53.

Zhang, X., Yana, V. B., Shawn, C., Matteo, P., & Steven, E. J., (2009). Genome-wide analysis of mono-, di-, and trimethylation of histone H3 lysine 4 in *Arabidopsis thaliana. Genome Biology, 10*(6), R62.

Zhang, Y., Dayong, L., Huijuan, Z., Yongbo, H., Lei, H., Shixia, L., Xiaohui, L., Zhigang, O., & Fengming, S., (2015). Tomato histone H2B monoubiquitination enzymes SlHUB1 and SlHUB2 contribute to disease resistance against *Botrytis cinerea* through modulating the balance between SA-and JA/ET-mediated signaling pathways. *BMC Plant Biology, 15*(1), 252.

Zhou, C., Lin, Z., Jun, D., Brian, M., & Keqiang, W., (2005). HISTONE DEACETYLASE19 is involved in jasmonic acid and ethylene signaling of pathogen response in *Arabidopsis*. *The Plant Cell, 17*(4), 1196–1204.

Zhou, J., Xiangfeng, W., Kun, H., Jean-Benoit, F. C., Axel, A. E., & Xing, W. D., (2010). Genome-wide profiling of histone H3 lysine 9 acetylation and dimethylation in *Arabidopsis* reveals correlation between multiple histone marks and gene expression. *Plant Molecular Biology, 72*(6), 585–595.

Zong, W., Xiaochao, Z., Jun, Y., & Lizhong, X., (2013). Genome-wide profiling of histone H3K4-tri-methylation and gene expression in rice under drought stress. *Plant Molecular Biology, 81*(1, 2), 175–188.

Zou, B., Dong-Lei, Y., Zhenying, S., Hansong, D., & Jian, H., (2014). Monoubiquitination of histone 2B at the disease resistance gene locus regulates its expression and impacts immune responses in *Arabidopsis*. *Plant Physiology,* 113.

CHAPTER 10

Understanding the Mechanism of Drought Tolerance in Pearl Millet

PRAVEEN SONI

Laboratory of Plant Molecular Biology, Department of Botany,
University of Rajasthan, Jaipur – 302004, India,
E-mail: praveen.soni15@gmail.com; praveensoni@uniraj.ac.in

ABSTRACT

Drought is one of the several abiotic factors which restrict yield by preventing the expression of the full genetic potential of crops. Pearl millet is a principal crop of the semi-arid regions of the earth. It can grow in harsh climatic conditions having extreme heat and recurrent droughts. This feature makes it an ideal model for studying plant stress biology. However, due to relatively large genome size and current unavailability of annotated genome, gene-mining for specific traits in pearl millet has been challenging. Nevertheless, in the last few years, the scientific community has gained success toward identification of quantitative trait locus (QTL) for drought tolerance in pearl millet using plant breeding approaches. It has been discovered that pearl millet genotypes having a QTL for terminal drought tolerance contain a high ABA level in leaves and a lower rate of water loss through transpiration. Using molecular approaches such as subtractive hybridization, a number of drought-inducible genes have been identified. Recently proteomics and transcriptomics studies have revealed tissue-specific drought responsive proteome and underlying gene regulatory network. This chapter summarizes the current investigations that have been done to unearth the complexity of drought tolerance in pearl millet.

10.1 INTRODUCTION

Pearl millet is the major coarse grain cereal and forage crop of the semi-arid tropical regions of Asia and Africa. Millet is a group of small-seeded crops which are largely cultivated in marginal environments. However, pearl millet is an exception in this respect as it produces large grains. Millets have been divided into two subfamilies, namely Panicoideae and Chloridoideae. Pearl millet belongs to the first subfamily. A few particulars of pearl millet have been mentioned in table 10.1.

TABLE 10.1 Facts about pearl millet

Characteristics	Description
Common name	Pearl millet
Other names	Spiked millet, Cattail millet, Bulrush millet, Bajra in Hindi, Kambu in Tamil
Botanical name	*Pennisetum glaucum* (L.) R. Br. Emend. Stuntz
Synonyms	*Pennisetum typhoides* (Burm.F) Stapf and C.E. Hubb, *Pennisetum americanum* (L.) Leeke, *Cenchrus americanus.* (L.) Morrone
Center of origin	Sahel zone of western Africa
Chromosome number	2n = 2x = 14
Genome size	2450 Mbp
Uses	Food, fodder, bird feed, fuel, fencing
Agronomic traits	Heat and drought tolerant
Nutritional facts	High energy, low starch, rich in mineral (iron and zinc)
Health benefits	Medium glycemic index, gluten-free

See FAO (1995); Martel et al. (1997); Nambiar et al. (2011) for more details.

10.1.1 MORPHOLOGY AND GROWTH STAGES OF PEARL MILLET

Pearl millet is a good tillering crop with a short life cycle. It can grow up to a height of 3 meters, though the high yielding varieties are shorter than this. Its shoots are 1–2 cm thick. Nodes bear axillary buds and alternately arranged leaves. The Kranz anatomy is found in the leaves. This distinctive feature of the C4 pathway not only increases photosynthetic efficiency of leaves but also makes them palatable to grazing cattle. It also contributes to drought tolerance in pearl millet. Leaves contain an equal number of stomata on both surfaces. The root system is the typical monocotyledonous type, consisting of a primary root, many adventitious, and crown roots.

The inflorescence is of panicle type. A spikelet generally contains two florets, one of which is hermaphrodite while the second one is staminate. Pearl millet is a highly cross-pollinated crop owing to its protogynous nature of flowering. Protogyny is a distinctive characteristic for preventing self-pollination in which the stigma come out earlier than anther dehiscence and receive pollen from other plants. In pearl millet, anthers emerge successively in two phases; first in the hermaphrodite florets and later on in the staminate florets. Pollination takes place by air. A panicle may contain 500 to 3,000 spikelets depending on the variety. Seeds vary in size (0.5–3 mm length and 0.003–0.02 g weight), shape (globular to conical), and color.

Pearl millet development occurs in three main stages, i.e., vegetative, reproductive, and grain filling (Maiti and Bidinger, 1981; Bidinger and Hash, 2004; Table 10.2). During the first stage seedling germination and subsequently root and shoot development takes place. This phase ends with panicle initiation. In the second stage, panicle elongation and development of spikelets take place. Elongation of the stem, enlargement of the leaves, and the emergence of all tillers also occur during this phase. The third and final stage starts with the fertilization of florets and ends with seed formation and maturity of the plant.

TABLE 10.2 Growth stages of pearl millet plant

Growth Stages (Sequential Identifying Phases)	Days After Emergence (DAE)*
Vegetative State (Germination and emergence-three leaf stage-Five leaf stage-panicle initiation)	0–21
Reproductive/Panicle Development Stage (Flag leaf stage-booting stage-half blooming)	21–42
Grain Filling Stage (Milk stage-dough stage-maturity)	42–77

See, Maiti and Bidinger (1981); Bidinger and Hash (2004) for more details.

*Vary with environmental conditions and genotypes.

10.1.2 IMPORTANCE OF PEARL MILLET

10.1.2.1 ECONOMIC SIGNIFICANCE

Pearl millet is the sixth most important cereal in terms of global production. It is the staple food of more than 90 million poor people living in the drought-prone regions of Africa and Asia. It plays a key role in the economy and food security of these regions of the world and accounts for over half of the total global millet production. The scenario of global millet production has been shown in Figure 10.1 (FAOSTAT, 2018). It is obvious from figure 10.1 that the area under millet cultivation has declined regularly while the production has shown an upward trend. An increase in yield (from 592 kg/ha in 1961 up to 894 kg/ha in 2016) has ensured that the production did not drop despite a continuous decrease in the cultivated area.

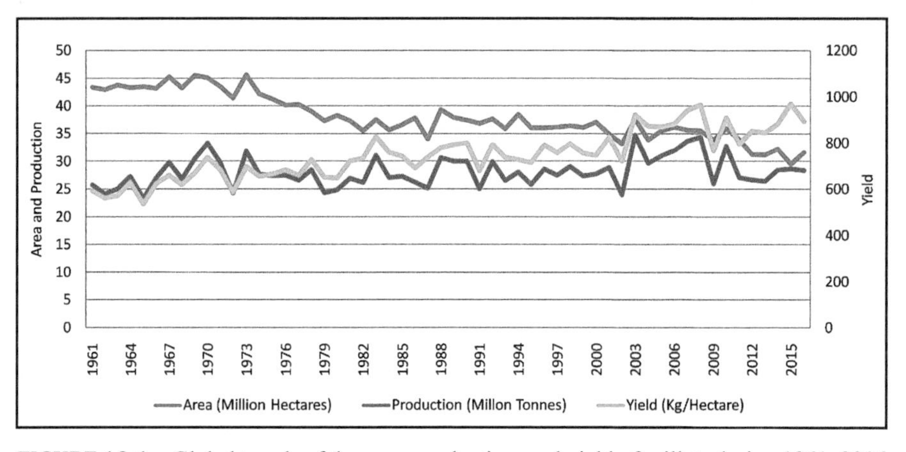

FIGURE 10.1 Global trends of the area, production, and yield of millets during 1961–2016 (*Source:* FAOSTAT, 2018, Food and Agriculture Organization of the United Nations).

India is the largest producer of pearl millet in the world with reference to the area as well as the production. Details of production of pearl millet from 1951 to 2014 at the all-India level are shown in figure 10.2 (Government of India, 2018). How the production of pearl millet in India has influenced and contributed to global millet production is clear by comparing figure 10.1 with figure 10.2. In 2014, globally 28.4 million tonnes of millet were produced from an area of 32.2 million hectares (Figure 10.1). In the same year, India (Figure 10.2) produced 9.2 million tonnes (32% of the global millet production) from only 7.3 million hectares (23% of the global millet area). The global yield of millet was only 882 kg /ha while it was 1255 kg/ha for pearl millet in India in the year 2014. The production in India is higher over other countries because of the introduction and adoption of improved varieties and hybrids (Figure 10.2) and use of better techniques on a regular basis. From 2011 to till now, more than 50 varieties/hybrids have been officially released in India (Singh et al., 2014). These hybrids/varieties are superior in respect of grain quality, harvest index, yield potential, disease resistance, stress tolerance, duration, etc. From the Figure 10.2, it is also apparent that the pearl millet area in India has also declined, but grain production has increased, due to an increase in productivity (from 246 kg/ha during 1951 to 1255 kg/ha during 2014). Pearl millet is mainly a rain-fed crop, excluding when cultivated as a summer irrigated crop. Less than 10% of the pearl millet area is irrigated in India.

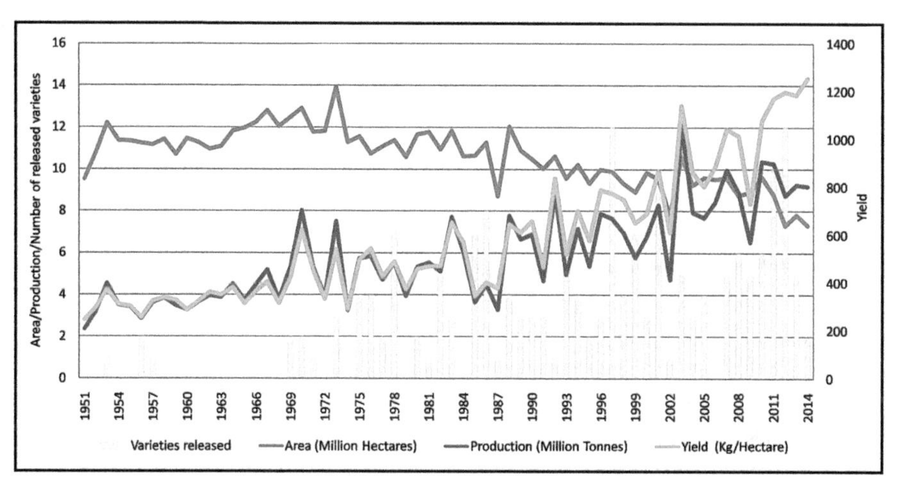

FIGURE 10.2 Trends of the area, production, yield, and release of varieties/hybrids of pearl millet in India during 1951–2014 (*Source*: Directorate of Economics & Statistics, Department of Agriculture, Cooperation & Farmers Welfare, Ministry of Agriculture & Farmers Welfare, Government of India).

10.1.2.2 AGRONOMIC ADVANTAGES

Pearl millet is well adapted to extreme environmental conditions like drought, heat, saline soil, high soil pH, high soil Al^{3+} ions and low soil fertility. In these conditions, other crops fail to survive. Pearl millet can grow and produce grain in regions where average annual rainfall is as low as 250 mm. This crop can tolerate more than 42°C temperatures during the reproductive phase (Gupta et al., 2015). Like other millets, it is also a C4 crop; therefore, it has an excellent photosynthetic efficiency. C4 pathway avoids photorespiration, hence, enhances the efficiency of utilization of the scant moisture available in the dry regions. As C4 plant can keep the stomata closed for a long time, it prevents water loss through the leaves. Being C4 plant, pearl millet has high biomass production potential even with less water requirement. These are the reasons that pearl millet is cultivated with minimal use of agricultural inputs like irrigation, pesticides, and fertilizers in marginal production environments. All these facts make pearl millet a dominant crop of semi-arid tropics.

10.1.2.3 NUTRITIONAL QUALITIES

Millets are known as poor man's crops as they significantly contribute to the diet of low-income consumers and poor rural populations of developing countries. Among them, pearl millet is also a nutrient-rich crop for both humans and livestock. Its grains contain more protein, micronutrients (iron and zinc), folic acid and higher energy (owing to oil content) than those of sorghum, maize, wheat, and rice (Léder, 2004; Nambiar et al., 2011). They are also rich in fibers (1.2 g/100 g) (Nambiar et al., 2011) and antioxidants (FAO 1995; Ragaee et al., 2006). Pearl millet grains are also used for bird feed. After harvesting, the crop remains, and straws are used for cattle feed by farmers.

10.1.2.4 HEALTH BENEFITS

Pearl millet has many potential health benefits. High-calorie pearl millet grains are considered to be a healthy food for growing children and pregnant women. Wheat and other common grains, which contain gluten cause celiac disease or other types of allergies. Pearl millet is best for gluten intolerant people as its grains are gluten-free. Pearl millet has a low glycemic index; therefore, it is useful in the dietary management of diabetes (Mani et al., 1993). Cookies and bread are made from pearl millet grain flour. High nutritional value of pearl millet has prompted researchers to formulate pearl millet-based 'ready-to-reconstitute' Kheer mix powder (Bunkar et al., 2014). It is also used for ethanol production in industries and as biofuel crops (Dahlberg et al., 2003). Iron deficiency is a recurring nutritional deficiency worldwide. The high amount of Iron (8 mg/100 g) and Zinc (3.1 mg/100 g) (Léder, 2004) can help in increasing the hemoglobin level in blood. Tako et al., 2015 have reported some pearl millet varieties with higher iron contents. But they also contain elevated polyphenols and phytates which reduce the bioavailability of iron; therefore, these anti-nutrients need to be reduced to accelerate mineral biofortification. Pearl millet grains contain a high level of antioxidants, especially flavonoids (Sharma and Kapoor, 1996;

Chandrasekara and Shahidi, 2011; Bhatt et al., 2011) which have the anti-tumor property (Huang and Ferraro, 1992).

10.2 DROUGHT: A MAJOR THREAT TO PEARL MILLET

There are some constraints to pearl millet production. Diseases caused by fungi (downy mildew, ergot, blast, smut, rust), bacteria (bacterial leaf streak, bacterial spot) and nematode (root-knot) result in huge yield loss and inferior fodder quality in pearl millet (Timper et al., 2002). Striga is a competitor and a problematic weed for pearl millet cultivation in Africa.

Natural selection driven adaptive evolution has made pearl millet, one of the most tolerant crops to heat and drought. Although pearl millet performs well in comparison to other crops in semi-arid regions, harsh climatic conditions prevailing in these areas are not at all suitable for its farming. Traditional landraces of pearl millet cultivated in drier areas are drought tolerant (Kusaka et al., 2005; Yadav OP, 2010) but their yield potential is lower in comparison to the improved cultivars. In semi-arid regions where it is the principal crop, drought is the major abiotic stress that reduces yield. Drought and other abiotic stresses such as soil salinity, soil acidity, and high temperatures become more severe threats to pearl millet production when they strike during vulnerable stages (crop establishment and reproductive growth).

10.2.1 CAUSES AND DEFINITIONS OF DROUGHT

In general, drought is a period of dry conditions unusually prolonged to cause crop damage, water supply shortage, and other problems. Not only high temperatures and lack of precipitation but also overpopulation and overuse are the reasons behind frequent drought occurrence. More than 150 published definitions of drought are available. Broadly they have been grouped into four basic categories: meteorological, hydrological, agricultural, and socio-economic. These categories define drought by comparing the current precipitation with that of the historical average. Different parameters such as the beginning, end, and degree of severity of drought are also included for comparison. The first three categories reflect drought in terms of physical phenomena. The last one specifies drought as a supply and demand problem caused by water shortfalls.

From an agricultural perspective, drought accounts for inadequate soil moisture, which fails to fulfill the requirements of a specific crop at a specific time of its growth. For instance, scarce moisture during plantation may hamper seed germination, causing low plant populations and low production. An absence or shortage of rainfall resulting in a loss of rain-fed agriculture is also drought. In another term, drought is a decline in water inputs into an agro-ecosystem eventually causing soil water deficit (SWD) over a period (Kramer, 1983). This definition includes various forms of drought, for instance, annual dry seasons, rainfall inconsistency or irrigation failure. SWD is a decline in the available water in the soil when water losses are greater than water inputs.

10.2.2 IMPACT OF AGRICULTURAL DROUGHT ON THE YIELD OF PEARL MILLET

So far, a lot of investment has been made in scientific research in plants with a focus on drought tolerance; nevertheless, the end results of these hard works have not met the increasing agricultural demands. This could be due to the high complexity and variability of drought. Firstly, drought can affect plants during different growth phases; either in the beginning during plant establishment (early drought), in the middle stage during vegetative development (intermittent drought) or late during the phase of reproduction (terminal drought). Secondly, droughts vary in their intensities and duration. Moreover, the effects of drought on a plant are also determined by the plant characteristics. To cope with drought, plants have developed various mechanisms at anatomical, morphological and phenological levels. In addition to these, they have also evolved different physiological and biochemical strategies. Because of the diverse drought patterns and different fine-tuning mechanisms adopted by plants to neutralize the drought effects, it is essential for scientists to specify the drought stress when they do such kind of studies.

Regarding pearl millet Bidinger et al., (1987) studied the effects of mid-season (panicle initiation to flowering; 20 to 50 days after emergence – DAE) and terminal (flowering to maturity; 50 DAE to maturity) drought stress. They conducted field trials in dry seasons at ICRISAT, India using advanced breeding lines of pearl millet. The yield declined by 60% due to the terminal drought, but it was only a little affected (7% loss) by the mid-season stress.

Mahalakshmi et al., 1987 also checked the vulnerability of pearl millet to SWD at different times of its growth. They also discovered interesting findings. When SWD was forced during early growth phases and relieved afterward, the degree of yield loss was determined by the timing of the withdrawal of stress with respect to flowering. Stress had no significant effect on yield when terminated at anthesis or the beginning of grain filling stage. But it caused a significant reduction in yield when it was not relieved until after flowering, thus, the duration of stress in relation to flowering is an important factor in determining the reduction in pearl millet grain yield. When stress was initiated during late developmental stages and was not withdrawn, the timing of imposing stress was straightforwardly correlated with the amount of yield loss. These observations indicate that the development stage at which stress is imposed and stage to which stress is continued both are determining factors regarding the final impact of drought on pearl millet yield.

Winkel et al., 1997 also evaluated (field trials done in dry seasons in Niger, Africa using a landrace Ankoutess which flowers in about 60 days) pearl millet's susceptibility to SWD during three growth stages: prior to flowering (from 30 to 45 DAE), at the beginning of flowering (from 30 to 45 DAE) and at the end of flowering (from 30 to 45 DAE). They found that the yield was severely reduced by 72% and 61% when drought was imposed prior to and at the flowering stage, respectively. Drought imposed at the end of flowering didn't have a significant effect on yield.

In a recent study, terminal drought was imposed on two pairs of genotypes (H77/833–2 and PRLT2/89–33; 841B and 863B) that are contrasting for terminal drought (Aparna et al., 2014). The sensitive genotype H77/833–2 showed higher yield under well-watered conditions in comparison to the tolerant genotype PRLT2/89–33. Under terminal drought stress, genotype H77/833–2 suffered 60% loss in seed number, whereas it was reduced little by 25% in the genotype PRLT2/89–33. Similarly, tolerant genotype 863B sustained a higher yield than the sensitive one 841B under terminal drought.

10.2.3 PLANTS' APPROACHES TO COPE WITH DROUGHT

Plants deal with drought by three main approaches which are drought escape, drought avoidance, and drought tolerance, though one more approach, drought recovery, has also been recognized (Table 10.3).

TABLE 10.3 General mechanisms evolved by plants to overcome drought stress

Approach	Lifecycle	Example	Characters	Degree of Drought Severity
Drought escape	Annual	Ephemeral plants	Rapid growth and early flowering	Plants do not face drought
Drought avoidance	Annual/ Biennial/ Perennial	Succulents, C4 and CAM plants	High water use efficiency, Optimized water uptake	Mild/moderate drought
Drought tolerance	Biennial/ Perennial	True xerophytes	Osmotic adjustment	Mild/moderate drought
Drought recovery	Perennial	Resurrection plants	Reduce damage, Repair damage, Restore growth	Extreme drought

see Levitt (1972, 1980); Kooyers (2015) for more details.

10.2.3.1 DROUGHT ESCAPE

It refers to the state in which a plant completes its life cycle before the occurrence of drought. Such plants are known as drought evaders. Plant characteristics such as fast development, early flowering, efficient photosynthesis and high level of nitrogen in leaves are linked to drought escape (Kooyers, 2015).

10.2.3.2 DROUGHT AVOIDANCE

It is related to the ability of a plant to avoid soil water deficiency. In this case, the plant maintains good water status in its tissues. Two kinds of drought avoidance strategies, namely, water-saving and water spending, have been recognized:

First mechanism involves conservation of water through low transpiration rate (stomatal closure), reduction of leaf-area (leaf rolling or leaf shedding) or slow growth of roots to leave water for later use during critical stages such as flowering and grain filling (Passioura, 1972; Sinclair et al., 2010). This approach may also involve specific permanent adaptation, such as succulence (fleshy, water-storing stems or leaves) (Nobel and Jordan, 1983).

In the second mechanism, plants are able to extract water from deeper soils (high root: shoot ratio) (Ho et al., 2005; Kooyers, 2015; Fang and Xiong, 2015). In place of the term 'drought avoidance,' 'drought postponement' is also preferred for this category to avert perplexity between 'escape' and 'avoidance.' In fact, in annual plants, these strategies do, indeed, simply postpone rather than avoid drought stress (Berger et al., 2016).

10.2.3.3 DROUGHT TOLERANCE

It applies to the ability of a plant to encounter SWD and avoid physiological drought by lowering its water potential by osmotic adjustment (Serraj and Sinclair, 2002). Compatible solutes, inorganic ions, organic compounds like polyols and sugars play an important role in drought tolerance (Blum, 2005; Kooyers, 2015). Other factors such as antioxidant defense, avoidance of photorespiration (Haupt-Herting and Fock, 2000; Ripley et al., 2007), alteration of leaf orientation away from the incident light (Kao and Forseth, 1992), thick cuticle, sunken stomata to reduce water loss and dissipation of excess energy absorbed by photosystem II through heat (Demmig-Adams et al., 2012) also contribute to drought tolerance.

10.2.3.4 DROUGHT RECOVERY

It refers to the capability of a plant to recover from the adverse effects of drought and resume growth. Plants are able to prevent the production of reactive oxygen species, stabilize proteins and preserve cellular membranes to survive extreme desiccation (Vander Willigen et al., 2004). Resurrection plants are the best examples of this approach that can survive near total desiccation (<5% relative water content) for several months or even years. Their branches curl inwards and form a dead-appearing ball. By curling up, they effectively shelter their tissues from the sun. It is a dormant and ametabolic state induced by the extreme dehydration. Plants come out of this phase and bloom under favorable conditions.

The above mentioned different strategies to cope with drought are interrelated. Readers are advised to go through the critical reviews by Gilbert and Medina, 2016 and Berger et al., 2016 in which authors have redefined the plant mechanisms for coping drought. More than one mechanism may operate simultaneously in plants for defense against drought stress (Mitra, 2001). Plants' capability to deal with drought is a complex polygenic trait and is also affected extensively by environments. This complexity may be resolved by a comprehensive study using integrated 'omics' technologies and by applying a system biology approach (Weckwerth, 2003, 2008, 2011a, b).

10.2.4 CHARACTERISTICS OF PEARL MILLET FOR DEALING WITH DROUGHT

Pearl millet is often called the 'Camel' of the crops, owing to its outstanding skill to tolerate drought. Plant traits such as yield, stover biomass, transpiration rate, photosynthesis parameters, osmolytes, pigment content and membrane integrity that are assessed in relation to drought stress are depicted in figure 10.3. Furthermore, different traits that have been studied in pearl millet in this respect are summarized in table 10.4.

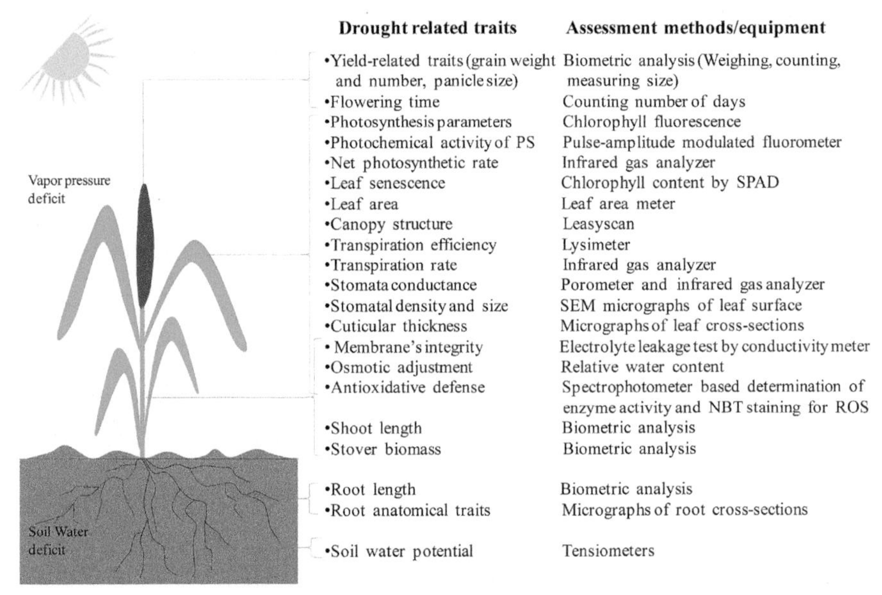

FIGURE 10.3 Important drought-related traits and methods/equipment used for their assessment.

10.2.4.1 AGRONOMIC TRAITS

Tiller numbers, panicle length, panicle diameter, grains number per panicle, grain weight, harvest index, grain yield, stover yield and time to flowering are important agronomic traits. Terminal drought stress limits the yield of pearl millet as it affects seed setting and grain filling. However, contrasting pearl millet genotypes differ in their sensitivities regarding these two factors (Bidinger et al., 1987; Aparna et al., 2014). The tolerant genotype is able to retain a higher number of seeds and more 100-seed weight in comparison to the sensitive one when drought occurs after flowering (Aparna et al., 2014). Contrasting genotypes also differ in their capacity to maintain the normal grain number per panicle and grain yield per panicle in response to terminal stress (Bidinger et al., 1987). Panicle harvest index is a good indicator of the success of spikelet fertility and the degree of grain filling (Bidinger et al., 1987). Furthermore, the tolerant genotype is also able to maintain a higher biomass than the sensitive one (Aparna et al., 2014). Quantitative trait loci (QTLs) associated with yield (Yadav et al., 2002, 2003; Serraj et al., 2005) and quality (Nepolean et al., 2006) of stover have been identified in pearl millet. A number of studies have also identified QTLs for grain yield under terminal drought (Yadav et al., 2002, 2003, 2004; Serraj et al., 2005; Bidinger et al., 2007; Tharanya et al., 2018a).

TABLE 10.4 Different traits of pearl millet studied in relation to drought tolerance

Traits	References
Agronomic Traits	
Grain yield and quality	Bidinger et al., 1987, 2007; van Oosterom et al., 1996; Yadav et al., 1999a, b, 2002, 2003, 2004; Serraj et al., 2005; Aparna et al., 2014; Tharanya et al., 2018a
Stover yield and quality	Nepolean et al., 2006; Yadav et al., 2002, 2003; Serraj et al., 2005; Aparna et al., 2014; Tharanya et al., 2018a
Flowering time	Bidinger et al., 1987
Physiological Traits	
Osmotic potential	Henson, 1982; Henson et al., 1982
Osmolytes	Patil et al., 2005; Kholová et al., 2008a
Transpiration related traits	Henson et al., 1981; Henson and Mahalakshmi, 1985; Ibrahim et al., 1995; Kholová et al., 2008b, 2010a, b, c; Vadez et al., 2013
Water transport	Tharanya et al., 2008b
Biochemical Traits	
ABA accumulation	Henson et al., 1981, 1983; Kholová et al., 2008b, 2010a, b
Photosynthetic pigments	Ibrahim et al., 1995; Ashraf et al., 2001; Kholová, 2008b
Antioxidative enzymes	Patil et al., 2005; Kholová et al., 2008b

10.2.4.2 PHYSIOLOGICAL TRAITS

Osmotic adjustment, water transport pathway and transpiration related traits such as transpiration rate, transpiration efficiency as well as transpiration response to the vapor pressure deficit are some of the important physiological traits. Compatible solutes help in maintaining turgor by lowering osmotic potential. Pearl millet genotypes differ in their capacity of osmoregulation during drought (Henson, 1982; Henson et al., 1982). Drought induces proline accumulation in pearl millet (Patil et al., 2005). Tolerant genotypes accumulate proline faster in compared to sensitive ones (Kholová, 2008a). Increased levels of proline might contribute to the plant's survival rather than to the yield enhancement (Kholová, 2008a). However, according to current opinion, traits enhancing osmotic adjustment are of little benefit for yield in the field conditions. Furthermore, drought-tolerant and drought-sensitive genotypes have different water use efficiencies (Vadez et al., 2013). Under conditions of high vapor pressure deficit, tolerant genotypes extract a smaller quantity of water before flowering and more water after flowering. Water conserved prior to flowering supports the tillers later on and increases yield in tolerant genotypes under terminal drought (Vadez et al., 2013). This water saving mechanism includes restriction of transpiration rate, which may be related to the higher level of ABA in leaves of tolerant genotypes even in well-watered conditions (Kholová et al., 2010b). Moreover, a recent study regarding water use efficiency has revealed that under non-stress or mild stress conditions, pearl millet uses more water to support increasing biomass and tiller production, which in turn results in increased yield (Tharanya et al., 2018a). In case of severe stress conditions, pearl millet saves water at vegetative stage (from lower plant vigor and fewer tillers) which leads to more water available during grain filling, expression of stay-green phenotypes and higher yield (Tharanya et al., 2018a).

Contrasting genotypes also differ in their degrees of dependence on the apoplastic (spaces between cells) and symplastic (through cells via aquaporin-facilitated transport) water transport pathways (Tharanya et al., 2018b). Tolerant genotype (having low transpiration rate) relies more on the apoplastic pathway for water transport, whereas the sensitive one (having a high transpiration rate) depends on both pathways. As mentioned earlier pearl millet is a C4 crop; therefore, it is able to avoid photorespiration, hence, it has an excellent photosynthesis capacity.

10.2.4.3 *BIOCHEMICAL TRAITS*

Stresses induce production of reactive oxygen species (ROS) which causes oxidative damage to cellular components. To minimize the destructive effects of ROS, plants produce antioxidants (carotenoids, flavonoids, ascorbate, glutathione, α-tocopherol) and ROS-scavenging enzymes (superoxide dismutases, catalases, peroxidases, reductases) (Ajithkumar and Panneerselvam, 2014). In pearl millet, the activity of superoxide dismutase, ascorbate peroxidase, nonspecific peroxidase and catalase gets elevated under drought conditions (Patil et al., 2005). In a study, biochemical traits related to isozyme analysis of antioxidative enzymes, chlorophyll, carotenoids, and ABA contents were measured in pearl millet genotypes contrasting in terminal drought tolerance (Kholová, 2008b). These genotypes were found to differ in APX5 isoenzyme activity and ABA content (Kholová, 2008b). However, the photosynthetic pigment content and most of the anti-oxidative isoenzymatic activities were found to be hardly related to the differences in drought tolerance between contrasting genotypes (Kholová, 2008b). Drought induces accumulation of ABA, which in turn regulates stomatal conductance. A study was conducted in another pair of contrasting pearl millet genotypes which differ in the amounts of endogenous ABA accumulating during drought (Henson et al., 1981). It was found that genotype with a high capacity to accumulate ABA showed a higher stomatal sensitivity to water stress than the low ABA accumulator.

10.2.4.4 MORPHOLOGICAL AND ANATOMICAL TRAITS

Leaf area, length of root and shoot are important morphological traits in relation to drought tolerance (Shao et al., 2008). In pearl millet, primary roots develop rapidly during the early growth stage, and they quickly colonize deeper soil horizon (Passot et al., 2016). Characterization of root anatomy and its architecture can contribute to the current understanding of drought tolerance in pearl millet. However, neither the genes related to such root characteristics nor the QTLs for such traits have been discovered in pearl millet. Furthermore, any anatomical character regarding the internal structures of leaf and shoot and its correlation with drought tolerance has not been reported yet.

10.3 'OMICS' STUDIES FOR GENE-MINING FOR DROUGHT TOLERANCE IN PEARL MILLET

Identification and characterization of drought-responsive genes in pearl millet are important for understanding drought-regulated pathways and for designing strategy to increase its production even in the condition of drought occurrence.

10.3.1 GENOMICS

This approach is of two types- structural genomics and functional genomics. The first one attempts to determine the physical structure of the genome with an aim to identify, locate, and arrange genomic characteristics. While the second one focuses on the discovery of the functions of genes. Regarding the first one, the genome of pearl millet has been sequenced recently (Varshney et al., 2017). This new wealth of 1.79 GB genome has revealed interesting features. First, pearl millet genome contains a high GC content (≈48%). This characteristic of the genome is correlated with drought tolerance (Šmarda et al., 2014) and has contributed to the adaptation of pearl millet to arid regions. Second, pearl millet genome contains 80% repetitive elements a proportion comparable to that found in the maize genome (85%) but greater than sorghum (61%), foxtail millet (46%) and rice (42%). These repetitive elements are dominated by transposable elements which play a major role in driving genome evolution. The high content of transposable elements has been proposed to confer a better adaptive potential in pearl millet. Transposable elements cause mutations resulting in altered gene expressions and phenotypes. These genetic variations undergo natural selection. It ultimately leads to better adaptation to the harsh environment. According to the current estimation, the pearl millet genome consists of 38,579 genes. Interestingly, analysis of pearl millet genome indicates a significant enrichment of gene families associated with the biosynthesis of wax, cutin, and suberin. The cuticle not only acts as a protective barrier to pathogens but also prevents excessive water loss (Seo et al., 2011). In the roots, thickening of Casparian strips of endodermis through suberization help in regulation of water movement and prevention of desiccation of internal tissues. Moreover, expansion of ABC transporter gene family also occurred in pearl millet genome. ABC transporters are involved in the transport of diverse secondary metabolites and auxin as well as ABA phytohormones (Kang et al., 2011).

Regarding the functional genomics approach, transgenic and breeding methods have been used to investigate the function of genes of pearl millet. They have been described later in the upcoming sections in this chapter.

10.3.2 TRANSCRIPTOMICS

So far only four reports have been published regarding the transcriptomic analysis of pearl millet in relation to drought (Table 10.5). Mishra et al., 2007 constructed a subtractive cDNA library of the drought-tolerant 863B genotype under abiotic stresses. This study showed the existence of a complex regulatory system (Mishra et al., 2007). A total of 1850 differentially expressed genes were identified in leaves of 7 days old seedlings which were subjected to desiccation, salinity (250 mM NaCl) and cold (4°C) stress for 4h and 24h in greenhouse condition. Broadly these genes belonged to two groups. Genes of the first group were involved in regulatory functions and signaling such as transcription factors (DREB, EREB, zinc finger), protein kinases (serine/threonine kinases, MAP kinases, receptor-related kinases, Ca^{++} dependent protein kinases), phosphatases and secondary messengers (inositol phosphate, cyclic AMP, calmodulin, calcium). While genes of the second group encoded proteins were involved in protection of cellular structures (LEA, dehydrins, RAB, RD); protein structure maintenance (HSPs); protein turnover (proteases, ubiquitins), cellular homeostasis (aquaporins, different solute transporters, ABC transporters, ATPases and antiporters); detoxification (superoxide dismutase, peroxidase, ascorbate peroxidase, glutathione-S-transferase, thioredoxin H, thioredoxin peroxidase and glyoxalase 1) and genes related to cell division, cytoskeleton, and DNA methylation.

In a study carried out by Choudhary et al., (2015), drought stress was imposed on 22 days old pearl millet plants of drought-tolerant PPMI741 genotype for different time periods (30 min, 2 h, 4 h, 8 h, 16 h, 24 h, and 48 h) in a phytotron facility by using 30% PEG 6000. A total of 745 ESTs (assembled into a collection of 299 unigenes) were identified in the leaves in response to drought stress. On the basis of biological and molecular functions, they belonged to various categories such as transcription factors, kinases, signal transduction and cellular homeostasis. Some of the important differential regulated genes identified were those encoding abscisic stress ripening protein, ascorbate peroxidase, aspartic proteinase oryzasin-1, chloroplast 30s ribosomal protein, photosystem I P700 chlorophyll A apoprotein A2, magnesium-protoporphyrin IX monomethylestercyclase, NADPH-dependent oxidoreductase, ATP synthase, citrate synthase, transketolase, mitochondrial carrier protein, glyoxalase, Hsp70 proteins, inosine monophosphate dehydrogenase, metallothionein, β-1,3-glucanase, ubiquitin, cupin domain-containing protein and DnaK family protein.

TABLE 10.5 Transcriptomics studies to understand drought responsive mechanisms operating in pearl millet

Method	Cultivar (Drought Tolerant/ Sensitive)	Plant Material	Differentially Expressed Genes Identified*	References
Subtractive cDNA library	863B (Tolerant)	Leaves	1850	Mishra et al., 2007
Subtractive cDNA library	PPMI741 (Tolerant)	Leaves	299	Choudhary et al., 2015
Illumina sequencing and *de novo* assembly	J-2454 (Tolerant)	Roots	1919 ↑ 2874 ↓	Jaiswal et al., 2018
	J-2454 (Tolerant)	Leaves	1626 ↑ 782 ↓	
RNA-Seq	ICMB 843 (Tolerant)	Roots	3354 ↑ 3447 ↓	Dudhate et al., 2018
	ICMB 863 (Sensitive)	Roots	879 ↑ 374 ↓	

*Vertical arrows pointing upward and downward indicate differential up-regulation and down-regulation of genes respectively.

The above-mentioned two studies were restricted to leaves only. Roots are the primary organs to perceive the drought stress. The effect of drought on leaves and roots differs considerably. In recent investigations, roots' response has also been analyzed (Jaiswal et al., 2018; Dudhate et al., 2018). A transcriptomic study based on next-generation sequencing followed by *de novo* assembly revealed the importance of purine and tryptophan metabolism. It also highlighted the role of MAP kinases in signal transduction in response to drought (Jaiswal et al., 2018). Plants of drought tolerant variety J-2454 were subjected to drought by the withdrawal of irrigation for 6 days after the 23rd day of sowing in the greenhouse. Higher numbers of differentially expressed drought-responsive genes were expressed exclusively in root in comparison to that of the leaf. However, genes commonly expressed in both leaf and root tissues were related to metabolism of purine, thiamine, sucrose, starch, serine, threonine, and glycine metabolisms. This group also included genes related to the biosynthesis of antibiotics and phenylpropanoids (Jaiswal et al., 2018). Purine metabolism is related to ABA accumulation. Phenylpropanoids contribute to stress tolerance. Phenylpropanoids base polymers such as suberin, lignin, and tannins increase strength and vigor towards environmental stress and mechanical damage like drought and wounding (Vogt, 2010). In response to drought, root showed higher expression of genes related to signaling, growth, energy balance (glycolysis and Krebs cycle), immunity (detoxification and scavenging) and protection (dehydrins, LEA). Root also expressed a higher number of transcripts related to osmoregulation (proline, betaines, polyols, and sugars), transporter activity (nitrate, potassium, and sodium transport) and antioxidant activity in comparison to the leaves (Jaiswal et al., 2018). Gene regulatory network analysis revealed MAPK Kinase pathway as major signal transduction pathway in the root.

Root-specific transcriptomic signatures were also compared between drought tolerant (ICMB 843) and sensitive genotype (ICMB 863) of pearl millet (Dudhate et al., 2018). Drought stress was imposed on 21 days old plants by withholding water for 5 days in the greenhouse. Drought-induced transcripts showed more variations in the tolerant genotype. Under drought stress, 4 times more genes were up-regulated while 10 times more genes were down-regulated in the drought-tolerant genotype than the sensitive one. This study revealed that the drought response in pearl millet is largely regulated by pathways associated with photosynthesis, plant hormone signaling, and MAP kinase signal transduction (Dudhate et al., 2018). Around 25 genes encoding different components of photosynthesis (Photosystem II, cytochrome b6/f complex, Photosystem I, photosynthetic electron transport, and F-type ATPase/synthase) were up-regulated in the tolerant genotype. Similarly, 22 genes associated with the signaling of phytohormones such as auxin (*Aux/IAA, SAUR, GH3*), gibberellic acid (*GID1, PIF*), ethylene (*ETR, EIN3, RTE, MKK6*) and ABA (*PP2C, SnRK2, ABF)* showed increased transcript abundance. Drought-induced up-regulation of genes of MAPK cascades (such as *MAPK1, MAPK6, MAP3K17/18, CAT1*) was also observed. MAP kinases belong to the family of Ser/Thr kinases and some of them are also involved in ABA and ethylene signaling.

10.3.3 PROTEOMICS

Analysis of tissue-specific proteomes was carried out to know the translational signatures of drought stress in pearl millet. Ghatak et al., (2016) identified a total of 2281 proteins from different tissues of pearl millet genotype ICTP 8203. It was found that 120, 25 and 10 drought-responsive proteins showed significant changes under drought stress in leaves, roots, and seeds respectively. Molecular chaperones, heat shock proteins, storage proteins, and late embryogenesis abundant proteins increased in seeds. Signaling proteins such as leucine-rich transmembrane protein kinase, GTP binding protein, calnexin, calreticulin, phospholipase C, 14–3-3 protein also increased under stress. However, interestingly proteins related to the metabolism of ABA, auxin, and jasmonate were decreased in leaf and root under water deficit condition.

10.3.4 METABOLOMICS

A number of metabolomics studies have been done in staple cereals and model plants in relation to drought. However, metabolomic analysis in pearl millet is still unreported.

10.3.5 TRANSGENESIS

There are only a few reports regarding functional validations of some of the drought-responsive genes of pearl millet using transgenic approach (Table 10.6). Generation of stress-tolerant transgenic pearl millet is still unreported as improvement of transformation protocol of pearl millet is an essential prerequisite for this.

TABLE 10.6 List of genes from pearl millet functionally validated through transgenesis to confer drought tolerance

Molecular Function	Gene (Encoded Protein)	Transgenic Organism Tested	References
Signal transduction	*PgRab7* (A small GTP binding protein)	Tobacco	Agarwal et al., 2008
Transcription factor	*PgDREB2A* (Dehydration responsive element binding factor)	Tobacco	Agarwal et al., 2010
Molecular chaperones	*PgHsp90* (Heat shock protein)	*E. coli*	Reddy et al., 2011
	PgDHN (Dehydrin)	*E. coli* and yeast	Singh et al., 2015
	PgHSF4 (Heat shockfactor)	Groundnut	Ramu et al., 2016
Antioxidant activity	*PgGPx* (Glutathione peroxidase)	*E. coli* and rice	Islam et al., 2015
Transport	*PgRab7* (A vesicle trafficking protein)	Rice	Tripathy et al., 2017
Housekeeping function	*PgeIF4A* (Eukaryotic translational initiation factor 4A)	Groundnut	Bhadra et al., 2017

In Figure 10.4, a simplified schematic depiction of drought stress signal-transduction and responses has been shown. Important genes that function under drought conditions in pearl millet have also been shown in this figure.

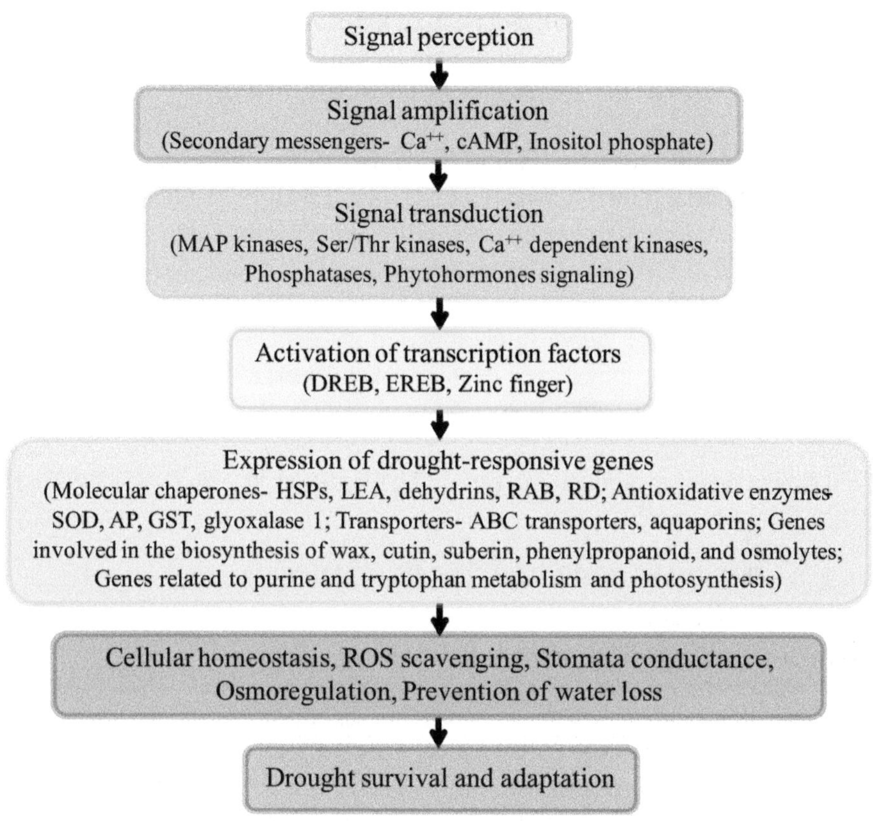

FIGURE 10.4 Drought signaling and response-related genes identified by 'omics' studies in pearl millet.

10.4 BREEDING FOR IMPROVEMENT OF PEARL MILLET GERMPLASM

10.4.1 BREEDING FOR DROUGHT ESCAPE

The strategy of drought escape is the decisive factor regarding crops' performance in harsh environments (Bidinger et al., 1987). Plant breeding approach for shortening of crops' life-cycle is particularly helpful in the regions where drought periods are highly predictable (Bernier et al., 2007). In such regions, natural selection has favored early flowering plants. Drought is a common phenomenon in the arid zone of northwestern India and West Africa. The landraces of pearl millet, which are preferentially cultivated in these areas, match their phenological stages to the regional precipitation pattern (Sivakumar, 1992). In this condition, the development of the panicles coincided with a period of rainfall to minimize the risks of drought incidences before or at the beginning of flowering. Therefore, such landraces have been utilized by plant breeders for reducing lifetime (Bidinger et al., 2009). For instance, CZP9802 is a high yielding and highly adaptive pearl millet variety derived from the landraces of Rajasthan (Yadav, 2004; Dwivedi et al., 2012). It escapes terminal drought on account of its exceptional characteristic of flowering within 48 days and maturity in 75 days (Yadav OP et al., 2004). Another example of an early maturing variety is Okashana 1 which is widely grown in Namibia (Daisuke, 2005). Similarly, West African landrace *Iniadi* (a photoperiod insensitive early maturing land-race) has greatly contributed to developing several pearl millet cultivars globally (Andrews and Kumar, 1996; Witcombe et al., 1997). Polymorphism in a gene *PgMADS11* belonging to MADS-box family has been found to be correlated with flowering time variation in pearl millet (Mariac et al., 2011).

10.4.2 BREEDING FOR DROUGHT AVOIDANCE

This is a useful strategy for the crop cultivation area where the drought pattern is variable. For achievement of drought avoidance, a comprehensive understanding of the intricate plant physiological processes is necessary. As proposed by Passioura (1977, 1996) crop yield is affected by three components and it is calculated as Yield = T X TE X HI; where T stands for the quantity of water transpired, TE for transpiration efficiency and HI for harvesting index. In this context, plant breeders try to increase crops' (1) ability to extract more water; (2) efficiency to produce more dry matter for each unit of absorbed water; and (3) capability to allocate more amount of the biomass into the grain. Apart from these factors, the timing of water consumption during all over plant development is also important for drought avoidance (Sinclair et al., 2005, Blum, 2009; Kholová et al., 2010a, b). As mentioned earlier pearl millet genotypes, carrying a terminal drought tolerance QTL, restrict the transpiration rate and use water economically in well-watered conditions to save water in the soil for the phase of grain filling (Kholová et al., 2010a) and tolerant genotypes maintain a significantly elevated level of ABA in leaves in well-watered conditions which probably lowers leaf conductance (Kholová et al., 2010b).

Drought avoidance strategies depend on the water use efficiency. Plant control over water use, in turn, depends on the balance between the water absorbed by the roots and that released from the leaves by transpiration. Root characteristics (length, conductivity, xylem diameter and chemical composition, symplastic versus apoplastic water movement) have been improved by breeding programs in many crops (Richards and Passioura, 1989; Umayal et al., 2001; Kashiwagi et al., 2006; Bernier et al., 2007; Hufstetler et al., 2007; Bramley et al., 2007, 2009). As mentioned earlier, tolerant and sensitive genotypes of pearl millet differ in water absorption through apoplastic and symplastic (aquaporins) routs (Tharanya et al., 2108b). Stomatal factors (stomata density, stomata conductivity and its sensitivity to soil drying) have been characterized in various crops (Henson et al., 1983; Muchow and Sinclair, 1989; Masle et al., 2005). Role of stomata is important in minimizing water use during pre-anthesis water scarcity in pearl millet (Winkel et al., 2001). Pearl millet and other C4 crops grown in semi-arid regions are sensitive to high vapor pressure deficits in the air. They tend to close their stomata to prevent water loss during the midday warm conditions (Sinclair et al., 2008, Kholová et al., 2010b). Restriction of daily transpiration rate certainly saves water in the soil and adds to transpiration efficiency (Sinclair et al., 2005, 2010).

10.4.3 BREEDING FOR DROUGHT TOLERANCE

It is a good approach in environments with highly unpredictable droughts. Potentials of osmolytes, photosynthetic pigments, and anti-oxidative enzymes have been utilized in breeding for drought tolerance strategies in various crops.

QTLs associated with osmolyte accumulation (Van Deynze et al., 1995, Teulat et al., 1998; Price and Courtis, 1999) have been identified in several crops. Moreover, plants having increased levels of osmolytes have also been developed through breeding (Nguyen et al., 1997) and transgenesis (De Ronde et al., 2000; Gao et al., 2001; Waditee et al., 2003; Majee et al., 2004). However, despite accumulating high concentration of osmolytes, these plants exhibit slight enhancement of drought tolerance. Indeed, higher level of osmolytes is expected to cause quick exhaustion of soil water (Serraj and Sinclair, 2002; Kholová et al., 2010a, b) and rapid shift of plant to the survival mode where even putative advantages are of little use for farmers (Serraj and Sinclair, 2002; Blum, 2005, 2009). As mentioned earlier pearl millet accumulates proline under drought, but it helps in its survival rather than contributing to the yield improvement (Kholová, 2008a). According to current opinion, traits enhancing osmotic adjustment are of little benefit for yield in the field conditions.

Drought causes a decrease in photosynthetic pigment content (Anjum et al., 2003; Massacci et al., 2008) which reduces photosynthesis and ultimately results in crop yield penalties. Moreover, maintenance of pigments in plants facing drought can induce elevated production of reactive oxygen species which can speed up the damage to photosynthesis machinery (Schmid, 2008). Basically, two detoxification mechanisms operate in plants to decrease ROS production: (i) Nonenzymatic antioxidant defense which involves free radical scavengers such as carotenoids, ascorbate, glutathione, tocopherol, mannitol, flavonoids, and some alkaloids. (ii) Enzymatic antioxidant defense which involves superoxide dismutase, ascorbic peroxidase, catalase, and glutathione reductase. However, in pearl millet, the contribution of maintenance of photosynthetic pigment contents, its connection to ROS detoxification systems and lastly their correlations to the yield stability under drought are still unclear. Therefore, there are limited plants breeding efforts in this direction. Though a gene *PgGPx* encoding glutathione peroxidase from pearl millet has been tested for its ability to enhance drought tolerance in rice and bacteria through transgenic approach (Islam et al., 2015).

10.4.4 BREEDING THE FOR DROUGHT RECOVERY

Plants can recover from drought by developing new leaves and producing tillers from buds which survive the dry period. In high drought-prone environments, farmers purposely grow high tillering landraces of pearl millet (van Oosterom et al., 1996). They are able to produce secondary tillers when the central culm is affected by drought. These landraces produce small panicles and small grains (Aparna et al., 2014) to minimize the chances of impaired grain-filling in drought conditions. Since yield is positively associated with grain number (Bidinger and Raju, 2000), the preferred cultivation of high tillering landraces having small grain and a short period of grain filling is a judicious strategy for coping drought and combating extreme aridity. However, any study related to the identification of QTL for drought recovery in pearl millet is still not reported.

10.5 CONCLUSION AND FUTURE PERSPECTIVE

Pearl millet is a major crop of semi-arid regions of the world where it provides an economical and nutritious food for a large population of poor people. Moreover, the livestock also depends on this crop for forage in these regions.

Pearl millet which was once considered 'orphan' in crop science research, is now receiving special attention. However, concrete efforts using both conventional and modern techniques, are still required for further improvement of our understanding of molecular mechanisms underlying drought stress to enhance the productivity of pearl millet. In the future, global climate changes will significantly affect crop cultivation. Important food crops such as rice and wheat that are currently widely cultivated are prone to the drastic effects of environmental stresses, particularly the global temperature rise and frequent droughts. In this scenario, pearl millet may prove as a climate-change resistant crop because of its better adaptations to challenging and harsh environments.

ACKNOWLEDGMENTS

I would like to thank the University Grant Commission (UGC), New Delhi for financial assistance through Start-up (Basic Science Research) grant and Departmental Research Support (DRS Phase-II), Department of Botany, University of Rajasthan. I would also like to acknowledge the University of Rajasthan, Jaipur for providing basic infrastructure facilities.

KEYWORDS

- **Breeding**
- **Drought**
- **Omics**
- **Pearl millet**
- **Signaling**
- **Tolerance**
- **Traits**

REFERENCES

Agarwal, P. K., Parinita, A., Parul, J., Bhavanath, J., Reddy, M. K., & Sopory, S. K., (2008). Constitutive over expression of a stress-inducible small GTP-binding protein PgRab7 from *Pennisetum glaucum* enhances abiotic stress tolerance in transgenic tobacco. *Plant Cell Reports, 27*(1), 105–115.

Agarwal, P., Pradeep, K. A., Arvind, J. J., Sudhir, K. S., & Malireddy, K. R., (2010). Over expression of PgDREB2A transcription factor enhances abiotic stress tolerance and activates downstream stress-responsive genes. *Molecular Biology Reports, 37*(2), 1125.

Ajithkumar, I. P., & Panneerselvam, R., (2014). ROS scavenging system, osmotic maintenance, pigment and growth status of *Panicum sumatrense* roth. Under drought stress. *Cell Biochemistry and Biophysics, 68*(3), 587–595.

Andrews, D. J., & Anand, K. K., (1996). Use of the West African pearl millet landrace India in cultivar development. *Plant Genetic Resources Newsletter, 105*, 15–22.

Anjum, F., Yaseen, M., Rasul, E., Wahid, A., & Anjum, S., (2003). Water stress in barley (Hordeum vulgare L.). II. Effect on chemical composition and chlorophyll contents. *Pak. J. Agric. Sci., 40*, 45–49.

Aparna, K., Hash, C. T., Yadav, R. S., & Vadez, V., (2014). Seed number and 100-seed weight of pearl Millet (*Pennisetum glaucum* L.) respond differently to low soil moisture in genotypes contrasting for drought tolerance. *Journal of Agronomy and Crop Science, 200*(2), 119–131.

Ashraf, M., Ashfaq, A., & McNeilly, T., (2001). Growth and photosynthetic characteristics in pearl millet under water stress and different potassium supply. *Photosynthetica, 39*(3), 389–394.

Berger, J., Jairo, P., & Vincent, V., (2016). An integrated framework for crop adaptation to dry environments: Responses to transient and terminal drought. *Plant Science, 253*, 58–67.

Bernier, J., Arvind, K., Venuprasad, R., Dean, S., & Gary, A., (2007). A large-effect QTL for grain yield under reproductive-stage drought stress in upland rice. *Crop Science, 47*(2), 507–516.

Bhadra, R. T. S. R., Naresh, J. V., Reddy, P. S., Reddy, M. K., & Mallikarjuna, G., (2017). Expression of *Pennisetum glaucum* eukaryotic translational initiation factor 4A (PgeIF4A) confers improved drought, salinity, and oxidative stress tolerance in groundnut. *Frontiers in Plant Science, 8*(453), 1–15.

Bhatt, D., Manisha, N., Priyadarshini, S., Saurabh, C. S., Anoop, K. D., & Sandeep, A., (2011). Responses to drought induced oxidative stress in five finger millet varieties differing in their geographical distribution. *Physiology and Molecular Biology of Plants, 17*(4), 347.

Bidinger, F. R., & Raju, D. S., (2000). Response to selection for increased individual grain mass in pearl millet. *Crop Science, 40*(1), 68–71.

Bidinger, F. R., & Thomas, H. C., (2004). Pearl millet. In: *Physiology and Biotechnology Integration for Plant Breeding*, (pp. 205–242). CRC Press.

Bidinger, F. R., Mahalakshmi, V., & Durga, P. R. G, (1987). Assessment of drought resistance in pearl millet (*Pennisetum americanum* (L.) Leeke). II. Estimation of genotype response to stress. *Australian Journal of Agricultural Research, 38*(1), 49–59.

Bidinger, F. R., Nepolean, T., Hash, C. T., Yadav, R. S., & Howarth, C. J., (2007). Quantitative trait loci for grain yield in pearl millet under variable post flowering moisture conditions. *Crop Science, 47*(3), 969–980.

Bidinger, F. R., Yadav, O. P., & Weltzien, R. E., (2009). Genetic improvement of pearl millet for the arid zone of northwestern India: Lessons from two decades of collaborative ICRISAT-ICAR research. *Experimental Agriculture, 45*(1), 107–115.

Blum, A., (2005). Drought resistance, water-use efficiency, and yield potential—are they compatible, dissonant, or mutually exclusive? *Australian Journal of Agricultural Research, 56*(11), 1159–1168.

Blum, A., (2009). Effective use of water (EUW) and not water-use efficiency (WUE) is the target of crop yield improvement under drought stress. *Field Crops Research, 112*(2, 3), 119–123.

Bramley, H., Neil, C. T., David, W. T., & Stephen, D. T., (2009). Roles of morphology, anatomy, and aquaporins in determining contrasting hydraulic behavior of roots. *Plant Physiology, 150*(1), 348–364.

Bramley, H., Turner, D. W., Tyerman, S. D., & Turner, N. C., (2007). Water flow in the roots of crop species: The influence of root structure, aquaporin activity, and water logging. *Advances in Agronomy, 96*, 133–196.

Bunkar, D. S., Alok, J., & Ankur, M., (2014). Optimization of the formulation and technology of pearl millet based 'ready-to-reconstitute' kheer mix powder. *Journal of Food Science and Technology, 51*(10), 2404–2414.

Chandrasekara, A., & Fereidoon, S., (2011). Antiproliferative potential and DNA scission inhibitory activity of phenolics from whole millet grains. *Journal of Functional Foods, 3*(3), 159–170.

Choudhary, M., & Jasdeep, C. P., (2015). Transcriptional profiling in pearl millet (*Pennisetum glaucum* LR, B.,) for identification of differentially expressed drought responsive genes. *Physiology and Molecular Biology of Plants, 21*(2), 187–196.

Dahlberg, J. A., Wilson, J. P., & Synder, T., (2003). Alternative uses of sorghum and pearl millet in Asia. In: *Proceedings of an Expert Meeting* (pp. 1–4). *ICRISAT, Patancheru, Andhra Pradesh, India.*

De Ronde, J. A., Spreeth, M. H., & Cress, W. A., (2000). Effect of antisense L-Δ1-pyrroline-5-carboxylate reductase transgenic soybean plants subjected to osmotic and drought stress. *Plant Growth Regulation, 32*(1), 13–26.

Demmig-Adams, B., Christopher, M. C., Onno, M., & William, W. A., (2012). Modulation of photosynthetic energy conversion efficiency in nature: From seconds to seasons. *Photosynthesis Research, 113*(1–3), 75–88.

Dudhate, A., Harshraj, S., Daisuke, T., Shenkui, L., & Tetsuo, T., (2018). Transcriptomic analysis reveals the differentially expressed genes and pathways involved in drought tolerance in pearl millet [*Pennisetum glaucum* (L.) R. Br]. *PloS One, 13*(4), e0195908.

Dwivedi, S. L., Hari, D. U., Senapathy, S., Tom, H. C., Kenji, F., Xiamin, D., Dipak, S. D. B., & Manoj, P., (2012). *Millets: Genetic and Genomic Resources* (pp. 247–375).

Fang, Y., & Lizhong, X., (2015). General mechanisms of drought response and their application in drought resistance improvement in plants. *Cellular and Molecular Life Sciences, 72*(4), 673–689.

FAO, (1995). *Sorghum and Millets in Human Nutrition (FAO Food and Nutrition Series No 27)*, p. 198.

FAOSTAT, (2018). www.faostat.org (accessed on 17 July 2018).

Gao, M., Ryutaro, T., Keisuke, M., Abhaya, M. D., & Akira, S., (2001). Transformation of Japanese persimmon (*Diospyros kaki* Thunb.) with apple cDNA encoding NADP-dependent sorbitol-6-phosphate dehydrogenase. *Plant Science, 160*(5), 837–845.

Ghatak, A., Palak, C., Matthias, N., Valentin, R., David, L., Gert, B., Wolfgang, P., et al., (2016). Comprehensive tissue-specific proteome analysis of drought stress responses in *Pennisetum glaucum* (L.) R. Br. (Pearl millet). *Journal of Proteomics, 143*, 122–135.

Gilbert, M. E., & Viviana, M., (2016). Drought adaptation mechanisms should guide experimental design. *Trends in Plant Science, 21*(8), 639–647.

Government of India, (2018). *Pocket Book of Agricultural Statistics* (pp. 1–50). Directorate of Economics and Statistics, Government of India.

Gupta, S. K., Rai, K. N., Piara, S., Ameta, V. L., Suresh, K. G., Jayalekha, A. K., Mahala, R. S., et al., (2015). Seed set variability under high temperatures during flowering period in pearl millet (*Pennisetum glaucum* L. (R.) Br.). *Field Crops Research, 171*, 41–53.

Haupt-Herting, S., & Heinrich, P. F., (2000). Exchange of oxygen and its role in energy dissipation during drought stress in tomato plants. *Physiologia plantarum, 110*(4), 489–495.

Hension, I. E., & Mahalaksshmi, V., (1985). Evidence for panicle control of stomatal behavior in water-stressed plants of pearl millet. *Field Crops Research, 11*, 281–290.

Henson, I. E., (1982). Osmotic adjustment to water stress in pearl millet (*Pennisetum americanum* (L.) Leeke) in a controlled environment. *Journal of Experimental Botany, 33*(1), 78–87

Henson, I. E., Mahalakshmi, V., Alagarswamy, G., & Bidinger, F. R., (1983). An association between flowering and reduced stomatal sensitivity to water stress in pearl millet [*Pennisetum americanum* (L.) Leeke]. *Annals of Botany, 52*(5), 641–648.

Henson, I. E., Mahalakshmi, V., Bidinger, F. R., & Alagarswamy, G., (1981). Stomatal responses of pearl millet (*Pennisetum americanum* (L.) Leeke) genotypes, in relation to abscisic acid and water stress. *Journal of Experimental Botany, 32*(6), 1211–1221.

Henson, I. E., Mahalakshmi, V., Bidinger, F. R., & Alagarswamy, G., (1982). Osmotic adjustment to water stress in pearl millet (*Pennisetum americanum* [L.] Leeke) under field conditions. *Plant, Cell and Environment, 5*(2), 147–154.

Ho, M. D., Juan, C. R., Kathleen, M. B., & Jonathan, P. L., (2005). Root architectural tradeoffs for water and phosphorus acquisition. *Functional Plant Biology, 32*(8), 737–748.

Hong-Bo, S., Li-Ye, C., Cheruth, A. J., & Chang-Xing, Z., (2008). Water-deficit stress-induced anatomical changes in higher plants. *Comptes Rendus Biologies* 331(3), 215–225.

Hufstetler, E. V., Roger, B. H., Thomas, E. C., & Hugh, J. E., (2007). Genotypic variation for three physiological traits affecting drought tolerance in soybean. *Crop Science, 47*(1), 25–35.

Ibrahim, Y. M., Marcarian, V., & Dobrenz, A. K., (1995). Pearl millet response to different irrigation water levels: II. Porometer parameters, photosynthesis, and water use efficiency. *Emirates Journal of Food and Agriculture*, 20–38.

Islam, T., Mrinalini, M., & Malireddy, K. R., (2015). Glutathione peroxidase of *Pennisetum glaucum* (PgGPx) is a functional Cd^{2+} dependent peroxiredoxin that enhances tolerance against salinity and drought stress. *PLoS One, 10*(11), e0143344.

Jaiswal, S., Tushar, J. A., Mandavia, M. K., Meenu, C., Rahul, S. J., Rukam, S. T., Jashminkumar, K., et al., (2018). Transcriptomic signature of drought response in pearl millet (*Pennisetum glaucum* (L.) and development of web-genomic resources. *Scientific Reports, 8*(1), 3382.

Kang, J., Jiyoung, P., Hyunju, C., Bo, B., Tobias, K., Youngsook, L., & Enrico, M., (2011). Plant ABC transporters. *The Arabidopsis Book/American Society of Plant Biologists,* 9.

Kashiwagi, J., Krishnamurthy, L., Crouch, J. H., & Serraj, R., (2006). Variability of root length density and its contributions to seed yield in chickpea (Cicer arietinum L.) under terminal drought stress. *Field Crops Research, 95*(2, 3), 171–181.

Kholová, J., Vadez, V., & Hash, C. T., (2008a). Proline- any perspective for pearl millet (*Pennisetum americanum* L.) drought tolerance improvement? In: *Golden Jubilee Conference on Challenges and Emerging Strategies for Improving Plant Productivity* (p. 61) New Delhi, India. Book of abstracts.

Kholová, J., Vadez, V., & Hash, C. T., (2008b). Mechanisms underlying drought tolerance in pearl millet (*Pennisetum americanum* L.). In: *5th International Crop Science Congress* (p. 188).

Kholová, J., Tom, H. C., Aparna, K., Marie, K., & Vincent, V., (2010a). Constitutive water-conserving mechanisms are correlated with the terminal drought tolerance of pearl millet [*Pennisetum glaucum* (L.) R. Br.]. *Journal of Experimental Botany, 61*(2), 369–377.

Kholová, J., Hash, C. T., Lava, K. P., Rattan, S. Y., Marie, K., & Vincent, V., (2010b). Terminal drought-tolerant pearl millet [*Pennisetum glaucum* (L.) R. Br.] have high leaf ABA and limit transpiration at high vapor pressure deficit. *Journal of Experimental Botany, 61*(5), 1431–1440.

Kholová, J., Zindy, P., Hash, C. T., Kočová, M., & Vadez, V., (2010c). *Genotypes Contrasting for Terminal Drought Tolerance Contrast for the Developmental Pattern of Water Use in Varying Environmental Conditions*. Australian summer grain conference, Gold coast, Australia.

Kooyers, N. J., (2015). The evolution of drought escape and avoidance in natural herbaceous populations. *Plant Science, 234*, 155–162.

Kramer, P. J., (1983). *Water Relations of Plants*. Academic Pres. Inc. New York.

Kusaka, M., Antonio, G. L., & Tatsuhito, F., (2005). The maintenance of growth and turgor in pearl millet (*Pennisetum glaucum* [L.] Leeke) cultivars with different root structures and osmo-regulation under drought stress. *Plant Science, 168*(1), 1–14.

Léder, I., (2004). Sorghum and millets. *Cultivated Plants, Primarily as Food Sources, 1*, 66–84.

Levitt, J., (1972). Water deficit (or drought) stress. In: Levitt, J., (ed.), '*Responses of Plants to Environmental Stresses'* (pp. 322–352).

Levitt, J., (1980). *Responses of Plants to Environmental Stress*. Water, radiation, salt, and other stresses.

Mahalakshmi, V., Bidinger, F. R., & Raju, D. S., (1987). Effect of timing of water deficit on pearl millet (*Pennisetum americanum*). *Field Crops Research, 15*(3, 4), 327–339.

Maiti, R. K., & Bidinger, F. R., (1981). *Growth and Development of the Pearl Millet Plant Research Bulletin no. 6.* ICRISAT, Hyderabad.

Majee, M., Maitra, S., Dastidar, K. G., Pattnaik, S., Chatterjee, A., Hait, N. C., Das, K. P., & Majumder, A. L., (2004). A novel salt-tolerant L-myo-inositol-1-phosphate synthase from *Porteresia* coarctata (Roxb.) tateoka, a halophytic wild rice: Molecular cloning, bacterial over expression, characterization, and functional introgression into tobacco-conferring salt tolerance phenotype. *The Journal of Biological Chemistry, 279*(27), 28539–28552.

Mani, U. V., Prabhu, B. M., Damle, S. S., & Mani, I., (1993). Glycemic index of some commonly consumed foods in Western India. *Asia Pac. J. Clin. Nutr., 2*, 111–114.

Mariac, C., Jehin, L., Saïdou, A. A., Thuillet, A. C., Couderc, M., Sire, P., Jugdé, H., et al., (2011). Genetic basis of pearl millet adaptation along an environmental gradient investigated by a combination of genome scan and association mapping. *Molecular Ecology, 20*(1), 80–91.

Martel, E., Di, D. N., Siljak-Yakoviev, S., Brown, S., & Sarr, A., (1997). Genome size variation and basic chromosome number in pearl millet and fourteen related *Pennisetum* species. *Journal of Heredity, 88*(2), 139–143.

Masle, J., Scott, R. G., & Graham, D. F., (2005). The ERECTA gene regulates plant transpiration efficiency in *Arabidopsis*. *Nature, 436*(7052), 866.

Massacci, A., Nabiev, S. M., Pietrosanti, L., Nematov, S. K., Chernikova, T. N., Thor, K., & Leipner, J., (2008). Response of the photosynthetic apparatus of cotton (Gossypium hirsutum) to the onset of drought stress under field conditions studied by gas-exchange analysis and chlorophyll fluorescence imaging. *Plant Physiology and Biochemistry, 46*(2), 189–195.

Mishra, R. N., Palakolanu, S. R., Suresh, N., Gorantla, M., Arjula, R. R., Sudhir, K. S., & Malireddy, K. R., (2007). Isolation and characterization of expressed sequence tags (ESTs) from subtracted cDNA libraries of *Pennisetum glaucum* seedlings. *Plant Molecular Biology, 64*(6), 713–732.

Mitra, J., (2001). Genetics and genetic improvement of drought resistance in crop plants. *Current Science*, 758–763.

Monneveux, P., & Jean-Marcel, R., (2006). Secondary traits for drought tolerance improvement in cereals. *Drought Adaptation in Cereals*, 97–143.

Mou-Tuan, H., & Ferraro, T., (1992). Phenolic compounds in food and cancer prevention. In: *ACS Symposium Series* (*USA*).

Muchow, R. C., & Sinclair, T. R., (1989). Epidermal conductance, stomatal density and stomatal size among genotypes of *Sorghum bicolor* (L.) moench. *Plant, Cell and Environment, 12*(4), 425–431.

Nambiar, V. S., Dhaduk, J. J., Neha, S., Tosha, S., & Rujuta, D., (2011). Potential functional implications of pearl millet (*Pennisetum glaucum*) in health and disease. *Journal of Applied Pharmaceutical Science, 1*(10), 62.

Nepolean, T., Blummel, M., Bhasker, R. A. G., Rajaram, V., Senthilvel, S., & Hash, C. T., (2006). QTLs controlling yield and stover quality traits in pearl millet. *Journal of SAT Agricultural Research, 2*(1), 4.

Nguyen, H. T., Chandra, B. R., & Blum, A., (1997). Breeding for drought resistance in rice: Physiology and molecular genetics considerations. *Crop Science, 37*(5), 1426–1434.

Nobel, P. S., & Peter, W. J., (1983). Transpiration stream of desert species: Resistances and capacitances for a C3, a C4, and a CAM plant. *Journal of Experimental Botany, 34*(10), 1379–1391.

Passioura, J. B., (1972). The effect of root geometry on the yield of wheat growing on stored water. *Australian Journal of Agricultural Research, 23*(5), 745–752.

Passioura, J. B., (1977). Grain yield, harvest index, and water use of wheat. *Journal of the Australian Institute of Agricultural Science*, 117–120.

Passioura, J. B., (1996). Drought and drought tolerance. *Plant Growth Regulation, 20*(2), 79–83.

Passot, S., Fatoumata, G., Daniel, M., Mikael, L., Soazig, G., Beatriz, M. O., Jonathan, A. A., et al., (2016). Characterization of pearl millet root architecture and anatomy reveals three types of lateral roots. *Frontiers in Plant Science, 7*, 829.

Patil, H. E., Mahatma, M. K., Patel, N. J., Bhatnagar, R., & Jadeja, G. C., (2005). Differential response of pearl millet hybrids to water stress in relation to antioxidant enzymes and proline. *Indian Journal of Plant Physiology, 10*(4), 344.

Price, A., & Brigitte, C., (1999). Mapping QTLs associated with drought resistance in rice: Progress, problems and prospects. *Plant Growth Regulation, 29*(1,2), 123–133.

Ragaee, S., El-Sayed, M. A., & Maher, N., (2006). Antioxidant activity and nutrient composition of selected cereals for food use. *Food Chemistry, 98*(1), 32–38.

Ramu, V. S., Thavarekere, N. S., Shekarappa, H. S., Chandrashekar, K. B., Sreevathsa, R., Malireddy, K. R., Narendra, T., et al., (2016). Simultaneous expression of regulatory genes associated with specific drought-adaptive traits improves drought adaptation in peanut. *Plant Biotechnology Journal, 14*(3), 1008–1020.

Reddy, P. S., Thirulogachandar, V., Vaishnavi, C. S., Aakrati, A., Sudhir, K. S., & Malireddy, K. R., (2011). Molecular characterization and expression of a gene encoding cytosolic Hsp90 from *Pennisetum glaucum* and its role in abiotic stress adaptation. *Gene., 474*(1), 29–38.

Richards, R. A., & Passioura, J. B., (1989). A breeding program to reduce the diameter of the major xylem vessel in the seminal roots of wheat and its effect on grain yield in rain-fed environments. *Australian Journal of Agricultural Research, 40*(5), 943–950.

Ripley, B. S., Matthew, E. G., Douglas, G. I., & Colin, P. O., (2007). Drought constraints on C4 photosynthesis: Stomatal and metabolic limitations in C3 and C4 subspecies of *Alloteropsis semialata. Journal of Experimental Botany, 58*(6), 1351–1363.

Schmid, V. H. R., (2008). Light-harvesting complexes of vascular plants. *Cellular and Molecular Life Sciences, 65*(22), 3619–3639.

Seo, P. J., Saet, B. L., Mi, C. S., Mi-Jeong, P., Young, S. G., & Chung-Mo, P., (2011). The MYB96 transcription factor regulates cuticular wax biosynthesis under drought conditions in *Arabidopsis. The Plant Cell*, tpc-111.

Serraj, R., & Sinclair, T. R., (2002). Osmolyte accumulation: Can it really help increase crop yield under drought conditions? *Plant, Cell and Environment, 25*(2), 333–341.

Serraj, R., Tom, H. C., Masood, H. R. S., Arun, S., Rattan, S. Y., & Fran, R. B., (2005). Recent advances in marker-assisted selection for drought tolerance in pearl millet. *Plant Production Science, 8*(3), 334–337.

Sharma, A., & Kapoor, A. C., (1996). Levels of antinutritional factors in pearl millet as affected by processing treatments and various types of fermentation. *Plant Foods for Human Nutrition, 49*(3), 241–252.

Sinclair, T. R., Carlos, D. M., Andy, B., & Mitch, S., (2010). Assessment across the United States of the benefits of altered soybean drought traits. *Agronomy Journal, 102*(2), 475–482.

Sinclair, T. R., Graeme, L. H., & Erik, J. V. O., (2005). Potential yield and water-use efficiency benefits in sorghum from limited maximum transpiration rate. *Functional Plant Biology, 32*(10), 945–952.

Sinclair, T. R., Maciej, A. Z., & Noel, M. H., (2008). Low leaf hydraulic conductance associated with drought tolerance in soybean. *Physiologia plantarum, 132*(4), 446–451.

Singh, J., Palakolanu, S. R., Chinreddy, S. R., & Malireddy, K. R., (2015). Molecular cloning and characterization of salt inducible dehydrin gene from the C4 plant *Pennisetum glaucum. Plant Gene., 4*, 55–63.

Singh, S. P., Tara, S. C., Mukesh, S. S., Chawla, H. S., Shrotria, P. K., & Jeena, A. S., (2014). *Hybrid Breeding in Pearl Millet: Past and Present Status.* New paradigms in heterosis breeding: Conventional and molecular approaches.

Sivakumar, M. V. K., (1992). Empirical analysis of dry spells for agricultural applications in West Africa. *Journal of Climate, 5*(5), 532–539.

Šmarda, P., Petr, B., Lucie, H., Ilia, J. L., Ladislav, M., Ettore, P., Lubomír, T., et al., (2014). Ecological and evolutionary significance of genomic GC content diversity in monocots. *Proceedings of the National Academy of Sciences, 111*(39), E4096-E4102.

Tako, E., Spenser, M. R., Jessica, B., Jonathan, J. H., & Raymond, P. G., (2015). Higher iron pearl millet (*Pennisetum glaucum* L.) provides more absorbable iron that is limited by increased polyphenolic content. *Nutrition Journal, 14*(1), 11.

Teulat, B., This, D., Khairallah, M., Borries, C., Ragot, C., Sourdille, P., Leroy, P., et al., (1998). Several QTLs involved in osmotic-adjustment trait variation in barley (Hordeum vulgare L.). *Theoretical and Applied Genetics, 96*(5), 688–698.

Tharanya, M., Jana, K., Kaliamoorthy, S., Deepmala, S., Charles, T. H., Basker, R., Rakesh, K. S., et al., (2018a). Quantitative trait loci (QTLs) for water use and crop production traits co-locate with major QTL for tolerance to water deficit in a fine-mapping population of pearl millet (*Pennisetum glaucum* LR B.). *Theoretical and Applied Genetics, 131*(7), 1509–1529.

Tharanya, M., Kaliamoorthy, S., Gloria, B., Jana, K., Thiyagarajan, T., & Vincent, V., (2018b). Pearl millet (*Pennisetum glaucum*) contrasting for the transpiration response to vapour pressure deficit also differ in their dependence on the symplastic and apoplastic water transport pathways. *Functional Plant Biology, 45*(7), 719–736.

Timper, P., Wilson, J. P., Johnson, A. W., & Hanna, W. W., (2002). Evaluation of pearl millet grain hybrids for resistance to *Meloidogyne spp.* and leaf blight caused by *Pyricularia grisea. Plant Disease, 86*(8), 909–914.

Tripathy, M. K., Budhi, S. T., Malireddy, K. R., Renu, D., & Sudhir, K. S., (2017). Ectopic expression of PgRab7 in rice plants (*Oryza sativa* L.) results in differential tolerance at the vegetative and seed setting stage during salinity and drought stress. *Protoplasma, 254*(1), 109–124.

Umayal, L., Babu, R. C., & Sadasivam, S., (2001). Water stress induced histological and enzymatic changes in roots of rice cultivars. *Plant Archives*, 1(1/2), 31–34.

Vadez, V., Jana, K., Rattan, S. Y., & Charles, T. H., (2013). Small temporal differences in water uptake among varieties of pearl millet (*Pennisetum glaucum* (L.) R. Br.) are critical for grain yield under terminal drought. *Plant and Soil, 371*(1, 2), 447–462.

Van, D., Allen, E., James, C. N., Eliana, S. Y., Sandra, E. H., Daniella, P. B., Susan, R. M., & Mark, E. S., (1995). Comparative mapping in grasses. Wheat relationships. *Molecular and General Genetics MGG, 248*(6), 744–754.

Van, O. E. J., Whitaker, M. L., & Weltzien, E., (1996). *Integrating Genotype by Environment Interaction Analysis, Characterization of Drought Patterns, and Farmer Preferences to Identify adaptive Plant Traits for Pearl Millet* (pp. 383–402).

Vander, W. C., Pammenter, N. W., Sagadevan, G. M., & Jill, M. F., (2004). Mechanical stabilization of desiccated vegetative tissues of the resurrection grass *Eragrostis nindensis*: Does a TIP 3; 1 and/or compartmentalization of sub cellular components and metabolites play a role? *Journal of Experimental Botany, 55*(397), 651–661.

Varshney, R. K., Chengcheng, S., Mahendar, T., Cedric, M., Jason, W., Peng, Q., He, Z., et al., (2017). Pearl millet genome sequence provides a resource to improve agronomic traits in arid environments. *Nature Biotechnology, 35*(10), 969.

Vogt, T., (2010). Phenylpropanoid biosynthesis. *Molecular Plant, 3*(1), 2–20.

Waditee, R., Yoshito, T., Kenji, A., Takashi, H., Hiroshi, J., Jun, T., Tetsuko, T., & Teruhiro, T., (2003). Isolation and functional characterization of N-methyltransferases that catalyze betaine synthesis from glycine in a halotolerant photosynthetic organism aphanothece halophytica. *Journal of Biological Chemistry, 278*(7), 4932–4942.

Weckwerth, W., (2003). Metabolomics in systems biology. *Annual Review of Plant Biology, 54*(1), 669–689.

Weckwerth, W., (2008). Integration of metabolomics and proteomics in molecular plant physiology–coping with the complexity by data-dimensionality reduction. *Physiologia plantarum., 132*(2), 176–189.

Weckwerth, W., (2011a). Green systems biology—from single genomes, proteomes and metabolomes to ecosystems research and biotechnology. *Journal of Proteomics, 75*(1), 284–305.

Weckwerth, W., (2011b). Unpredictability of metabolism—the key role of metabolomics science in combination with next-generation genome sequencing. *Analytical and Bioanalytical Chemistry 400*(7), 1967.

Wen-Yuan, K., & Forseth, I. N., (1992). Responses of gas exchange and phototropic leaf orientation in soybean to soil water availability, leaf water potential, air temperature, and photosynthetic photon flux. *Environmental and Experimental Botany, 32*(2), 153–161.

Winkel, T., Jean-François, R., & Payne, W. A., (1997). Effect of the timing of water deficit on growth, phenology and yield of pearl millet (*Pennisetum glaucum* (L.) R. Br.) grown in Sahelian conditions. *Journal of Experimental Botany, 48*(5), 1001–1009.

Winkel, T., Payne, W., & Renno, J. F., (2001). Ontogeny modifies the effects of water stress on stomatal control, leaf area duration and biomass partitioning of *Pennisetum glaucum. New Phytologist, 149*(1), 71–82.

Witcombe, J. R., Rao, M. N. V. R., Raj, A. G. B., & Hash, C. T., (1997). Registration of 'ICMV 88904'pearl millet. *Crop Science, 37*(3), 1022–1023.

Yadav, O. P., (2010). Drought response of pearl millet landrace-based populations and their crosses with elite composites. *Field Crops Research, 118*(1), 51–56.

Yadav, O. P., Weltzien-Rattunde, E., & Bidinger, F. R., (2004). Diversity among pearl millet landraces collected in north-western. *India Ann. Arid Zone, 43*, 45–53.

Yadav, R. S., Hash, C. T., Bidinger, F. R., & Howarth, C. J., (1999a). *QTL Analysis and Marker-Assisted Breeding of Traits Associated with Drought Tolerance in Pearl Millet.* Forthcoming from IRRI: Los Banos, Philipines.

Yadav, R. S., Hash, C. T., Bidinger, F. R., Cavan, G. P., & Howarth, C. J., (2002). Quantitative trait loci associated with traits determining grain and Stover yield in pearl millet under terminal drought-stress conditions. *Theoretical and Applied Genetics, 104*(1), 67–83.

Yadav, R. S., Hash, C. T., Bidinger, F. R., Devos, K. M., & Howarth, C. J., (2004). Genomic regions associated with grain yield and aspects of post-flowering drought tolerance in pearl millet across stress environments and tester background. *Euphytica, 136*(3), 265–277.

Yadav, R. S., Hash, C. T., Bidinger, F. R., Dhanoa, M. S., & Howarth, C. T., (1999b). *Identification and Utilization of Quantitative Trait Loci (QTLs) to Improve Drought Tolerance in Pearl Millet (Pennisetum glaucum (L.) R. Br.).* Forthcoming from CIMMYT: El Batan, Mexico.

Yadav, R., Bidinger, F., Hash. C., Yadav, Y., Yadav, O., Bhatnagar, S., & Howarth, C., (2003). Mapping and characterization of QTL× E interactions for traits determining grain and Stover yield in pearl millet. *Theoretical and Applied Genetics, 106*(3), 512–520.

CHAPTER 11

Biotic Stress on Wheat: An Overview

PANKAJ KUMAR SINGH[1,2], SUNITA SINGH[3], and
GIRISH CHANDRA PANDEY[2,4]

[1]*ICAR-Central Rainfed Upland Rice Research Station, Hazaribagh, Jharkhand, India, E-mail: singhpankajkumar8@gmail.com*

[2]*ICAR-Indian Institute of Wheat and Barley Research, Karnal, Haryana, India*

[3]*ICFRE-Institutes of Forest Productivity, Ranchi, Jharkhand, India*

[4]*Departments of Bioscience and Biotechnology, Banasthali Vidyapith, Rajasthan, India*

ABSTRACT

The wheat crop is exposed to several biotic and abiotic constraints. Apart from the biotic factors, including rusts, smuts, bunts, leaf blight, powdery mildew, head scab, etc., there are various abiotic stresses, which lower the yield, wherever the wheat is grown in the country. Biotic stresses refer to the type of stresses caused by the living organisms. Every year, diseases, insect pests, and weeds cause substantial yield loss to both major and orphan crops. The extent and severity of biotic stresses are more pronounced in tropical region than in the temperate region. This is mainly due to the presence of more conducive environment in the tropics throughout the year where pests and diseases are continuously feed on their host. Biotic stress which is often called decay is caused by infectious diseases that develop in crops and is usually caused by bacteria, fungi, or viruses. The importance of biotic stress factors to cause yield or quality loss depends on the environment and thus varies from different agro-climatic regions.

11.1 INTRODUCTION

Biotic and abiotic both stresses are the major factor of crop losses in agriculture. Wheat is affected by many biotic and abiotic stresses. So the term of biotic stress means in the plant body affected by many living organisms, such as microorganisms like fungi, bacteria, and viruses, some insect and pest nematodes, and weeds. It is different from abiotic stresses; the plants are also affected by natural factors such as sunlight, wind, temperature, heat, salinity, and drought and, flooding. The types of biotic stresses imposed on an organism depends the climate where it lives as well as the species' ability to resist particular stresses. Biotic stress remains a broadly defined term and those who study it face many challenges, such as the greater difficulty in controlling biotic stresses in an experimental context compared to abiotic stress. Plants respond to biotic stress through a defense system. The defense mechanism is classified as an innate and systemic response. This review aims to summarize the current status of knowledge on biotic stress on wheat and its implication on wheat. In the wheat crop, rust is a major disease which is caused by fungi. In our country, out of three types of rusts, brown rust which is known as leaf rust, yellow rust commonly known as stripe rust and the third one is black rust which called stem rust. The three types of host-related of rust disease are alternate host, collateral host, and primary host which are discussed in subsections.

11.1.1 ALTERNATE HOST

This type of host means, two kinds of plants on which a parasitic fungus must develop to complete the life cycle. In our country, the alternate host has not importance in the life cycle of rust fungus. Stripe rust (*Puccinia striiformisf. sp. tritici*) has no alternate host in India. Whereas the Alternate host of stem rust (*Puccinia graminisf. sp. tritici*) is Barberis and *Thalictrum* is the alternate host of leaf rust (*Puccinia reconditaf. sp. tritici,*).

11.1.2 COLLATERAL HOST

This host means a host of the same family in which he helps in surviving a pathogen. In some crop Uredial and conidial stages of rust pathogen survive on Grass hosts. Thus, *Agropyron* and *Bromus* species is the Collateral host of Stripe rust like this. In North India *Bromus* sp. are collateral host of Black rust and similarly In the southern part of India, Brija minor are collateral hosts for Black rust.

11.1.3 PRIMARY HOST

In these types of the host, the rust pathogen produces its telial and resting stages. Dr. K. C. Mehta and his team believed that three hosts on which uredial spores are produced like continued available host, self-growing wheat plants, and *Bromus*. (ex. Grass host). In India, wheat is a second-most staple cereals crop. Rust disease is one of the devastating diseases in all wheat-producing countries. Since the discovery of rust many studies, surveys, surveillance, workshop, and research have been done on rust pathogens and their management. Leaf and stripe rust is the most common disease of rust in the northwestern plain zone, therefore we are going to describe the details of Wheat rust disease symptoms, survival, and spread, favorable condition and control measures, or their details management.

11.2 LEAF RUST

This rust disease is also known as another name of brown rust which is airborne disease.

11.2.1 SURVIVAL AND SPREAD

Thalictrum species is an alternate host of this rust. Rust is an airborne disease which is commonly seen in the wheat-growing northern part of India like Haryana, Punjab, Jammu, Himachal Pradesh, and some part of Northern Uttar Pradesh like Muzzafarnagar, Saharanpur, and Meerut. The Rust Pathogen (*Puccinia* sp.) over-summers in mid and low altitudes of Southern hill (Nilgiris) and Northern Hill (Himalaya). In the part of eastern Indo-Gangetic plains, the Primary infections develop from wind deposited urediospores where it multiplies and moves westwards by the month of last February and the first week of March.

11.2.2 SYMPTOMS

The most common site of this disease appears on the leaf surface of wheat and barley crop, sheaths, glumes, and awns may be occasionally infected and display the symptoms like the tip of the needle in Brown color hence this disease called brown rust. The sporulation occurs 10–14 days after infection. The pustules are slightly elliptical or circular and are often not coalescing; in addition, they have contained masses of orange to orange-brown Urediospores.

11.2.3 FAVORABLE CONDITION

The most favorable condition of this disease the Temperature of 20–25°C with free moisture cause epidermics. Wherever the infection of this disease has been seen, there is a loss of 30%.

11.2.4 CONTROL MEASURES

Always use the resistant variety, Follow the crop rotation, and use the recommended fungicides. Destroy the volunteer wheat plants in the month of March, as they can provide a green bridge to carry the rust spores. During the crop season survey, surveillance, and monitoring the wheat crop it's very important activities for early detection of disease.

11.3 STRIPE RUST

Yellow rust is the second most popular disease of stripe rust. This is also an airborne disease.

11.3.1 SURVIVAL AND SPREAD

Stripe rust requires wet and cool weather to infect the wheat crop. During offseason in the zone of Northern hill, the inoculums survive in the form of uredospores or teliospores on self sown crop or volunteer hosts. So this rust only visible in Northern plain zone of India like Jammu, Himachal Pradesh, Punjab Haryana, and some northern part of Uttar Pradesh also. In the offseason, Northern hills provide an excellent source of inoculums and uredospores primary spread from hills. It is an excellent air traveler and can spread to long distances under favorable climatic conditions because of the proliferating attribute (Brown and Hovmoller, 2002; Chen, 2005).

11.3.2 SYMPTOMS

The most common site of infection on the leaves, stem, leaf sheaths, and awns uredia (bright Yellow color pustules) appear on host plant leaves at an early stage of the crop the yellow color pustules are arranged in a linear row as Stripes, therefore, this disease is also called stripe Rust of wheat. These stripes of yellow rust are orange-yellow in color or yellow like turmeric. Teliospores are also arranged on the surface of the leaves in the form of stripes and are dull black in color when the temperatures arise 25–30°C. The pustules of stripe rust, which, contain yellow to orange-yellow urediospores, usually form narrow stripes on the leaves. These pustules are also found on the leaf sheaths, necks, and glumes.

11.3.3 FAVORABLE CONDITIONS

The infection of this disease occurs only when there are free moisture and optimal temperatures between 10–15°C. Pustules erupt within 10–14 days after infection. It causes a loss of production of about 25% when the stripe rust weather is favorable.

11.3.4 CONTROL MEASURES

Many fungicides are recommended for the control of stripe rust. Identifying the correct disease is very important; we can consult the ICAR, State departments of agriculture, and Krishi Vigyan Kendra also some websites, Agriculture expert's fact sheets, and charts. Select a recommended resistant variety. The survey, surveillance, and crop monitoring are very important for the early detection of diseases. Adaptation of new varieties which are proven to be resistant to stripe rust and cultivation of these resistant varieties is the best approach to control wheat losses to stripe rust (Line, 1972; Konzak et al., 1977; Robbelen and Sharp, 1978; Line and Chen, 1995; Wan et al., 2007).

11.4 STEM RUST

The stem rust of disease also known as black rust of wheat.

11.4.1 SURVIVAL AND SPREAD

Berberis species is an alternate host of this rust. This rust is commonly known as Black rust of wheat which is usually found in the southern part of India where temperatures up to 20°C. In comparison to other rust diseases of wheat, the maximum temperature is required for the spread of this rust disease, the range of which is favorable from 14 to 20°C. This disease both survive on stubbles and volunteer crops. Spores require free moisture like dew, rain, or irrigation and take up to six hours to infect the plant and pustules can be seen after 10–20 days of infection.

11.4.2 SYMPTOMS

In this rust, important symptoms are uredial pustules that occur almost all aerial parts of the plant like stem, leaf sheaths, and upper and lower leaf surfaces. The pustules (containing masses of urediospores) of stem rust are dark reddish-brown color on both sides of the leaves, stems, spikes, and elongated in shapes.

11.4.3 FAVORABLE CONDITIONS

The temperature above 20° C and moisture favors the development of this rust disease. Stem rust is very common in the area of Tamil Nadu state where the wheat is grown. In the past, stem rust has had the ability to cause significant economic damage between 50–100% of yield.

11.4.4 CONTROL MEASURES

Always grow the recommended rust-resistant varieties such as HD 2733, UP 2445, HD 2189, PBW 343, HW 291, HS 240, etc. Avoid the use of excess Nitrogenous fertilizers in the field. Follow the crop rotation activities and crop monitoring also. For control, the stem rust use recommended foliar fungicide.

11.5 KARNAL BUNT (KB) DISEASE OF WHEAT

Karnal bunt (KB) or partial bunt is a fungal disease of Wheat. The disease has been named due to its origin in the Karnal City of Haryana State. In

1931 according to Mitra, the causal organism of this disease is *Tilletia indica* but in the year 1940 Mundkur renamed this fungus is *Neovossia indica Tilletia indica,* a basidiomycetous fungus causing KB disease of wheat, was first time reported from Karnal, Haryana, India by Mitra (1931), subsequently from Pakistan (Munjal, 1975), Nepal (Singh et al., 1989), Iraq (CM, 1989), Mexico (Duran, 1972), the USA (Ykema et al., 1996) and South Africa (Crous et al., 2001). This fungus is a seed and soil-borne the infection occurs at a flowering time by airborne spridia. The side effect of this disease of wheat is more on its quality and not on its reduction. Common wheat, durum wheat, triticale, and some other related species are mainly affected by this disease. This fungus *Tilletia indica,* a basidiomycete, invades the kernels and obtains nutrients from the endosperm, leaving behind waste products with a disagreeable odor that makes bunted kernels too unpalatable for use in flour or pasta. While KB generally does not lead to devastating crop losses, it has the potential to dramatically decrease yield and poses additional economic concerns through quarantines which limit the export of suspected infectious wheat. At the time of spike are emerge wheat plants are more Susceptible but infestation can occur in whole a thesis (Warham, 1984).

11.5.1 DISEASE SYMPTOMS

This disease appears when the wheat grains have developed. Mature ear heads in the field show spikelet's with wide-open glumes having bunted grains. In this disease the ear heads and grains are infected at random, i.e., neither all the ear heads nor all the grains in a spike get infected. In this disease, some wheat grains in the spike are fully or partially converted into black powder masses. The black powder gives is foul smell due to the presence of trimethylamine secreted by teliospores.

11.5.2 FAVORABLE CONDITIONS

Varietal susceptibility and excessive use of fertilizers and irrigation are a favorable condition of this disease. some weather parameters like weekly mean max temperature between 18–22°C. Weekly mean minimum temperature between 9 – 12°C and relative humidity >80% are favorable conditions for spread this disease.

11.5.3 CONTROL MEASURES

Always avoid sowing the susceptible variety and use the recommended resistance varieties. Crop monitoring, survey, surveillance, and post-harvest visual examination are using for detection the KB seeds. Diagnoses and test of seed health is another most important step for control this disease. Sodium hydroxide soak method can be used for seed health test. Excessive irrigation should be avoided at the time of flowering. Use the Vitavax @ 2–2.5 gram per kg seed for eliminating seed-borne infection.

11.6 BLACK POINT DISEASE OF WHEAT

Black point disease of wheat also known as kernel smudge of wheat which is caused mainly fungus by *Alternaria alternata*, *Bipolaris sorokiniana*, *Cladosporium cladosporioides*, *Curvularia lunata,* and *Fusarium* spp. is one of them (Fakir, 1998). *A. alternata*, and *C. sativus* are the most commonly isolated fungus (Fernandez et al., 1994b). The black point disease does not reduce the yield of wheat but gives the quality of the grain as bad. This disease is very common where the wheat grown (Mathur and Cunfer, 1993). The symptoms are visible only when the crop harvested and grain is threshed from the spike (Southwell et al., 1980b).The black point is an insidious type of disease. This is visible only when the grain starts losing moisture. Black point disease has an adverse effect on seed quality germination seed weight, and seedling emergence (Khanum et al., 1987; Rahman and Islam, 1998). Black point infection becomes severe when prolonged wet weather prevails during grain filling period of the crop. The disease is characterized by brown to black discoloration usually restricted to the embryonic end of the grain, but in case of severe infection, the whole grain may be discolored and shriveled (Hanson and Christensen, 1953; Adlakha and Joshi, 1974).

11.6.1 SYMPTOMS

The symptoms of black point or kernel smudge disease the embryo tip shows a black to brown discoloration that may extend into the crease of the kernel. Symptoms of this disease cannot be observed until the plants are harvested and the grain is threshed from the head (Southwell et al., 1980b). Apart from this, delay in maturity of the crop also promotes and favor this disease.

11.6.2 CONTROL MEASURES

To control this disease, irrigation should be reduced after flowering in wheat. Spray the foliar fungicides at the time of flowering and head emergence. The development of resistant cultivars is generally considered the most practical way to control BP, but no variety is fully resistant (Conner and Thomas, 1985; Conner and Davidson, 1988).

KEYWORDS

- **black point disease**
- **Karnal bunt disease**
- **kernel smudge disease**
- **leaf rust**
- **resistance varieties**
- **stem rust**
- **stripe rust**

REFERENCES

Adlakha, K. L., & Joshi, L. M., (1974). Black point of wheat. *Indian Phytopathology, 27*, 41–44.

Ahmed, S. M., & Meisner, C. A., (1996). *Wheat Research and Development in Bangladesh* (p. 201). Bangladesh-Australia Wheat Improvement Project and CIMMYT, Bangladesh.

Brown, J. K. M., & Hovmoller, M. S., (2002). Aerial dispersal of pathogens on the global and continental scales and its impact on plant disease. *Science, 297*, 537–541.

Conner, R. L., & Davidson, J. G. N., (1988). Resistance in wheat to black point caused by *Alternaria alternata* and *Cochliobolus sativus*. *Canadian Journal of Plant Science, 68*, 351–359.

Conner, R. L., & Thomas, J. B., (1985). Genetic variation and screening techniques for resistance to black point in soft white spring wheat. *Canadian Journal of Plant Pathology, 7*, 402–407.

Crous, P. W., Van, J., Castelebury, A. B., Carris, C. A., Frederick, L. M., Pretorious, R. D., & Z. A., (2001). Karnal bunt of wheat newly reported from the African Continent. *Plant Disease, 85*, 561.

Duran, (1972). Further aspects of teliospore germination in North American smut fungi. *Canadian Journal of Botany, 50*, 2569–2573.

Fakir, G. A., (1998). *Black Point Disease of Wheat in Bangladesh* (2nd edn., p. 81). Seed Pathology Laboratory, Bangladesh Agricultural University, Mymensingh.

Fernandez, M. R., Clarke, J. M., DePauw, J. M., Irvine, R. B., & Knox, R. E., (1994). Black point and red smudge in irrigated durum wheat in southern Saskatchewan in 1990–1992. *Can. J. Plant Pathol., 16,* 221–227.

Hanson, E. W., & Christensen, J. J., (1953). The black point disease of wheat in the United States. *Minnesota Agricultural Experiment Station Technical Bulletin, 206,* 30.

Konzak, C. F., Line, R. F., Allan, R. E., & Schafer, J. F., (1977). Guidelines for the production, evaluation, and use of induced resistance to stripe rust in wheat. *Proc. Induced Mutation to Plant Diseases* (pp. 437–460). International Atomic Energy, Vienna.

Line, R. F., & Chen, X. M., (1995). Successes in breeding for and managing durable resistance to wheat rusts. *Plant Dis., 79,* 1254–1255.

Line, R. F., (1972). Recording and processing data on foliar diseases of cereals. *Proc. European and Mediterranean Cereal Rusts Conference* (pp. 175–178). Prague.

Mathur, S. B., & Cunfer, B., (1993). Black point. In: Mathur, S. B., & Cunfer, B., (eds.), *Seed-Borne Diseases and Seed Health Testing of Wheat* (pp. 13–21). Danish Government Institute of Seed Pathology for Developing Countries, Copenhagen, Denmark.

Mitra, M., (1931). *A New Bunt of Wheat in India Annals of Allied Biology, 18,* 178–179.

Munjal, R. L., (1974). Technique for keeping the cultures of *Neovossia indica* in sporulating condition. *Indian Phytopathology, 27,* 248–249.

Robbelen, G., & Sharp, E. L., (1978). Mode of inheritance, interaction and alication of genes conditioning resistance to yellow rust. *Fortschr Pflanzenzucht, 9,* 88.

Singh, D. V., Aggarwal, R., Shreshtha, J. K., Thapa, B. R., & Dubin, H. J., (1989). First report of *Neovossia indica* on Wheat in Nepal. *Plant Disease, 73,* 277.

Southwell, R. J., Wong, P. T. W., & Brown, J. F., (1980b). Resistance of durum wheat cultivars to black point caused by *Alternaria alternata. Australian Journal of Agricultural Research, 31,* 1097–1101.

Wan, A. M., Chen, X. M., & He, Z. H., (2007). Wheat stripe rust in China. *Aust. J. Agric. Res.,58,* 605–619.

Warham, E. J., (1984). A comparison of inoculation methods for karnal bunt *Neovossia indica. Phytopathol, 74,* 856, 857.

Ykema, R. E., Floyd, J. P., Palm, M. E., & Peterson, G. L., (1996). *First Report of Karnal Bunt of Wheat in the United States.* Plant Disease, *80, DO – 10.1094/PD-80-1207B.*

CHAPTER 12

Effect of Water Scarcity and High Temperature on Wheat Productivity

VIDISHA THAKUR and GIRISH CHANDRA PANDEY

Department of Bioscience and Biotechnology, Banasthali Vidyapith, Rajasthan, India, E-mail: girish.dwr@gmail.com (G. C. Pandey)

ABSTRACT

In recent years, considerable efforts have been made to understand the factors controlling grain weight in wheat. Under stress such as heat and drought, reduction in grain weight contributes significantly to the loss of grain yield. The yield components related to yield stability under heat and drought stress have been evaluated and efforts have been made to reveal the mechanisms of tolerance under a stressed condition. The agronomic traits from vegetative to reproductive phase have been significantly correlated with grain filling in wheat. Several genes and QTLs have been identified that determine the grain size and weight. The heat and drought-tolerant wheat cultivars have a higher biomass accumulation, well-developed root system, longer duration, and rate of grain filling and stronger sink activity.

12.1 INTRODUCTION

Wheat is one of the most important crops as it forms a major part of the human diet. The world population derives 20% calories and nearly 23% protein from wheat (FAO, 2017). Wheat also provides other nutrients such as dietary fiber, vitamins, and minerals. Adaptation of wheat to a wide range of agro-ecologies is evident from the fact that it appears in sub-tropical to temperate agriculture from coast to the hills in a diverse type of soils which range from sandy loam to deep black as witnessed in India. Improving the yield of wheat can have a great impact on global nutritional security. The projected climate change will make this an increasingly important issue in the world. Global warming has caused an imbalance in the amount of precipitation worldwide. Drought occurs due to an extended period of below-average precipitation. Other factors such as soil conditions and erosion triggered by poorly planned agricultural endeavors cause a shortfall in water available to the crops. Today many areas of the world face water scarcity and high temperatures which severely affect agriculture.

12.2 GRAIN DEVELOPMENT AND POSITIONAL VARIATION IN SPIKE

The spike of wheat is composed of specialized branches, termed spikelets that are positioned alternatively on two sides of the main axis called rachis. There can be up to 20 spikelets in a spike and each has two outer coverings known as glumes. Initially 8–12 floret primiordia originate in each spikelet but only the first 4–6 are fertile. Florets consist of two sheaths, lemma, and palea that enclose one carpel and three stamens (Kirby and Appleyard, 1987; Evers and Millar, 2002). The grain of wheat is developed by the fertilization of carpels inside the florets.

Within a spike, different spikelets start growing at different times thus the development of all spikelets completes in a time span of several days but all the florets of a spike fully develop in a time-lapse of 2–6 days as they grow at different rates. Spikelet formation starts at the middle of spike and then advances both ways (upwards and downwards) in the spike (Bonnett, 1936; Kirby, 1974; Baillot et al., 2018). This can explain several observations made of higher grain weight in middle spikelets as compared to the apical and bottom spikelets. Floret development within the spikelets begins from the bottom which is floret 1. The two bottom florets develop at almost the same time but floret two gives maximum grain weight. Distal florets have grains of relatively lower weights (Calderini and Ortiz Monasterio, 2003; Liu et al., 2006; Lizana et al., 2010; Xie et al., 2015). Conversely, Herzog and Stamp found the proximal grain to be generally heavier than the one in the second position within a spikelet, while Feng et al. reported no significant weight differences between these positions (Herzog and Stamp, 1983; Feng et al., 2018). Rawson and Evans showed that florets in the center of the ear reach anthesis first, followed by the more distal florets, one day later, and then the more proximal florets, three days later. Moreover, within a spikelet, as many as four days can elapse between the anthesis of the first and last florets (Rawson and Evans, 1970). Many studies have found a strong correlation between the size of carpel and the final grain weight (Calderini et al., 1999). The genetic variability with respect to intra-spike variation in grain size can be one of the opportunities to improve grain weight per spike and hence to improve the productivity of wheat.

Miralles and Slafer stated that the rate of grain filling varies at different positions in the spike resulting in differences in grain weight (Miralles and Slafer, 1995). Bremmer and Herzog and Stamp found that position within spike influenced both rate and duration (Bremmer, 1972; Herzog and Stamp, 1983). However, the mechanism involved is not known. Specifically, the time of floret development and carpel size at anthesis was not considered as putative co-variables. Individual floret flowering time depends on the same two variables. Carpel weight is directly related to final grain weight in the case of central spikelets (Hasan et al., 2011). A study depicted that larger the size of carpel, the faster the rate of grain filling, resulting in larger grains (Xie et al., 2015). Although the spikelet position and carpel size have not been correlated, carpel size could be characteristic of spikelet positions within the spike, finally determining the intra-spike variation in grain weight. However, most of the above studies were largely descriptive and did not propose candidate processes that might explain the observed differences in grain weight. A comprehensive literature on mechanisms of intra-spike variation in grain size and weight is lacking. The comparison of kinetic parameters of grains at different positions within the spike is also required. Under stress, the individual grain weight is reduced and this reduction varies across the length of the spike.

12.3 INTERACTION BETWEEN GRAIN YIELD AND GRAIN NUTRIENT CONTENT

Wheat breeding has created a substantial reduction in nitrogen and phosphorus levels in grain as a result of biomass dilution (Calderini et al., 1995; Ortiz-Monasterio et al., 1997). Grain nutrient concentration tends to get diluted at the cost of improvement in weight. However, grain position in the spike has been correlated to nitrogen concentration. Within the spikelets, the distal lighter grains had low N content than the heavier proximal grains on central and basal spikelets (Simmons and Moss, 1978; Herzog and Stamp, 1983). This suggests that the nutrient content of the grains is affected by grain position in spike. Therefore, nutrient partitioning within spike and dilution affects both need to be considered in subsequent breeding programs for yield improvement. Previous efforts have focused on a higher grain number per square meter than enhancement in grain size. The distal grain setting in spikelets has contributed to an increase in the grain number (Slafer et al., 1994; Calderini et al., 1999). Miralles and Slafer (1995) have found a negative correlation between plant height and proportion of grains in the distal position. However, the selection for additional grains in spikelets is compromising the nutritional quality of grains. *Atauschii* has higher grain micronutrient content than *T. aestivum* (Ortiz-Monasterio and Graham, 2000). Daniel F. Calderini and Ivan Ortiz-Monasterio have observed genetic variation for micronutrient concentration in grains (Calderini and Ortiz-Monasterio, 2003).

12.4 GENETIC CONTROL OF GRAIN WEIGHT

Wheat can have multiple ploidy levels and the species that are presently cultivated worldwide include diploid, tetraploid, and hexaploid (Tadesse et al., 2015). Genes related to the transport of assimilates, protein, and carbohydrate metabolism, starch synthesis, and storage proteins are expressed at a high level during grain filling. The processes that occur during grain growth have been specified phenotypically, whereas genetic characterization of grain weight is limited. Transcriptomic and proteomic studies have been focused on middle and later stages of grain filling, however the genetic control of the important early phases of carpel and grain growth remains unknown (Zhang et al., 2018). Final grain weight showed few or no significant correlations with enzyme activities, sugar levels, or starch content during grain filling or with starch content at maturity. We conclude that neither sugar availability nor enzymatic capacity for starch synthesis during grain filling significantly influenced final grain weight in our field conditions. We suggest that final grain weight may be largely determined by developmental processes prior to grain filling (Fahy et al., 2018). Zhang et al. have concluded that cytokinin oxidase (CKX2)/dehydrogenase gene TaCKX6-D1 was significantly associated with higher grain weight on the basis of linkage mapping, association analysis, and gene expression analysis (Zhang et al., 2012). ABA regulates senescence and therefore controls the time available for nutrient mobilization to developing grains (Yu et al., 2015).

GRAINSIZE3 (*GS3*) is a major QTL for grain length and weight, functions as a negative regulator of grain size, and encodes a transmembrane protein (Fan et al., 2006; Mao et al., 2010). Syntenic genes were discovered in maize (Li et al., 2010) and wheat (Zhang et al., 2014), where a QTL for marker *TaGS-D1* was described in a recombinant in bred population. In wheat, a series of syntenic genes was described for all three genomes of the group six chromosomes (Su et al., 2011; Qin et al., 2014). Significant associations with grain weight were reported for haplotypes of the genes *TaGW2-6A* and *TaGW2-6B* (Qin et al., 2014). Grain filling is affected by the *GIF1* (*grainincomplete filling-1*) that encodes a cell-wall invertase required for carbon-partitioning during early grain-filling (Wang et al., 2008). A sequence-based GWAS (genome-wide association study) and functional genome annotation approach identified *OsGASR7*, a gibberellin-regulated gene that controls grain length in rice (Huang et al., 2012). The syntenic genes *TaGASR7* were discovered in *Triticumurartu* and hexaploid wheat and their natural variation could be linked to effects for grain length and grain weight (Ling et al., 2013; Dong et al., 2014).

Several candidate genes with synteny to known rice genes for grain weight were identified in wheat and their natural variation was associated to grain size-related traits in various wheat panels (Dong et al., 2014; Qin et al., 2014; Zhang et al., 2014). There exists a wealth of QTL for grain size and shape-related traits in wheat (Börner et al., 2002; Dholakia et al., 2003; Groos et al., 2003; Gupta et al., 2006; Huang et al., 2006; Gegas et al., 2010; Williams and Sorrells, 2014) mostly in bi-parental mapping populations. Also in advanced backcross populations QTL for TGW were described (Huang et al., 2003, 2004; Narasimhamoorthy et al., 2006) and a TGW-QTL on chromosome 7D stemming from a synthetic wheat was fine-mapped and further characterized (Röder et al., 2008).

However, all of these studies only reflect the genetic content of a limited number of wheat accessions. Therefore, in the recent years genome-wide association mapping (GWAS) has emerged as alternative strategy to linkage mapping in bi-parental populations. Association mapping is based on "meioses of the past," which occurred during the evolution or development of a line. The major advantages of GWAS are an increased resolution due to an increased number of recombination events compared to bi-parental mapping populations and especially the fact that larger germplasm panels can be surveyed (Hamblin et al., 2011). In specific cases, association mapping even led to the direct molecular identification of causal genes for a trait, such as a gene for spike architecture in barley (Ramsay et al., 2011). Recently, GWAS was also applied to analyze grain size-related traits in wheat (Breseghello and Sorrells, 2006; Neumann et al., 2011; Wang et al., 2012c; Rasheed et al., 2014). Overall, TGW was determined by many markers with small effects. Only three SNP-markers had R2 values above 6% (Christine et al., 2015).

Robust QTL that increases grain weight without reducing grain number has been identified (Griffiths et al., 2015).

12.5 APPLICATION OF PHENOMICS

Trait-based breeding is the need of the hour which can be met through genome aided complementation of conventional breeding approach. While genomics has become robust enough to meet these demands there has been no matching progress in generating phenotypic data at a level desired for identification of genes and QTLs with high power of predictions. Hence, the characterization of plant responses has now assumed high throughput mode with interventions of image-based tools and automation in the recently emerged area of science called phenomics. These image-based tools allow noninvasive quantification of crop growth, leaf senescence, morphological changes when cameras with a visible range of wavelength are employed. It is possible to monitor tissue water content, photosystem health, and canopy temperature dynamics through NIR, chlorophyll fluorescence, and infrared imaging systems. Further, it has been clearly shown that it is possible to employ CT techniques to observe root dynamics without disturbing the soil and metabolite changes through hyperspectral imaging. Surrogate traits provided by these phenomics tools can effectively differentiate the genotypes for potential adaptive traits in plants. Thus, they have the potential to accelerate trait identification and trait phenotyping for facilitating the detection of genes relevant to abiotic stress tolerance. However, it is essential to optimize the methods for evolving phenotyping solutions targeted for a particular crop and agro-ecological situations. With optimized methods, not only the mapping population aimed at genes and markers but also germplasm collected and preserved in germplasm banks can be screened for potential donors for desired traits. Such efforts for hunting genes of relevance to abiotic stress tolerance can begin with a small set of mini core or core set of genotypes representing the huge collection. This needs an inter-institutional and meaningful multidisciplinary approach involving crop scientists and even computational biologists which is possible with advances in these areas of science.

12.6 ABIOTIC STRESS IN WHEAT

The efforts to mitigate the agricultural losses due to abiotic stress include irrigation, soil reclamation, use of fertilizers, and others. Due to the economic and ecological limitations of these strategies, there is a need for exploring the plant's genetic potential for resistance (Baillot et al., 2018). Two main yield components are grain number per unit area and thousand-grain weight. The number of grains is fixed before anthesis and final grain weight is decided during grain filling. Yield stability mainly depends on grain number, however, predicted climate change means a rise in the occurrence of environmental stresses and necessitates the scrutinization of variation in thousand-grain weight (Gouache et al., 2012; Semenov et al., 2011). Thousand-grain weights are basically the average of individual grain weights in a plot. Grain development involves three subsequent phases of cell proliferation, grain assimilate accumulation and grain maturation/desiccation (Shewry et al., 2012). The duration and rate of grain filling are the two aspects that determine the final grain weight (Gebeyehou et al., 1982; Xie et al., 2015).

12.7 WATER STRESS

There are two main processes that control the water availability to plants: (i) water absorption depends on root system and soil properties; (ii) crop transpiration that involves relative humidity, radiation, canopy cover, and stomatal conductance. The evapotranspiration (ET) of crop is directly proportional to grain yield consequently water stress reduces yield. Grain number is determined between tillering and heading stage which is most sensitive to the yield loss due to drought stress.

12.7.1 GERMINATION TO EMERGENCE

Germination to emergence of wheat may experience water stress in the arid environments, which leads to diminished crop establishment through decreased seed reserves, low germination percentage, and high soil impedance (Bouaziz and Hicks, 1990). There is strong correlation between the parameters such as seed size, initial root and shoot biomass and protein content (Ries and Everson, 1973). Wheat germination and establishment is higher with larger seeds as concluded by several studies (Ries and Everson, 1973; Kalankanavar et al., 1989). Bigger seeds also provide higher tolerance against the harmful effects of early exposures to drought (Mian and Nafzigar, 1994). In case of early water stress due to deeper water level profile in the soil, larger root system developed from seedlings of bigger seeds also help to sustain plant water status. Coleoptile length is an important seedling trait for crop establishment.

The major variation in the coleoptile length is genetic (ICARDA, 1987). Genotypes with a long coleoptile allow sowings at greater soil depth avoiding a 'false start' by rainfall that is not of sufficient magnitude or frequent enough to assure the establishment of the crop. Early autumn planting has demonstrated clear advantages in rainfed Mediterranean environments (Acevedo et al., 1991a); the penalty in terms of crop yield associated with delayed planting is in the order of 1% yield loss per day (Acevedo et al., 1998a). Deep early sowing is required in this case to avoid a 'false start.' In soil that is dry at the beginning of the season, seeds should be planted at a depth that would not allow germination unless significant rainfall has occurred to wet the first 10 cm of soil.

12.7.2 EMERGENCE TO TILLERING

Water stress during emergence to tillering may increase the phyllochron of bread and durum wheat (Krenzer et al., 1991; Simane et al., 1993), but leaf expansion is most sensitive to water stress (Acevedo et al., 1971); and leaf growth can be drastically reduced at leaf water potentials of –0.7 to –1.2 MPa (Eastham et al., 1984). Tillering is also very sensitive to water stress, being almost halved if conditions are dry enough (Peterson et al., 1984). As a result, leaf area index development is the most affected physiological process during this stage. Water deficit just before flower initiation may also decrease the number of spikelet primordia at this stage (Oosterhuis and Cartwright, 1983).

12.7.3 TILLERING TO ANTHESIS

Wheat plant growth (roots, leaves, stems, and ears) continues up to approximately ten days after anthesis. Tillering to heading is therefore a period of very active plant growth. It follows that mild to moderate water deficits during this period will decrease cell growth and leaf area with a consequent decrease of photosynthesis per unit area. If the water deficit is more intense, net photosynthesis will decrease even more due to partial stomata closure (Acevedo, 1991a). Stomata start to close in wheat at leaf water potentials of-1.5 MPa (Kobata et al., 1992; Palta et al., 1994). Decreased leaf internal CO_2 (Ci) has the effect of reducing electron transport. Continued over-excitation of the light-harvesting system with no electron transport causes photoinhibition, thus damaging the system (Long et al., 1994). Maintenance of the plant's water status and open stomata is therefore important not only for cooling but also for maintaining a high conductance for CO_2, which keeps photo-synthetic dark reactions going and electron transport functioning (Loomis and Amthor, 1996). Chlorophyll fluorescence is observed when light-harvesting exceeds the capacity of the dark reactions; consequently, fluorescence measurements are now used widely for the detection of stress effects on crops (Seaton and Walker, 1990).

Yield reduction is at a maximum when water stress develops from ten days before spike emergence. Water stress during this stage also decreases the spikelets per spike of fertile tillers (Moustafa et al., 1996) and causes death of the distal and basal florets of the spikes (Oosterhuis and Cartwright, 1983). As mentioned earlier, carbon, and nitrogen availability for spike growth are critical at this stage of development; both are decreased by water stress.

12.8 EFFECT OF TERMINAL DROUGHT STRESS ON GRAIN YIELD

Drought stress can cause damage to wheat during any growth phase (Xie et al., 2015). Drought stress during any phase of the wheat growth cycle leads to a similar decline in agronomic traits (Zhang et al., 2003). The terminal drought stress is often experienced by wheat crop and the grain-filling is affected considerably. All the early events of grain filling such as carpel and grain development are severely impacted which restricts grain growth and reduces grain yield (Farooq et al., 2014). During the vegetative phase, exposure to drought negatively affects growth of roots consequently decreasing leaf area and deteriorating functioning of leaves. Drought stress at tillering stage has been found to reduce number of leaves and leaf area and also causes early senescence. Therefore, sensitivity of this stage is high with 47% yield loss as compared to booting stage that decreases yield by 21% (Boonjung et al., 1996). This is backed by results of an investigation that excluded the late tillers. In case drought occurs only during tillering initiation, wheat can grow late tillers that add to final yield (Daryanto et al., 2016). Drought applied from jointing stage to physiological maturity decreased grain number per spike, spike length and number of spikes per unit area. This resulted in significant decrease in grain yield (Zhang et al., 2003). The late tillers do not yield if the drought stress occurs during mid growth phase till anthesis as it impairs spikelet growth and reduces the number of spikelets per spike (Guan et al., 2010). Maximum loss of yield occurs when the flowering and grain filling stages are exposed to water scarcity as it bring about substantial decline in the rate of photosynthesis and poorly filled grains. This is the outcome of constrained metabolic performance (e.g., degradation of chloroplasts and closure of stomata) (Pinheiro et al., 2000). A mild drought (60% soil water content) post-anthesis reduces yield by 10–30% whereas when extending from flowering to maturity it decreases yield by 60–90% (Maitu et al., 2017). Severe stress (40% soil water content) during grain filling causes a yield reduction of 63–75% (Dixit et al., 2012).

Plant leaf gas exchange has been studied and a decrease of 64% in net photosynthesis has been reported under severe stress, whereas under mild stress the decrease is only 28% (Yan et al., 2016). Drought tolerance is significantly dependent on the root system of crops (Xie et al., 2015). Yield stability under drought can be achieved by well-developed root architecture (Dodd et al., 2011). A deep root system directly correlates to high yield as it ensures water availability from deeper layers of soil (Manschadi et al., 2006). Most of the studies have used artificial mediums for root system analysis which does not apply well to field-grown crops (Nakhforoosh et al., 2013). Wheat roots develop to nearly 113 cm and pots substantially reduce the root environment resulting in a decrease of yield (Hamblin, 1985). Thus, the results of pot studies cannot be extrapolated to field scale.

Photosynthetic efficiency is reduced by drought stress and simultaneously causes accelerated leaf senescence which decreases the duration of grain filling as a result due to elevated ABA levels and transport of carbon reserves from stem to leaves (Yang et al., 2003; Yang and Zhang, 2006). Stay green characteristic is greatly beneficial to grain growth under terminal drought due to delayed leaf senescence (Jordan et al., 2012). Grain dehydration, less endosperm cell proliferation, small grain size (Nicolas et al., 1985), lower amylase content (Singh et al., 2008) are consequences of water stress and lead to decreased transport of assimilates and adverse effect on starch synthesis mechanisms, finally causing a huge decline in starch content of grains (Ahmdi and Baker, 2001).

In wheat grain, expression of 121 proteins was considerably altered under water stress and 57 of these have been identified (Hajheidari et al., 2007). Expression of sucrose transporter genes (SUTs) is modified by mild to severe drought (Xu et al., 2018), and according to some reports; it is downregulated (Xue et al., 2016). Several enzymes including AGPase, starch synthase, and starch branching enzymes play important roles in the accumulation of starch in grains (Morell et al., 2001). Under water stress, the reduction in transcript level of SSS is more than that of other enzymes involved in starch synthesis (Hurkman et al., 2003). The concentrations of glucose, fructose, and sucrose in grains fall significantly in sensitive genotypes which coincided with an abrupt decline in cell wall invertase and soluble invertase activities. Sucrose synthase activity positively correlated to grain size in drought-tolerant cultivar due to an increase in assimilates supply to grains (Saeedipour, 2011). Water stress increased the albumin and gliadin concentration, however, globulin, and glutenins were unaffected (Zhang et al., 2014).

12.9 DROUGHT TOLERANCE

The quantification of drought resistance has also been approached by a yield stability index across environments (Eberhart and Russell, 1966), as well as by drought susceptibility indexes (Fischer and Maurer, 1978). These indexes are highly dependent on yield potential and crop phenology, which are characters with a high genotype x environment interaction (Acevedo, 1991b). To avoid these effects, Bidinger et al. (1987a, 1987b) proposed a drought resistance index (DRI) equivalent to the residual effect of yield under stress once the effects of yield potential, phenology, and experimental error had been removed. The DRI is a criterion to be used to select drought-resistant genotypes or genotypic traits related to drought resistance that could be manipulated as independent genetic characters (Acevedo and Ceccarelli, 1989).

Physiological and morphological characters that confer drought resistance can be classed according to their association to water absorption or water loss by the crop. Morphological and physiological traits related to an increase in water absorption include: root growth, osmotic adjustment, and related solutes and membrane stability (Acevedo et al., 1998a). Morphological and physiological traits related to a decrease in transpiration include: leaf color (van Oosterom and Acevedo, 1992), leaf movements, epicuticular waxes and trichomes on leaf surfaces (Upadhayaya and Furnes, 1994), stomatal behavior (Venora and Calcagno, 1991), transpiration efficiency (Farquhar and Richards, 1984; Acevedo, 1993) and air to canopy temperature difference (Rees et al., 1993). Morgan and Condon (1986) demonstrated that segregating lines of bread wheat and durum wheat with a high capacity for osmotic adjustment had a yield advantage (11 to 17% in bread wheat and 7% in durum wheat) when compared to near-isogenic lines without this character.

12.10 HIGH-TEMPERATURE STRESS

Wheat can be grown not only in the tropical and sub-tropical zones, but also in the temperate zone and the cold tracts of the far north, beyond even the 60°

North altitude. Wheat can tolerate severe cold and snow and resume growth with the setting in warm weather in spring. It can be cultivated from sea level to an altitude of 3300 meters. The most favorable climatic condition for wheat cultivation is cool and moist weather during the vegetative growth period followed by dry, warm weather for the grain to mature and ripe. The optimum temperature range for ideal germination of wheat seed is 20–25°C. Warm and damp climatic conditions are not suited for wheat growing. During the heading and flowering stages, excessively high or low temperatures and drought are harmful to wheat. Cloudy weather, with high humidity and low temperatures is conducive for rust attack. Wheat plant requires about 25–30°C optimum average temperature at the time of ripening. The temperature at the time of grain filling and development are very crucial for yield. Temperatures above 25°C during this period tend to depress grain weight.

Temperature requirements may slightly differ from one variety to another at the time of germination. The critical minimum temperature for wheat crop is from 3.5 to 5.5°C, optimum 20–25°C and the maximum is around 35°C. If temperature is more than 30°C at the time of maturity, it leads to forced maturity and yield loss. An UN report (2011) stated that the earth will be warmer by 2.4°C by the year 2020 and the crop yield in India may fall by up to 30% by then. Instances in India are there indicating that the terminal heat creates a significant yield reduction in wheat. Therefore, maintaining the optimum sowing time and growing ideal cultivars may manage the problem to some extent. Conventional breeding processes generally aimed at disease resistance, quality improvement, and ultimately yield enhancement. In view of the climate change impact particularly that of temperature rise; these breeding programs need to be focused to develop heat-tolerant varieties. Low temperature during the initial stage and high temperature at the later maturity stage lead to the completion of a major part of the wheat growth cycle.

In the northwestern India, it has been observed that considering every grain weighs 40 mg, every degree centigrade in mean temperature during the terminal reproductive phase beyond 17°C causes a loss of yield up to 2.5 q/ha. This is due to the fact that the crucial advanced reproductive phase is affected by rising temperature results in poor grain filling, shorter earhead, lesser 1000 grain weight, and ultimately lower production.

High temperatures accelerate plant development and specifically affect the floral organs, fruit formation, and the functioning of the photosynthetic apparatus. Although recognizing the fundamental linkage between water and heat stresses in plants, attention here will focus on one of them, heat stress, and assume that the wheat plants do not suffer water shortages. For breeding purposes, however, resistance to these two stresses usually has to be combined.

Transpiration, a mechanism of heat avoidance, is the primary agent for energy dissipation. A crop that maintains transpirational cooling may be a good heat avoider. The temperature of plant organs in the field may differ from air temperature by several degrees. This difference increases with a greater rate of transpiration. For wheat with no shortage of soil water, the leaf to air temperature difference increases linearly with vapor pressure difference (Idso et al., 1984). If water shortage arises and stomata begin to close, leaf temperature rises and may exceed air temperature. Leaf photosynthesis is negatively affected as leaf temperature rises above 25°C in cool-grown wheat leaves, but leaves acclimated to warm temperature start to show a similar decline as temperatures exceed 35°C. At 45°C leaf, photosynthesis may be halved.

Heat stress decreases total above-ground biomass and grain yield in wheat. Temperature has a differential effect on each of these phases (Shpiler and Blum, 1986; O'Toole and Stockle, 1991). The most thermosensitive stage of wheat grain yield is reproductive stage, when kernel number is being determined.

12.10.1 GERMINATION TO SEEDLING

From sowing to emergence, seedling mortality, and hence crop establishment, is a problem when soil temperatures are high. Plant emergence and population establishment are the starting points of crop growth. In hot environments, however, the maximum soil temperature in the top centimeters may exceed maximum air temperature by 10–15°C if the soil surface is bare and dry and radiation intensity is high. Under such conditions, maximum soil temperature may reach 40–45°C with serious effects on seedling emergence. The initial plant population may fall below 100 plants/m_2 considered to be deleterious to crop yield. Significant differences in crop establishment, genotypes, and genotype x environment interactions were found under heat stress by Acevedo et al. (1991). Angus et al. (1981) also found that the effect of temperature on emergence varied among wheat genotypes. If seedlings emerge satisfactorily, brief exposures to extreme soil temperatures may inhibit crown root growth and tiller initiation (Fischer, 1985).

12.10.2 EMERGENCE TO TILLERING

Sensitivity to high temperature increases as vegetative growth develops and tillering proceeds towards the end of vegetative stage (O'Toole and Stockle, 1991). The sensitivity to high temperature during this phase is expressed as a decreased duration of vegetative (Shpiler and Blum, 1986) and reduced leaf area and growth. A reduction in total number of leaves and spike-bearing tillers is also an effect of high temperature during this phase (Mid-more et al., 1984). Acevedo et al. (1991b) exemplify these effects. The phyllochron increases when the growth temperature increases (Cao and Moss, 1994), reducing the number of leaves.

12.10.3 TILLERING TO ANTHESIS

The presence of double ridges marks the beginning of the reproductive growth stage. By the end of this stage, the potential number of grains, kernel number, has been determined. The reproductive stage is therefore critical in setting the extent to which the grain yield potential is realized. The main effect of heat stress after floral initiation is observed on kernel number. The number of kernels per unit area decreases at a rate of 4% for each degree increase in means temperature during the 30 days preceding anthesis (Fischer, 1985). The effect of temperature on grains per unit area may be attributed to a decreased number of fertile spikes or to fewer grains per year. In a controlled experiment, Warrington et al. showed that wheat grown at 25°C during reproductive stage had only 40% of the kernel number in the main spike when compared with plants grown at 15°C during this period (Warrington et al., 1977).

The decrease in duration of reproductive stage at high temperatures is affected by genotypic variation in photoperiod sensitivity, vernalization response and optimum temperature for spikelet formation (Blum, 1988). If genotypes are able to maintain high carbon exchange rates at high temperatures, the decrease in reproductive phase duration and spike weight is smaller (Blum, 1986). High temperatures affect the capacity of the chloroplast membranes for electron transport (Berry and Rawson, 1981). An increase in chlorophyll fluorescence at temperatures where CO_2 fixation begins to be affected indicates heat damage of photosystem II (Seeman et al., 1984).

12.10.4 ANTHESIS TO MATURITY

Heat stress during grain filling mainly affects assimilates availability, translocation of photosynthates to the grain and starch synthesis and deposition in the developing grain. The net result is a lower kernel weight. Over the range of 12–26°C increase in mean temperature during grain filling, grain weight is reduced at a rate of 4–8%/°C (Wardlaw et al., 1980; Wiegand and Cuellar, 1981). Acevedo et al. reported a mean reduction of 4% in grain weight per degree increase in mean temperature during grainfilling (Acevedo et al., 1991). Shortened grainfilling duration is partially offset by increased grain-filling rate (Sofield et al., 1977), but the effects are much more complex. Hastened senescence, on the other hand, reduces assimilate supply to the grain. Also high temperature reduces final cell number in the endosperm, reducing grain weight. The results, however, are usually confounded with water stress.

Heat stress tolerance has been related to membrane stability, increased compatible solutes, increased protein stability, and the synthesis of heat shock proteins (HSP). Plant response to heat shock is characterized by a rapid production of a specific set of proteins for supra-optimal temperatures. This happens when plant cells are abruptly exposed to temperatures that are about 5–10°C above their normal physiological growth temperature. The HSP synthesis is also induced by other stresses, such as drought and salinity (El Madidi and Zivy, 1993). These proteins are presumably involved in repairing and/or protecting structures that have been damaged by an increase in temperature or other stress. The presence of denatured proteins inside the cell is enough to induce the synthesis of HSPs, which may have a 'chaperone' role interacting with other proteins protecting their structure and conformation (Ellis, 1990). Genotypes having higher heat tolerance appear to synthesize HSPs in a higher concentration.

Temperatures greater than 24°C for few days at heading lead to great loss of floret fertility whereas a consistent heat stress of 35°C resulted in complete floret sterility. There is a linear decrease in grain weight with increasing duration of heat stress at heading stage (Prasad and Djanaguiraman, 2014). Heat stress during grain growth severely inhibits carbon assimilation in the chloroplasts by reducing the activities of PEP carboxylase and RuBP carboxylase (Xu et al., 2004). Heat stress breaks down chlorophyll, impedes photosynthesis rate and induces leaf senescence thereby affecting grain weight and quality. The rate of grain filling increases along with reduction in its duration, decrease in starch synthesis and loss of sink activity (Hurkman et al., 2003). Decreased starch synthesis is due to lower expression of starch-related genes which results in reduction in seed size (Du Pont and Altenbach, 2003). The concentration of soluble sugars is lower under high temperature stress as compared to optimum temperature conditions (Thomas et al., 2003). Jenner (1991) observed that soluble sugars, sugar nucleotides, and hexose phosphate levels decreased due to high temperature stress. The decline in soluble sugar concentration may be associated to higher assimilate utilization instead of production (Asthir et al., 2012). Thus, heat stress decreases accumulation of carbohydrates by affecting metabolic pathways. High temperature during grain filling has been reported to reveal significant increase in level of proteins involved in photosynthesis, signal transduction, antioxidant activity, ATP synthase, HSPs and other nitrogen metabolism-related proteins in tolerant genotypes implying their critical role in tolerance (Wang et al., 2015).

Quality of storage proteins is changed due to alteration in accumulation of N under heat stress (Triboi et al., 2003; Barnabas et al., 2008). Gliadin accumulation is increased and high molecular weight glutenins decreased followed by reduction in aggregation properties leading to poor dough quality (Stone et al., 1997). Panozzo and Eagles (2000); DuPont and Atlenbach (2003) observed that the concentration of large polymers decreased and the concentration of gliadins and glutenins increased in wheat grain flour.

The differential gene expression among the normal and heat stressed condition has been correlated to the biochemical differences in grains which are mainly lower amylose content, side chain elongation of amylopectin and smaller grain size. The expression of Amy 3 genes which encodes an alpha-glucosidase and alpha-amylase which catalyze the hydrolysis of the raw starch granules is altered under drought stress at grain filling stage (Ma et al., 2017).

12.11 COMBINED HEAT AND DROUGHT STRESS

Both heat and drought stress bring about reduction in grain size due to lower cell number in endosperm, changes in the grain filling rate and different biochemical reactions (Nicolas, 1985). Combined effects of heat and drought stress are not known in detail. The data generated from limited studies on combined heat and drought suggests that effects on crop growth and productivity are severe (Cairns et al., 2013; Awasthi et al., 2014; Seghal et al., 2017). The early events of reproductive phase are most often exposed to simultaneous effect of high temperature and water scarcity, substantially decreasing the rate of fertilization leading to lower grain number (Prasad et al., 2008). Shortening of the grain development phase in crop growth cycle is directly responsible for yield loss (Reddy et al., 2004).

12.12 CONCLUDING REMARKS

Much is known about the phenotypic, physiological, and molecular biology of wheat that can be of direct use to agronomists and breeders. The generation of potential grain yield in wheat is quite well understood in addition to the avenues for improving yield potential. Yield under stress is generally less understood, but available physiological knowledge should allow better and more rapid progress in the future. Important aspects of wheat physiology, such as lodging resistance, the use of growth regulators for wheat growth, weed competition, soil mechanical impedance, and nutrient toxicities/deficiencies, were not discussed here since a priority was given to yield and yield-forming processes with the idea that the application of these concepts would have a higher impact on wheat production around the world.

KEYWORDS

- **drought resistance index**
- **drought stress**
- **evapotranspiration**
- **genome-wide association study**
- **heat shock proteins**
- **sucrose transporters**

REFERENCES

Acevedo, E., & Ceccarelli, S., (1989). Role of physiologist-breeder in a breeding program for drought resistance conditions. In: Baker, G. W. F., (ed.), *Drought Resistance in Cereals* (pp. 117–139). Wallingford UK. CAB International.

Acevedo, E., (1993). Potential of 13C discrimination as a selection in barley breeding. In: Ehleringer, J., Hall, A. E., & Farqu-har, D. G., (eds.), *Stable Isotopes in Agriculture* (pp. 399–417). San Diego CA, USA, Academic Press.

Acevedo, E., Harris, H., & Cooper, P. J. M., (1991a). Crop architecture and water use efficiency in Mediterranean environments. In: Harris, H., Cooper, P. J. M., & Pala, M., (eds.), *Soil and Crop Management for Improved Water Use Efficiency in Rain Fed Areas* (pp. 106–118). Ankara, ICARDA.

Acevedo, E., Hsiao, T. C., & Henderson, D. W., (1971). Immediate and subsequent growth responses of maize leaves to changes in water status. *Plant Physiol., 48*, 631–636.

Acevedo, E., Nachit, M., & Ortiz-Ferrara, G., (1991b). Effects of heat stress on wheat and possible selection tools for use in breeding for tolerance. In: Saunders, D. A., (ed.), *Wheat for the Nontraditional Warm Areas* (pp. 401–421). Mexico, F. D., CIMMYT.

Acevedo, E., Silva, H., & Silva, P., (1998a). Current trends in research on resistance to water stress of cultivated plants. *Bol. Technician. This. Exp. Agron., 49*(1/2), 1–28.

Ahmadi, A., & Baker, A. D., (2001). The effect of water stress on the sucrose to starch pathway. *Plant Growth Regul., 35*, 81–91.

Angus, F. J., Cunningham, B. R., Moncur, M. W., & Mackenzie, D. H., (1981). Phasic development in field crops. I. Thermal response in the seedling phase. *Field Crops Res., 3*, 365–378.

Asthir, B., Koundal, A., & Bains, S. N., (2012). Putrescine modulates antioxidant defense response in wheat under high temperature stress. *Biol. Plant, 56*, 757–761. doi: 10.1007/s10535-012-0209-1.

Awasthi, R., Kaushal, N., Vadez, V., Turner, C. N., Berger, J., Siddique, H. K., et al., (2014). Individual and combined effects of transient drought and heat stress on carbon assimilation and seed filling in chickpea. *Funct. Plant Biol., 41*, 1148–1167.

Baillot, N., Girousse, C., Allard, V., Piquet-Pissaloux, A., & Le Gouis, J., (2018). Different grain-filling rates explain grain-weight differences along the wheat ear. *PLoS One, 13*(12), e0209597.

Barnabás, B., Jäger, K., & Fehér, A., (2008). The effect of drought and heat stress on reproductive processes in cereals. *Plant Cell Environ., 31*, 11–38.

Berry, J. A., & Rawson, J. K., (1981). Response of macrophytes to temperature. In: Lange, L. O., Nobel, S. P., Osmond, C. B., & Ziegler, H., (eds.), *Physiological Plant Ecology* (p. 278). I. Responses to physical environment. Berlin, Springer-Verlag.

Bidinger, R. F., Mahalakshmi, V., & Rao, G. D., (1987a). Assessment of drought resistance in pearl millet (*Pennisetum americanum* (L. Leeke). I. Factors affecting yields under stress. *Austr. Agric. J. Res., 38*, 37–48.

Bidinger, R. F., Mahalakshmi, V., & Rao, G. D., (1987b). Assessment of drought resistance in pearl millet (*Pennisetum americanum* (L. Leeke). II. Estimation of genotype response to stress. *Austr. Agric. J. Res., 38*, 49–59.

Blum, A., (1986). The effect of heat stress on wheat leaf and ear photosynthesis. *Exp. J. Bot., 37*, 111–118.

Blum, A., (1988). *Plant Breeding for Stress Environments* (p. 223). Boca Raton FL, USA, CRC Press.

Bonnett, O., (1936). The development of the wheat spike. *J. Agric. Res., 53*, 445–451.

Boonjung, H., & Fukai, S., (1996). Effects of soil water deficit at different growth stages on rice growth and yield under upland conditions. 1. Growth during drought. *Field Crop Res., 48*, 37–45.

Bouaziz, A., & Hicks, D. R., (1990). Consumption of wheat seed reserves during and during and early growth as affected by soil water potential. *Plant Soil, 128*, 161–165.

Bremner, M. P., (1972). Accumulation of dry matter and nitrogen by grains in different positions of the wheat ear as influenced by shading and defoliation. *Australian Journal of Biological Sciences, 25*(4), 657.

Brendan, F., Hamad, S., Laure, C. D., Stephen, J. P., Philippa, B., Cristobal, U., & Alison, M. S., (2018). Final grain weight is not limited by the activity of key starch-synthesizing enzymes during grain filling in wheat. *Journal of Experimental Botany, 69*(22), 5461–5475.

Breseghello, F., & Sorrells, M. E., (2006). Association mapping of kernel size and milling quality in wheat (*Triticum aestivum* L.) cultivars. *Genetics, 172*, 1165–1177.

Breseghello, F., Sorrells, M. E., Dholakia, B. D. M., Röder, S. M., Rao, V. S., et al., (2003). Molecular marker analysis of kernel size and shape in bread wheat. *Plant Breed*, pp. 392–395.

Cairns, E. J., Crossa, C., Zaidi, H. P., Grudloyma, P., Sanchez, C., & Araus, L. J., (2013). Identification of drought, heat, and combined drought and heat tolerance donors in maize (*Zea mays* L.). *Crop Sci., 53*, 1335–1346.

Calderini, D., Abeledo, L., Savin, R., & Slafer, A. G., (1999). Effect of temperature and carpel size during pre-anthesis on potential grain weight in wheat. *J. Agric. Sci., 132*, 453–459.

Calderini, F. D., & Ortiz-Monasterio, I., (2003). Grain position affects grain macronutrient and micronutrient concentrations in wheat. *Crop Sci., 43*, 141–151.

Calderini, F. D., & Slafer, A. G., (1999). Has yield stability changed with genetic improvement of wheat yield? *Euphytica, 107*, 51–59.

Calderini, F. D., Torres, L. S., & Slafer, A. G., (1995). Consequences of wheat breeding on nitrogen and phosphorus yield, grain nitrogen and phosphorus concentration and associated traits. *Ann. Bot. (London), 76*, 315–322.

Cao, W., & Moss, D. N., (1994). Sensitivity of winter wheat phyllochron to environmental changes. *Agron. J., 86*, 63–66.

Daryanto, S., Wang, L., & Jacinthe, P. A., (2016). Global synthesis of drought effects on maize and wheat production. *PLoS One, 11*, e0156362.

Dixit, S., Swamy, B. P. M., Vikram, P., Ahmed, H. U., Sta, C. M. T., Amante, M., Atri, D., Leung, H., & Kumar, A., (2012). Fine mapping of QTLs for rice grain yield under drought reveals sub-QTLs conferring a response to variable drought severities. *Theor. Appl. Genet., 125*, 155–169.

Dodd, I. C., Whalley, W. R., Ober, E. S., & Parry, M. A., (2011). Genetic and management approaches to boost UK wheat yields by ameliorating water deficits. *Exp. J. Bot., 62*, 5241–5248.

Dong, L., Wang, F., Liu, T., Dong, Z., Li, A. G. C., Robert, N., Bervas, E., & Charmet, G., (2003). Genetic analysis of grain protein-content, grain yield and thousand kernel weight in bread wheat. *Theor. Appl. Genet., 106*, 1032–1040.

DuPont, M. F., & Altenbach, B. S., (2003). Molecular and biochemical impacts of environmental factors on wheat grain development and protein synthesis. *J. Cereal Sci., 38*, 133–146. doi: 10.1016/S0733-5210(03)00030-4.

Eastham, J., Oosterhuis, D. M., & Walker, S., (1984). Leaf water and turgor potential threshold values for leaf growth of wheat. *Agron. J., 76*, 841–847.

Eberhart, S., & Russell, W., (1966). Stability parameters for comparing varieties. *Crop Sci., 6*, 36–40.

El Madidi, S., & Zivy, M., (1993). Genetic variability of heat shock proteins and thermotolerance in wheat. Does genetic progress go through the identification and inventory of genes? In: Chlyah, H., & Demarly, Y., (eds.), *Actualités scientifiques* (pp. 173–181). Paris, John Libbey Eurotext.

Ellis, H., Spielmeyer, W., Gale, K. R., Rebetzke, G. J., Richards, R. B. A. E., Fürste, A., Cöster, H., Leithold, B., Röder, M. S., et al., (2002). Mapping of quantitative trait loci for agronomic important characters in hexaploid wheat (*Triticum aestivum* L.). *Theor. Appl. Genet., 105*, 921–936.

Ellis, R. J., (1990). Molecular chaperones. *Semin. Cell Biol.,* 1–72.

Evers, T., & Millar, S., (2002). Cereal grain structure and development: Some implications for quality. *J. Cereal Sci., 36*, 261–284.
Fan, C., Xing, Y., Maom, H., Lu, T., Han, B., Xu, C., et al., (2006). GS3, a major QTL for grain length and weight and minor QTL for grain width and thickness in rice, encodes a putative transmembrane protein. *Theor. Appl. Genet., 112*, 1164–1171.
Farooq, M., Hussain, M., & Siddique, K. H. M., (2014). Drought stress in wheat during flowering and grain-filling periods. *Crit. Rev. Plant Sci., 33*, 331–349.
Farquhar, G. D., & Richards, R. A., (1984). Isotopic composition of plant carbon correlates with water-use efficiency of wheat genotypes. *Austr. J. Plant Physiol., 11*, 539–552.
Feng, F., Han, Y., Wang, S., Yin, S., Peng, Z., Zhou, M., et al., (2018). The effect of grain position on genetic improvement of grain number and thousand grain weight in winter wheat in North China. *Frontiers in Plant Science, 9*, 129.
Fischer, R. A., & Maurer, R., (1978). Drought resistance in spring wheat cultivars. I. Grain yield responses. *Austr. Agric. J. Res., 29*, 897–912.
Fischer, R. A., (1985b). Physiological limitation to producing wheat in semitropical and tropical environments and possible selection criteria. In: *Proc. Int. Symp. Wheats for More Tropical Environments* (pp. 209–230). Mexico, F. D., CIMMYT.
Gebeyehou, G., Knott, R. D., & Baker, J. R., (1982). Rate and duration of grain filling in durum wheat cultivars 1. *Crop Science, 22*(2), 337–340.
Gegas, V. C., Nazari, A., Griffiths, S., Simmonds, J., Fish, L., Orford, S., et al., (2010). A genetic framework for grain size and shape in wheat. *Plant Cell, 22*, 1046–1056.
Gouache, D., Le, B. X., Bogard, M., Deudon, O., Page, C., & Gate, P., (2012). Evaluating agronomic adaptation options to increasing heat stress under climate change during wheat grain filling in france. *European Journal of Agronomy, 39*, 62–70.
Griffiths, S., Wingen, L., Pietragalla, J., et al., (2015). Genetic dissection of grain size and grain number trade-offs in CIMMYT wheat germ plasm. *PLoS One, 10*, e0118847.
Guan, Y. S., Serraj, R., Liu, S. H., Xu, J. L., Ali, J., Wang, W. S., Venus, E., Zhu, L. H., & Li, Z. K., (2010). Simultaneously improving yield under drought stress and non-stress conditions: A case study of rice (*Oryza sativa* L.). *Exp. J. Bot., 61*, 4145–4156.
Gupta, P. K., Rustgi, S., & Kumar, N., (2006). Genetic and molecular basis of grain size and grain number and its relevance to grain productivity in higher plants. *Genome, 49*, 565–571.
Hajheidari, M., Eivazi, A., Buchanan, B. B., Wong, H. J., Majidi, I., & Salekdeh, H. G., (2007). Proteomics uncovers a role for redox in drought tolerance in wheat. *J. Proteome Res., 6*, 1451–1460.
Hamblin, A. P., (1985). Root characteristics of some temperate legume species and varieties on deep, free-draining entisols. *Aust. Agric. J. Res., 36*, 63–72.
Hamblin, M. T., Buckler, E. S., & Jannink, J. L., (2011). Population genetics of genomics-based crop improvement methods. *Trends Genet.,27*, 98–106.
Hasan, K. A., Herrera, J., Lizana, C., & Calderini, F. D., (2011). Carpel weight, grain length, and stabilized grain water content are physiological drivers of grain weight determination of wheat. *Field Crops Research, 123*(3), 241–247.
Herzog, H., & Stamp, P., (1983). Dry matter and nitrogen accumulation in grains at different ear positions in 'gigas,' semi-dwarf and normal spring wheats. *Euphytica, 32*(2), 511–520.
Herzog, H., & Stamp, P., (1983). Dry matter and nitrogen accumulation in grains at different ear positions in 'gigas' semi dwarf and normal spring wheat's. *Euphytica, 32*, 511–520.
Huang, X. Q., Cloutier, S., Lycar, L., Radovanovic, N., Humphreys, G. D., Noll, S. J., et al., (2006). Molecular detection of QTLs for agronomic and quality traits in a doubled haploid

population derived from two Canadian wheat (*Triticum aestivum* L.). *Theor. Appl. Genet., 113*, 753–766.

Huang, X. Q., Cöster, H., Ganal, M. W., & Röder, M. S., (2003). Advanced back cross QTL analysis for the identification of quantitative trait loci alleles from wild relatives of wheat (*Triticum aestivum* L.). *Theor. Appl. Genet., 106*, 1379–1389.

Huang, X. Q., Kempf, H., Ganal, M. W., & Röder, M. S., (2004). Advanced back cross QTL analysis in progenies derived from a cross between a German elite winter wheat variety and asynthetic wheat (*Triticum aestivum* L.). *Theor. Appl. Genet., 109*, 933–943.

Huang, X., Zhao, Y., Wie, X., Li, C., Wang, A., Zhao, Q., et al., (2012). Genome-wide association study of flowering time and grain yield traits in a worldwide collection of rice germplasm. *Nat. Genet., 44*, 32–39.

Hurkman, J. W., McCue, F. K., Altenbach, B. S., Korn, A., Tanaka, K. C., Kothari, M. K., et al., (2003). Effect of temperature on expression of genes encoding enzymes for starch biosynthesis in developing wheat endosperm. *Plant Sci., 164*, 873–881.

Idso, S. B., & Brazel, J. A., (1984). Rising atmospheric carbon dioxide concentrations may increase stream flow. *Nature, 312*, 51–52.

Jenner, F. C., (1991). Effects of exposure of wheat ears to high temperature on dry matter accumulation and carbohydrate metabolism in the grain of two cultivars. *Funct. Plant Biol., 18*, 165–177. doi: 10.1071/PP9910165.

Jing, R., et al., (2014). Natural variation of TaGASR7-A1 affects grain length in common wheat under multiple cultivation conditions. *Mol. Breed., 34*, 937–947.

Jordan, R. D., Hunt, H. C., Cruickshank, W. A., Borrell, K. A., & Henzell, G. R., (2012). The relationship between the stay green trait and grain yield in the elite *Sorghum* hybrids grown in a range of environments. *Crop Sci., 52*, 1153–1161.

Kalakanavar, M. R., Shashidhara, S. D., & Kulkarni, G. N., (1989). Effect of grading on quality of wheat seeds. *Seed Res., 17*(2), 182–185.

Kirby, E., (1974). Ear development in spring wheat. *J. Agric. Sci., 82*, 437–447.

Kirby, J. E. M., & Appleyard, M., (1987). *Cereal Development Guide* (2nd edn.), Arable Unit, National Agricultural Centre, Coventry, UK.

Kobata, T., Palta, J. A., & Turner, N. C., (1992). Rate of development of post anthesis water deficits and grain filling of spring wheat. *Crop Sci., 32*, 1238–1242.

Krenzer, G. E., Nipp, T. L., & McNew, R. W., (1991). Winter wheat main stem leaf appearance and tiller formation vs. moisture treatment. *Agron. J., 83*, 663–667.

Li, Q., Yang, X., Bai, G., Warburton, M. L., Mahuku, G., Gore, M., et al., (2010a). Cloning and characterization of a putative GS3 ortholog involved in maize kernel development. *Theor. Appl. Genet., 120*, 753–763.

Ling, H. Q., Zhao, S., Liu, D., Wang, J., Sun, H., Zhang, C., et al., (2013). Draft genome of the wheat A-genome progenitor *Triticum urartu*. *Nature, 496*, 87–90.

Liu, H. Z., Wang, Y. H., Wang, E. X., Zhang, P. G., Chen, D. P., & Liu, J. D., (2006). Genotypic and spike positional difference in grain phytase activity, phytate, inorganic phosphorus, iron, and zinc contents in wheat (*Triticum aestivum* L.). *J. Cereal Sci., 44*, 212–219.

Lizana, C. X., Riegel, R., Gomez, D. L., Herrera, J., Isla, A., Mcqueen-Mason, J. S., & Calderini, F. D., (2010). Expansin expression is associated with grain size dynamics in wheat (*Triticum aestivum* L.). *J. Exp. Bot., 61*, 1147–1157.

Long, P. S., Humphries, S., & Falkowski, P. G., (1994). Photo inhibition of photosynthesis in nature. *Ann. Rev. Plant Physiol. Plant Mol. Biol., 45*, 633–664.

Loomis, R. S., & Amthor, J. S., (1996). Limits of yield revisited. In: Reynolds, P. M., Rajaram, S., & McNab, A., (eds.), *Increasing Yield Potential in Wheat: Breaking the Barriers* (pp. 76–89). Mexico, F. D., CIMMYT.

Ma, J., Li, R., Wang, H., Li, D., Wang, X., Zhang, Y., et al., (2017). Transcriptomics analyses reveal wheat responses to drought stress during reproductive stages under field conditions. *Front. Plant Sci., 8*, 592.

Manschadi, A. M., Christopher, J., Devoil, P., & Hammer, G. L., (2006). The role of root architectural traits in adaptation of wheat to water-limited environments. *Funct. Plant Biol., 33*, 823–837.

Mao, H., Sun, S., Yao, J., Wang, C., Yu, S., Xu, C., et al., (2010). Linking differential domain functions of the GS3 protein to natural variation in grain size in rice. *Proc. Natl. Acad. Sci. U.S.A., 107*, 19579–19584.

Matiu, M., Ankerst, D. P., & Menzel, A., (2017). Interactions between temperature and drought in global and regional crop yield variability during 1961–2014. *PLoS One, 12*, e0178339.

Mian, M. A. R., & Nafziger, E. D., (1994). Seed size and water potential effects on germination and seedling growth of winter wheat. *Crop Sci., 34*, 169–171.

Midmore, J. D., Cartwright, P. M., & Fischer, R. A., (1984). Wheat in tropical environments. II. Crop growth and grain yield. *Field Crops Res., 8*, 207–227.

Miralles, J. D., & Slafer, A. G., (1995). Individual grain weight responses to genetic reduction in culm length in wheat as affected by source-sink manipulations. *Field Crops Research, 43*(2), 55–66.

Morell, K. M., Rahman, S., Regina, A., Appels, R., & Li, Z., (2001). Wheat starch biosynthesis. *Euphytica, 119*, 55–58.

Morgan, J., & Condon, A. G., (1986). Water use, grain yield and osmoregulation in wheat. *Austr. J. Plant Physiol., 13*, 523–532.

Moustafa, A. M., Boersma, L., & Kronstad, W. E., (1996). Response of four spring wheat cultivars to drought stress. *Crop Sci., 36*, 982–986.

Nakhforoosh, A., (2013). Recent approaches in screening methodology for drought resistance. *Agric. Vet. Sci. Nutr. Nat. Resour., 8*, 1–14.

Narasimhamoorthy, B., Gill, B. S., Fritz, A. K., Nelson, J. C., & andBrown-Guedira, L. G., (2006). Advanced backcross QTL analysis of a hard winter wheat × synthetic wheat population. *Theor. Appl. Genet., 112*, 787–796.

Neumann, K., Kobiljski, B., Dencic, S., Varshney, R. K., & Börner, A., (2011). Genome-wide association mapping, a case study in bread wheat (*Triticum aestivum* L.). *Mol. Breed., 27*, 37–58.

Nicolas, E. M., Gleadow, M. R., & Dalling, J. M., (1985). Effect of postanthesis drought on cell division and starch accumulation in developing wheat grains. *Ann. Bot., 55*, 433–444.

O'Toole, J. C., & Stockle, C. D., (1991). The role of conceptual and simulation modeling in plant breeding. In: Acevedo, E., Fereres, E., Gimenez, C., & Srivastava, P. J., (eds.), *Improvement and Management of Winter Cereals under Temperature, Drought and Salinity Stresses, Proc. ICARDA-INIA Symp.* (pp. 205–225). Cordoba, Spain.

Oosterhuis, D. M., & Cartwright, P. M., (1983). Spike differentiation and floret survival in semidwarf spring wheat as affected by water stress and photo-period. *Crop Sci., 23*, 711–716.

Ortiz-Monasterio, I. J., & Graham, D. R., (2000). Breeding for trace minerals in wheat. *UNU Food Nutr. Bull., 21*(4), 392–396.

Ortiz-Monasterio, I. J., Sayre, D. K., Rajaram, S., & McMahon, M., (1997). Genetic progress in wheat yield and nitrogen use efficiency under four nitrogen rates. *Crop Sci., 37*, 898–904.

Palta, A. J.,Kobata, T., Turner, N. C., & Fillery, I. R., (1994). Remobilization of carbon and nitrogen in wheat as influenced by post-anthesis water deficits. *Crop Sci., 34*, 118–124.

Panozzo, F. J., & Eagles, A. H., (2000). Cultivar and environmental effects on quality characters in wheat II. *Protein Aust. Agric. J. Res., 51*, 629–636.

Peterson, M. C., Klepper, B., Pumphrey, F. B., & Rickman, R. W., (1984). Restricted rooting decreases tillering and growth of winter wheat. *Agron. J., 76*, 861–863.

Pinheiro, B. D. S., Austin, R. B., Carmo, M. P. D., & Hall, M. A., (2000). Carbon isotope discrimination and yield of upland rice as affected by drought at flowering. *Pesqui. Agropecu. Brasil., 35*, 1939–1947.

Prasad, V. P., V., & Djanaguiraman, M., (2014). Response of floret fertility and individual grain weight of wheat to high temperature stress: Sensitive stages and thresholds for temperature and duration. *Funct. Plant Biol., 41*, 1261–1269.

Prasad, V. P., V., Pisipati, R. S., Ristic, Z., Bukovnik, U., & Fritz, K. A., (2008b). Impact of nighttime temperature on physiology and growth of spring wheat. *Crop Sci., 48*, 2372–2380.

Qin, L., Hao, C., Hou, J., Wang, Y., Li, T., Wang, L., et al., (2014). Homologous haplotypes, expression, genetic effects and geographic distribution of the yield gene TaGW2. *BMC Plant Biol., 14*, 107.

Ramsay, L., Comadran, J., Druka, A., Marshall, D. F., Thomas, W. T. B., Macauly, M., et al., (2011). Intermedium-C, a modifier of lateral spikelet fertility in barley,is an ortholog of the maize domestication gene teosinte branched 1. *Nat. Genet., 43*, 169–173.

Rasheed, A., Xia, X., Ogbonnaya, F., Mahmood, T., Zhang, Z., Mujeeb-Kazi, A., et al., (2014). Genome-wide association for grain morphology in synthetic hexaploid wheat's using digital imaging analysis. *BMC Plant Biol., 14*, 128.

Rawson, M. H., & Evans, L. T., (1970). The pattern of grain growth within the ear of wheat. *Australian Journal of Biological Sciences, 23*(4), 753.

Reddy, R. A., Chaitanya, V. K., & Vivekanandan, M., (2004). Drought-induced responses of photosynthesis and antioxidant metabolism in higher plants. *J. Plant Physiol., 161*, 1189–1202.

Rees, D., Sayre, K., Acevedo, E., Nava, E., Lu, Z., Zeiger, E., & Limon, A., (1993). *Canopy Temperatures of Wheat: Relationship with Yield and Potential as a Technique for Early Generation Selection.* Wheat Special Report No. 10. Mexico DF, CIMMYT.

Ries, S. K., & Everson, E. H., (1973). Protein content and seed size relationships with seedlings vigor of wheat cultivars. *Agron. J., 65*, 884–886.

Röder, M. S., Huang, X. Q., & Börner, A., (2008). Fine mapping of the region on wheat chromosome 7D controlling grain weight. *Funct. Integr. Gen., 8*, 79–86.

Saeedipour, S., (2011). Activities of sucrose-metabolizing enzymes in grains of two wheat (*Triticum aestivum* L.) cultivars subjected to water stress during grain filling. *J. Plant Breed. Crop Sci., 3*, 106–113.

Seaton, G. R., & Walker, D. A., (1990). Chlorophyll fluorescence as a measure of photosynthetic carbon assimilation. *Proc. Royal Soc., Lond. B, 242*, 29–35.

Seeman, R. J., Berry, J. A., & Downton, W. J. S., (1984). Photosynthetic response and adaptation to high temperature in desert plants: A comparison of gas exchange and fluorescence methods for studies of thermal tolerance. *Plant Physiol., 75*, 364–368.

Sehgal, A., Sita, K., Kumar, J., Kumar, S., Singh, S., Siddique, H. K., M., et al., (2017). Effects of drought, heat and their interaction on the growth, yield, and photosynthetic function of lentil (*Lens culinaris* Medikus) genotypes varying in heat and drought sensitivity. *Front. Plant Sci., 8*, 1776.

Semenov, A. M., & Shewry, R. P., (2011). Modeling predicts that heat stress, not drought, will increase vulnerability of wheat in Europe. *Scientific Reports, 1*, 66.

Shewry, R. P., Mitchell, A. R. C., Tosi, P., Wan, Y., Underwood, C., Lovegrove, A., et al., (2012). An integrated study of grain development of wheat (Cv. Hereward). *Journal of Cereal Science, Cereal Grain Development: Molecular Mechanisms and Impacts on Grain Composition and Functionality, 56*(1), 21–30.

Shpiler, L., & Blum, A., (1986). Differential reaction of wheat cultivars to hot environments. *Euphytica, 35*, 483–492.

Simane, B., Peacock, J. M., & Struik, P. C., (1993). Differences in development and growth rate among drought-resistant and susceptible cultivars of durum wheat (*Triticum turgidum* L. var. durum). *Plant Soil, 157*, 155–166.

Simmons, R., & Moss, N., (1978). Nitrogen and dry matter accumulation by kernels formed at specific florets in spikelets of spring wheat. *Crop Sci., 18*, 139–143.

Singh, S., Singh, G., Singh, P., & Singh, N., (2008). Effect of water stress at different stages of grain development on the characteristics of starch and protein of different wheat varieties. *Food Chem., 108*, 130–139.

Slafer, A. G., Satorre, H. E., & Andrade, H. F., (1994). Increases in grain yield in bread wheat from breeding and associated physiological changes. In: Slafer, G. A., (ed.), *Genetic Improvement of Field Crops* (pp. 1–68). Marcel Dekker, New York.

Sofield, I., Wardlaw, F. I., Evans, L. T., & Lee, S. Y., (1977). Nitrogen, phosphorus, and water contents during grain development and maturation in wheat. *Austr. J. Plant Physiol., 4*, 799–810.

Stone, J. P., Gras, W. P., & Nicolas, E. M., (1997). The influence of recovery temperature on the effects of brief heat shock on wheat. III. Grain protein and dough properties. *J. Cereal Sci., 25*, 129–141.

Su, Z., Hao, C., Wang, L., Dong, Y., & Zhang, X., (2011). Identification and development of a functional marker of TaGW2 associated with grain weight in bread wheat (*Triticum aestivum* L.). *Theor. Appl. Genet., 122*, 211–223.

Tadesse, W., Ogbonnaya, C. F., Jighly, A., Sanchez-Garcia, M., Sohail, Q., Rajaram, S., & Baum, M., (2015). Genome-wide association mapping of yield and grain quality traits in elite winter wheat genotypes. *PLoS One,10*, e0141339.

Thomas, M. J., G., Boote, J. K., Allen, H. L., Gallo-Meagher, M., & Davis, M. J., (2003). Elevated temperature and carbon dioxide effects on soybean seed germination and transcript abundance. *Crop Sci., 43*, 1548–1557.

Triboï, E., Martre, P., & Triboï-Blondel, M. A., (2003). Environmentally-induced changes of protein composition for developing grains of wheat are related to changes in total protein content. *Exp. J. Bot., 54*, 1731–1742.

Upadhayaya, M. K., & Furnes, N. H., (1994). Influence of light intensity and water stress on leaf surface characteristics of *Cynoglossum* officinale, *Centaurea* spp. and *Tragopogon* spp. *Can. Bot. J., 72*, 1379–1386.

Van, O. E. J., & Acevedo, E., (1992). Adaptation of barley (*Hodeum vulgare*L.) to harsh Mediterranean environments. I. Morphological traits. *Euphytica, 62*, 1–14.

Venora, G., & Calcagno, F., (1991). Study of stomatal parameters for selection of drought resistance varieties in *Triticum durum* desf. *Euphytica, 57*, 275–283.

Wang, E., Wang, J., Zhu, X., Hao, W., Wang, L., Li, Q., et al., (2008). Control of rice grain-filling and yield by a gene with a potential signature of domestication. *Nat. Genet., 40*, 1370–1374.

Wang, L., Ge, H., Hao, C., Dong, Y., & Zhang, X., (2012c). Identifying loci influencing 1,000-kernel weight in wheat by microsatellite screening for evidence of selection during breeding. *PLoS One, 2*, e29432.

Wang, X., Dinler, S. B., Vignjevic, M., Jacobsen, S., & Wollenweber, B., (2015). Physiological and proteome studies of responses to heat stress during grain filling in contrasting wheat cultivars. *Plant Sci., 230*, 33–50.

Warrington, J. I., Dunstone, R. L., & Green, L. M., (1977). Temperature effects at three developmental stages on the yield of the wheat ear. *Austr. Agric. J. Res., 28*, 11–27.

Wiegand, C. L., & Cuellar, J. A., (1981). Duration of grain filling and kernel weight of wheat as affected by temperature. *Crop Sci., 21*, 95–101.

Williams, K., & Sorrells, M. E., (2014). Three-dimensional seed size and shape QTL in hexaploid wheat (*Triticum aestivum* L.) populations. *Crop. Sci., 54*, 98–110.

Xie, Q., Mayes, S., & Sparkes, L. D., (2015). Carpel size, grain filling, and morphology determine individual grain weight in wheat. *J. Exp. Bot., 66*, 6715–6730, erv378.

Xu, L. X., Zhang, H. Y., & Wang, M. Z., (2004). Effect of heat stress during grain filling on phosphoenolpyruvate carboxylase and ribulose-1,5-bisphosphate carboxylase/oxygenase activities of various green organs in winter wheat. *Photosynthetica, 42*, 317–320.

Xu, Q., Chen, S., Yunjuan, R., Chen, S., & Liesche, J., (2018). Regulation of sucrose transporters and phloem loading in response to environmental cues. *Plant Physiol., 176*, 930–945.

Xue, J. L., Frost, J. C., Tsai, J. C., & Harding, A. S., (2016). Drought response transcriptomes are altered in poplar with reduced tonoplast sucrose transporter expression. *Sci. Rep., 6*, 33655.

Yan, W., Zhong, Y., & Shangguan, Z., (2016). A meta-analysis of leaf gas exchange and water status responses to drought. *Sci. Rep., 6*, 20917.

Yang, C. J., Zhang, H. J., Wang, Q. Z., Zhu, S. Q., & Liu, J. L., (2003). Involvement of abscisic acid and cytokinins in the senescence and remobilization of carbon reserves in wheat subjected to water stress during grain filling. *Plant Cell Environ., 26*, 1621–1631.

Yang, J., & Zhang, J., (2006). Grain filling of cereal under soil drying. *New Phytol., 169*, 223–236.

Yu, M. S., Lo, F. S., & Ho, H. T. D., (2015). Source-sink communication, regulated by hormone, nutrient, and stress cross-signaling. *Trends Plant Sci., 20*, 844–857.

Zanke, C., Ling, J., Plieske, J., Kollers, S., Ebmeyer, E., Korzun, V., Argillier, O., et al., (2015). Analysis of main effect QTL for thousand grain weight in European winter wheat (*Triticum aestivum* L.). by genome-wide association mapping. *Frontiers in Plant Science, 6*.

Zhang, F. Y., Huang, W. X., Wang, L. L., Wei, L., Wu, H. Z., You, S. M., et al., (2014). Proteomic analysis of wheat seed in response to drought stress. *Integr. J. Agric., 13*, 919–925.

Zhang, J., Zhang, S., Cheng, M., Jiang, H., Zhang, X., Peng, C., Lu, X., Zhang, M., & Jin, J., (2018). Effect of drought on agronomic traits of rice and wheat, a meta-analysis. *Int. Environ. J. Res. Public Health, 15*, 839.

Zhang, L., Zhao, L. Y., Gao, F. L., Zhao, Y. G., Zhou, H. R., Zhang, S. B., & Jia, Z. J., (2012). TaCKX6-D1, the ortholog of rice OsCKX2, is associated with grain weight in hexaploid wheat. *New Phytol., 195*, 574–584.

Zhang, Y., Liu, J., Xia, X., & He, Z., (2014). TaGS-D1, an ortholog of rice OsGS3, is associated with grain weight and grain length in common wheat. *Mol. Breed., 34*, 1097–1107. doi: 10.1007/s11032-014-0102-7.

CHAPTER 13

Insect Pest Management of Agricultural Crops

RITIKA, NIKITA BHATI, SHREYA, and ARUN KUMAR SHARMA

Department of Bioscience and Biotechnology, Banasthali Vidyapith, Rajasthan, India, E-mail: arun.k.sharma84@gmail.com (A. K. Sharma)

ABSTRACT

Among the major global concern, the protection of crops from insect pests is an important issue. Continuous use of chemical pesticides results in toxicity which causes harmful effects on the animals and food consumers. The increasing social concern on the usage of chemical pesticide led their replacement with other alternatives which are less harmful to humans and crops leading to the development of biological agents that acts as pesticides. New researches in pesticide chemistry, genetics, population dynamics, and public opinion resulted in the involvement of beneficial insects into Integrated Pest Management programs. IPM is an eco-friendly method to reduce the pest population causing minimal disturbance to the crops.

13.1 INTRODUCTION

When an insect causes harm to crops, animals, humans, and properties it is defined as pest. Insects can be classified as pests when they cause direct damage to harvestable products, e.g., codling moth larvae causing damage to apple, e.g., by being a vector for several plants and livestock diseases. The knowledge of pest control adds an important dimension in studying pest management because it provides insights into the driving forces (economical, technical, and social) of current pest management practices which in turn will provide some idea of the forces likely to be acting in the future (Norton and Mumford, 1993). While pests have been a major problem in agriculture since the beginning, many of today's serious pest problems are the direct consequence of actions taken to improve crop production (Waage, 1993). The growing demand for agriculture has created greater pest problems in a number of ways:

1. When the concentration of a single plant species and monocultures increases, it leads to rise in the number of pest species which colonize it (Strong, Lawton, and Southwood, 1984).
2. High yielding crop cultivars can provide improved conditions for pest colonization, spread, and rapid growth.
3. Reduced number of indigenous natural enemies because of unavailability of food and shelter during harsh conditions render them to move out of their habitat resulting in pest outbreak.
4. The trading of plant species around the world also carries the pest associated with it. Plant breeders, commercial importers, distributors of food aid and general commerce inadvertently introduce pest species.
5. Increased dependency of farmers on chemical pesticides leads to an increasing pest problem.

The changes in agricultural practices as the introduction of new crop species or enlargement and aggregation of fields, use of monocrops, and plant density have been held responsible for causing many pest problems (Table 13.1).

13.2 INTEGRATED PEST MANAGEMENT

According to Alastair (2003), IPM is a strategy of justifying pesticide use to reduce or prevent the re-emergence of pest populations, which have become resistant to pesticides, and to save useful insects. Recently, concerns about pesticide residues penetrating in the food chain and in the environment have led to elimination of traditional use of pesticides.

However, there are wide agreements on the basic idea of IPM which include:

- An integrated scheme which involves methods of control that are regarded as component technologies rather than alternatives.

TABLE 13.1 Pests of Major Economically Important Crops in India

Crop	Common Name	Scientific Name	Economic Threshold Level	Existing Control Methods
Legumes				
Pigeonpea	Pod fly	*Melanagromyza obtuse* (malloach)	In all endemic locations	Chemical
	Pod borer	*Helicoverpa armigera* (hubner)	5 eggs or 3 small larvae per plant	IPM
Chickpea	Cutworm	*Agrotis ipsilon* (hufnagel)	5% plant mortality	Chemical
Soybean	Stem fly	*Ophiomyia phasioli* (tryon)	5% plant infestation	Chemical
Cereals				
Wheat	Aphid	*Schizaphis graminum* (rondani)	5–10% of plants with infestation	Host Plant Resistance (HPR)
Rice	Stem borer	*Scirpopha gaincertulus* Walker	5% white ears/ One egg mass sqm^{-1}	IPM
Maize	Earworm	*Helicoverpa armigera* hubner	25–30% damage to cobs	Chemical
Fruits				
Apple	Codling moth	*Cydia pomonella* (L.)	1–2% incidence	IPM
Mango	Hopper	*Amritodes atkinsoni* leth.	2–5 hoppers per inflorescence	Chemical
Grapes	Thrips	*Retithrips syriacus* (mayet)	5 thrips/young leaf	Chemical
Oranges	Defoliators	*Papilio demoleus* L.	20–30% foliar damage	Chemical
Vegetables				
Tomato	Fruit worm	*Helicoverpa armigera* hubner	1–5% fruit damage	IPM
Brinjal	Fruit and stem borer	*Leucinodes orbanalis*	1–5% shoot/fruit infestation	IPM
Cabbage and cauliflower	Tobacco caterpillar	*Spodoptera litura* (fab)	1–5% incidence	IPM

TABLE 13.1 *(Continued)*

Crop	Common Name	Scientific Name	Economic Threshold Level	Existing Control Methods
Oil Seeds				
Sunflower	Gram pod borer	*Helicoverpa armigera* hubner	One larva per head	Chemical
Groundnut	Leaf miner	*Aproaerema modicella* deventer	5 mines per plant at 30 days of crop age	IPM
Sesame	Leaf webber	*Antigastra catalaunalis* dub	2–5 webbs per plant	Chemical
Cash Crops				
Cotton	American bollworm	*Helicoverpa armigera* hub.	5–10% boll infestation	IPM
Tobacco	Tobacco caterpillar	*Spodoptera litura* fab.	5–10% leaves with damage	IPM
Sugarcane	Scale insect	*Melanapsis glomerata* (green)	20–30% canes with scale incidence	IPM

Source: Ranga Rao et al. (2009).

- It focuses on pest management within ecologically balanced system whereas control strategy suggests direct involvement with little concern for sustainability (Alastair, 2003).

IPM is believed to be a flexible and comprehensive system. It provides an insight of agro-ecosystem as an interconnected whole, which utilizes a variety of biological, genetic, cultural physical, and chemical techniques that maintain pests below economically harmful levels with a little disturbance to the crops and the nature (Malena, 1994).

13.2.1 STRATEGIES FOR IPM IMPLEMENTATION

IPM strategy aids in reducing the usage of chemical sprays. It also results in upsurge of natural enemies by three-fold, reducing the usage of insecticide and environmental pollution (Dhaliwal and Ramesh, 1996).

The integrated strategy for regulation of major pests and diseases is possible due to:

1. New varieties are crossed with in-built resistance;
2. Development of effective methods of pest control through monitoring and pest surveys; and
3. Using conservation and augmentation as a biological tool to control pests and protect natural enemies like predators, parasites, and insect pathogens.

IPM strategies that are economically viable have been developed for the control of major pests in cotton, sugarcane, rice, pulses, etc. Control of mealy bug of coffee, pyrilla, and top borer of sugarcane, lepidopterous pests affecting tobacco, sugarcane, cotton, coconut, etc., are few examples where success has been attained through the release of biocontrol agents.

13.2.2 MAJOR OBSTACLES

- Limited implementation at the farmers level;
- Complex nature of IPM vs. simplicity of chemical pesticides;
- For a large number of crops there is a lack of location-specific IPM modules.

13.3 PEST MANAGEMENT STRATEGY

13.3.1 CULTURAL PRACTICES

Here, manipulation of the environment is done to make it unfavorable for the growth of the pests. It involves many methods that hinder with the pests ability to colonize a crop, promoting dispersal, reducing reproduction and its existence. It is attained through techniques such as crop rotation, intercropping, hermetic storage systems, and manipulation of planting periods (Coaker, 1987).

13.3.2 CROP ROTATION

Crop rotation serves as a way to maintain soil fertility which results in better yields than cultivating the same crop every year. In the rotation, method the type of crop used is very important to maximize the yield. There is a benefit in using crops having diverse rooting habits and thus have varying demand between different soil layers, leguminous plants can also be used to fix soil nitrogen. Crop type in a rotation is selected on the basis of combination that reduces pest damage efficiently. It reduces the accumulation of insect pests along with weeds and pathogens. This method is most useful against pest species having a narrow host range and a limited range of dispersal. The next generation of an insect pest which has inhabited the host crop will face a non-host crop in the following season. It will lead to dispersal of the insects, and the low dispersal power has a low probability of finding appropriate host. The result will be reduced colonization on host crop. Here, the choice of crop should be appropriate and it should not lead to introduction of new pests into the system. Lately, this technique has become less popular as more intensive farming methods are used but they do not provide an efficient means of pest suppression (Coaker, 1987).

13.3.3 PEST RESISTANT VARIETIES

Breeding is done to develop resistance in the plants. But, meanwhile the pests, especially plant pathogens, co-evolve with their hosts. The gene transfer technology is used to develop cultivars resistant to plant pathogens, insects, and herbicides, e.g., incorporation of genetic material from *Bacillus thuringiensis* (Bt), a bacterium present in cotton, potato, and corn, making plant tissues toxic to insects. This technique contains huge potential in managing pests, but it also contains probability of increased selection pressure for development of resistance against it and its impact on other crops in environment. Thus, due to ethical and social concern, this potential technology is surrounded by many disputes (Khader, 1996).

13.3.4 PLANT SANITATION

In IPM, preventive practices play an important role. It consists of planting quality seeds, regular cleaning and maintenance of field equipment (e.g., haying equipment, tillage equipment, etc.), and quarantine of infected crops or fields. These methods are helpful in preventing introduction of pests into the farmland (Robert and William, 1994).

13.3.5 MIXED CROPPING

The most important step in crop plantation is selection of the type of crop to be grown. In one study, dryland rice intercropped with pigeon pea or cotton resulted in low numbers of green leafhopper and planthopper. On mixing dryland rice cultivars of various maturities and harvesting individual panicles at maturity reduces storage time and as a result of storage pest problem (Raymundo, 1986).

13.3.6 PLANTING METHOD

The influence of planting method on pest numbers are the result of variation in planting time, plant density, or water level which occur due to altering planting methods (Israel, 1969).

13.3.7 TILLAGE

In this method turning over the soil crushes the insects, buries them, destroys their nests, and exposes them to desiccation and predators. Tillage can be frequently performed in the dryland areas during land preparation or inter-row weed cultivation of rice or crops preceding rice. Insects in the soil prefer moderately moist soil. More insects will be in the plow zone if the soil is wet. It allows insect pests to carryover between crops. During land preparation, tillage can control dormant stem borer larvae before they emerge as adults, but the timing of these operations on a community-wide basis is highly critical. If any delay in tillage occurs, pest problems can worsen (Khan, 1967).

13.4 PHYSICAL CONTROL

The growth of the insect pest populations depends on its physical environment. To regulate the level of pest in the field; the physical surrounding of the pest should be manipulated. The physical environment is defined as any abiotic variable that affects the pest in terms of temperature (Yokoyama and Miller, 1996), light intensity, humidity (Santoso et al., 1996), soil composition and structure (Blackshaw and Thompson, 1993; Leather, 1993) and atmospheric conditions (Hodges and Surendro, 1996).

Irrigation is used to physically change crop environment according to human needs and it also influences pest population development (Tabashnik and Mau, 1986; Wheatley et al., 1989). Another method used is the modification of mosquito breeding grounds by drainage to control malaria.

13.4.1 STORED GRAINS

13.4.1.1 TEMPERATURE

In storage systems insect development occurs within the range of 10–40°C of temperature; its optimal range is 25–30°C (Evans, 1987). Temperature and humidity are modified above or below the tolerance limits of the insect or to the levels that kill the pest. To control the pest population temperature can be reduced to 15–17°C. Cowpea beetle *Callosobruchus maculates* and other beetles such as *Tribolium castaneum* require temperatures above 20°C for normal growth (Benz, 1987), hence any fall in temperature will impact pest population. The rise in temperature (35–40°C) is also utilized to control pests of stored product.

13.4.1.2 CONTROLLED ATMOSPHERE

In hermetic storage method changes in the interstitial atmosphere of storage, pits are done to regulate the pest population. Due to modified storage conditions respiration of the storage product results in an increase in CO_2 level and a decrease in the level of O_2 to the extent that kills any insect present in the store. The modified form of the hermetic storage system is the use of gas-tight storage facilities which provide oxygen-deficient, nitrogen, and carbon dioxide enriched atmosphere to regulate pest population (Soderstrom and Brandl, 1982).

13.4.1.3 AERATION

Control of storage pests can also be attained by maintaining suitable timing of aeration (Hagstrum and Flinn, 1990; Flinn, Hagstrum, and Muir, 1997), by compression of air or by a mixture of fumigant and aeration (Yokoyama and Miller, 1996). The controlled environment of storage systems are economically competitive with chemical treatments such as phosphine. On a long term basis, this method requires a protective tight sealing of the system and there is a restriction to access for inspection and handling. Still, a controlled environment remains less harmful than chemical fumigants and there is the least probability of leaving any residue in the stored products (Soderstrom and Brandl, 1982).

13.4.1.4 IRRADIATION

Gamma radiation and micro radiation (use of microwaves) are effectively used against stored grain pests. Infrared radiation can be treated dually to the insects or to the stored grain infested with insects. At lower doses, ionizing radiation (x-rays, gamma rays) have sterilizing effect but at higher doses have lethal effects. The major limitation of this method is the acceptability of the irradiated food by consumers.

13.5 MECHANICAL BARRIERS

In this method, the repression or control of insect population is done by using manual devices. Generally, manual methods of pest control are beneficial for high-value crops where cheap labor is available, pests are collected and can be used as food or have an economic importance (Isley, 1951).

13.5.1 HAND PICKING

The most commonly used manual method is elimination of insects by hand. Camouflaged or fast moving insects are hard to capture. Sluggish and aggregated large worms and larvae can be easily picked by hands from foliage. Adult stage of grasshopper is sluggish at night and can be directly picked from foliage and is either used as food or is destroyed by crushing, burning, burial, or placing in oil or kerosene (Jones and Mackie, 1913).

Eggs are stationary and thus are attractive stages for handpicking. Species laying eggs in masses are selected. In India, China, and Japan, armyworm, and stem borer egg masses plus moths were traditionally removed by hand from seedbeds (Ramakrishna, 1933). Hand picking of moths is difficult so they are captured by net.

13.5.2 ROGUING

Removal of pest infected plants or parts of plant using hand or hand tools. This method is used to prevent the damage from growing. Roguing is helpful in controlling root aphids, stem borers, and leaf folders (Maxwell-Lefroy, 1907).

13.5.3 BARRIERS

Nursery boxes are used in temperate areas, to incubate plantlets for machine transplantation in the early spring as this method prevents early invasion of insect vectors. Seedbeds are covered with mesh cloth in warmer climates to prevent virus transmission by leafhoppers. But then seedbeds developed high stem maggot invasion (Kuwayama, 1963).

13.5.4 LIGHT TRAPS

Long ago farmers used fire and established light traps as a control method because many insects were attracted to light. Light traps controls army-worms, leafhoppers, stem borers, seed bugs, caterpillars, and root weevils. Shortcomings of light trap are their high cost, killing of useful insects, theft, erratic attraction, all the species are not attracted and trapped, and among the collected insects the sex ratio of males is high (Stewart, 1934).

13.5.5 ELECTROCUTING TRAP

In this method, insects are electrocuted on live metal screens. This type of traps is used to reduce housefly populations.

13.5.6 COLORED TRAPS

Insects show diverse responses to different colors. Colors are selected on the basis of physiological stage of insect, position of traps, and quality of the incident wavelengths striking the traps.

13.6 BIOLOGICAL CONTROL

Biological control is the advantageous action of predators, parasites, pathogens, and competitors in controlling pests and their damages. These active organisms provide biocontrol action and are particularly important for decreasing the number of mites, pests, and insects. The major means of decreasing pest population and pest effects by using natural enemies and this act become an independent goal in the 1990s. The biological control strategies are Classical, Augmentation, and Conservation. In the last 3–4 decades, the use of biological control agents have been increased remarkably (Van Driesche et al., 2008).

13.6.1 IMPORTATION (CLASSICAL BIOLOGICAL CONTROL)

Classical biological control involves the import and release of exotic/alien natural enemies to a novel environment for their permanent establishment in order to target the pest in that area. The major goal of importation includes selecting a significant natural enemy that can establish a stable interaction with pest, that can co-exist and grow on pest (as an alternative food source) to keep the pest population below the economic threshold level without any extra human involvement. The importation process engages in determining the origin of the intruder crop pest then collecting the suitable natural enemy related to that introduced pest followed by selection and testing of natural enemies to ensure that no contagious organisms are selected or introduced. The selected natural enemies are mass-reared and released into the fields to study long term effects of natural enemies in pest control. The classical biological control method is critically examined because of its crucial roles in ecological restoration and suppressing the invasion of alien insects and plant species into natural ecosystems.

13.6.2 AUGMENTATION AND INOCULATION WITH NATURAL ENEMIES

In a normal environment where natural enemies are either absent or their number is too small to target the pest and keep their number below the economic injury level, consequently, their number is augmented by releasing the mass-reared natural enemies from companies or laboratories. Augmentation is a temporary biological control and multiple episodic releases of natural enemies might be required. The major constrain in the augmentation approach is the mass rearing of natural enemies, appropriate packaging, and storage methods for mass-reared organisms.

The method of mass rearing of natural enemies on their natural hosts involves a triple-phase development program in which the host plant is propagated for the insect pest then natural enemies are developed on to the insect pest. But the contamination of the natural host with pest or other diseases makes mass rearing a very expensive and unfeasible method. Under these circumstances, using synthetic diets may cut down the production cost but the unnatural host used in insectary can be disadvantageous because it can result into altered host preference by natural enemies making augmentation process futile. The cost of mass rearing of natural enemies is evaluated by comparing its value with the cost of other pest-controlling chemical agents. Augmentation of natural enemies appears to be a more viable and economic method for the chemical-free production of agricultural or horticultural crops. Augmentation involves two major approaches as inoculative and inundative releases (Nafiu, Dong, and Cong, 2014).

13.6.2.1 INUNDATIVE RELEASES

It is a corrective and immediate pest control measure which involves multiple release of huge number of insectary-reared natural enemies for instantly reducing the damaging pest population. The release of these insects is repeated because the persistence of later generations in ample amount to control the pest is not evident. When the crop is long lived, a new pest insect appears into the crop or the complete population of pest was not there during initial susceptible stages results in repeated release of natural enemies (Nafiu, Dong, and Cong, 2014).

13.6.2.2 INOCULATIVE RELEASES

It is a chief preventive and long term measure, applied when the population of pest is very small hence require natural enemies in small quantity at the prescribed time. The major aim is to keep the number of pests below an economic injury level and targeting the pest after the multiplication of natural enemies. The natural enemies(predators and parasitoids) are generally released during the early season of crops in the glasshouse with an alternative food source and the natural enemies are released again in the next generation. This method of controlling insect pests is strongly density-dependent.

13.6.3 CONSERVATION BIOLOGICAL CONTROL

The use of severe crop intensification techniques has reduced the activity and efficiency of indigenous natural enemies to target the pest. This can be remunerated by manipulating their natural habitat or environment rendering them effective biological control agents. In habitat manipulation, habitat refuges are created like grassbanks, headlands, or hedgerows for recolonization of natural enemies or a chief food sources is provided to attract them to crop field. The act of providing food source and shelter during harsh conditions might operate as micro-ecotone zones for achieving appropriate floral and physical diversity and pest suppression (Smith, 1991).

13.6.3.1 HABITAT REFUGES

After harvesting the crop the natural enemies like invertebrate parasitoids and predators, are unable to leave arable land and they will be exposed to harsh conditions of burning, drilling, plowing, and a range applied pesticides in the field with few food resources during autumn. And during wintering, the predator densities are higher in uncultivated grasses and herbaceous plants at the field boundaries than the crop itself because of lower soil moisture and warmer temperature during the daytime. Thus providing sufficient food sources and refuges, form extreme perturbations can raise the number and effectiveness of natural enemies in crop fields (Nafiu, Dong, and Cong, 2014).

13.6.3.2 FOOD SOURCES

A variety of natural enemies are not host-specific and they will search for a substitute food resource in the absence of their prey. If they found the alternative prey, they will tend to go away from their indigenous crop habitat resulting increase in pest population and their late return to crop might not considerably target the pest population. Hence, the presence of alternative or secondary food source can solve this problem. Some parasitoid species of hoverflies and wasps require pollen or nectar as alternative food and energy source for maturation of egg and presence of apt flower in the field or nearby the crop can enhance their efficiency. *Orius majuscules* are anthrocorid bugs target thrips *Frankliniella occidentalis* of western flowers and they can feed on alternative preys when thrips are sparse (Nafiu, Dong, and Cong, 2014; Wakeil et al., 2017).

13.6.4 PREDATORS

A predator depends upon predation or sustains their life by consuming animals of other species as its food. This behavior is prevalent among mites, spiders, and insects. Around 40 families of predators are important for the suppression of pest in forestry and agriculture of these the *Anthocoridae, Coccinellidae, Reduviidae, Pentatomidae, Gerridae, Formicidae, Ceci-domyiidae, Cecidomyiidae, Staphylinidae,* and *Miridae* are mainly found preying on pest. Insect predators may be vertebrates or entomophagous insects. In every five groups of animals, natural pest's enemies can be found, while mammals and birds have concerned for future perspective. The amphibians and reptiles have small consumption rates therefore have less potential as biological control agents. Though, fish play a vital role in controlling rice stem borers and mosquitoes. Birds have also been acknowledged as predators of pest insect.

Mammals and birds are recognized to control the insect pests in ecosystem; birds feed on adult insects (mainly free flying) whereas the mammals are limited mostly to pupae stages of insects. Feeding of predators on each host stages, i.e., larvae, pupae, and adults and every predator needs many individual preys to attain maturity but only one individual required by parasitoids. However, the biological control of these agents is small in few circumstances like masked shrew, which controls the *Pristiphoraerich sonii* and sawfly.

Predators are mainly divided into two types, i.e., with sucking and chewing mouthparts but usually lack extreme adaptations related with parasitism. Carabidae and Coccinellidae are important families for biological control within the Coleoptera. Spiders and predatory mites are predators for arthropod and show an important role as controlling agents in glasshouse process and orchards by degrading phytophagous mites (Ribaand Silvy, 1989).

13.6.5 PARASITE

A parasite is an organism living in or on another living organism, obtaining nutrients from their host, resulting altered growth, development, and reproduction or death for the host.

Out of all species of insects, approx. 10% are classified as parasitoids. The important difference between parasitoids and parasite is that parasitoids kill their hosts but parasites do not. Approximately 75% of the parasitoids belong to Hymenoptera and rest 25% consists of Neuroptera, Lepidoptera, Strepsiptera, Coleoptera, and Diptera.

The main difference between parasites and parasitoids is that parasites may not kill their hosts but parasitoids do. About 75% of the parasitoids are Hymenoptera; the remaining 25% is composed of Diptera, Strepsiptera, Neuroptera, Coleoptera, and Lepidoptera. Diptera families are parasitoids of sanils and arthropods. The frequently used groups for biological control in the Hymenoptera are *Ichneumonidae, Pteromalidae, Braconidae, Encrytidae, Aphelinidae*, and *Eulophidae* (Ribaand Silvy, 1989).

13.6.6 PATHOGEN

13.6.6.1 FUNGAL PATHOGEN

Fungi are heterotrophic, multicellular eukaryotic organism and play a vital role in nutrient cycle. It shows symbiotic relationship with bacteria and plants. Entomopathogenic fungi are natural parasites and are generally pathogenic to arthropods and bacteria. These include various ecologically, morphologically, and phylogenetically multiple species of fungi, these organisms grow at the expense of other insects. Among 31 insect orders, approximately 20 orders are infected in growth and developmental levels by these fungi (Butt, Jackson, and Magan, 2001).

Attachment and penetrability degree of fungi inside the exoskeleton of host, demonstrates the infection success rate. During the insect integument degradation by entomopathogenic fungi, production of various extracellular enzymes occurred. Microbial pathogenic agents are used over chemical pesticides because of the following purposes: reducing the residues of pesticides in biodiversity and food, protection of nontarget microorganism, and for the human safety.

Unregulated and undefined use of chemical insecticides has led to improvement in resistance development towards various chemicals present in insects and plant protection products. Above 500 arthropods, species have developed resistance to several chemical pesticides. Hence, novel substitute strategies are required to overcome the pest outbreaks (Butt, Jackson, and Magan, 2001). The biological control models are:

1. ***Metarhizium anisopliae*:** First biocontrol agent recognized in 1880's and used as a biological control agent against various pests and insects involving locusts, beetles, and spittle bugs. Several conidial and spore formulation of *Metarhizium anisopliae* are made and applied. After the commencement of epizootic spread, production of spores and vegetative cells occur in affected insect. These spores infect healthy insect population and enhance continual control (Zimmermann, 1993).
2. ***Beauveria bassiana*:** The entomopathogenic deuteromycete. *Beauveria bassiana* is a filamentous fungus grows in soil naturally and acts as a parasite on several species of arthropod. This fungus has high host specificity and causes white muscardine disease. Vectors like *Glossina morsitans*, *Rhodinus chagas*, and *Phlebotomus* are the hosts of efficient transporter of leishmania are the hosts of medical significance. Consequently, codling moths and colorado potato beetle are classified as the hosts of agricultural importance. Entomopathogenic fungi are vastly enduring in the atmosphere therefore provide long-standing inhibition of the pest and enhanced control. *B. bassiana* is used as a biological control agent against the *Ostrinia nubilalis, Dendrolimus superans,* and *Nephotettix virescens* (Zimmermann, 1993).
3. ***Trichoderma*:** Various species linked to *Trichoderma* genus are known for their capability to generate industrially useable enzymes. This species have potent role in the naturally controlling pathogen of plants. *Trichoderma* species uses an approach of biocontrol called mycoparasitism, during this process activation of signaling cascades occurred against a fungal pathogen. *Trichoderma* spp. are recognized for nonribosomal peptides production like siderophores and epipoly-thiodioxo-piperazines which are antimicrobial in nature and improve the lysis of cell wall by substituting in a collaborative manner with hydrolytic enzymes that are included in breakdown of cell wall.
4. ***Metarhizium anisopliae*:** It spreads in soil, where wet conditions led to filamentous growth and conidial spores (infectious) production;

these spores when contact with soil dwelling insects infect them. First defined by Metschnikoff in 1879, as a pathogen of wheat cockchafer and also various fungal pathogens are used as pesticides for microbes. This species consists of diverse isolates and strains of topographic area and from many host types (Roberts and Leger, 2004). This species potentially used as biological controlling agent against malaria vector species and is a model organism for future research and development (Mnyone et al., 2010).

Genetic Modifications. The use of fungal pathogens depends on genetic modification of wild type strains and results in a different strain. Enhancement in the competence of the insecticide, decrease in the dose required for pest killing or to decrease the crop damage by decreasing the pests feeding time and increasing the range of host are two strategies, out of various types of enhancements that may be included. Genetic modification methods are important for current fungal research and the selectable markers accessibility determines the procedure success rate. Genetic modification techniques have been utilized for isolation of particular virulent gene. For entomopathogenic fungi, studying the fungal pathogenesis mechanism in insects will give a base for genetic engineering (Domsch, Gams, and Anderson, 1980).

13.6.6.2 BACTERIAL PATHOGENS

Bacteria are moderately unicellular microbes which lack organelles like mitochondria and nucleus, they reproduce by binary fission. Most of the bacteria used as insecticides grow voluntarily on a broad diversity of low priced substrates, a feature that deeply facilitate their production in bulk. Currently, majority of the spore forming members of Bacillaceae family are used as biocontrol agents and belongs to the *Bacillus* genus. These pathogenic bacilli enter in unhealthy and healthy insects, and isolated from several habitats consisting plants, soil, insect frass, aquatic environments, and granaries. Several bacterial subspecies and species, mainly *Pseudomonas*, Bacillus, etc., are mainly used to control insect and plant diseases (Shelton, Zhao, and Roush, 2002).

Bacillus thuringiensis. The *Bacillus thuringiensis* species is mainly composed of all bacterial species and distinguished via fabrication of parasporal body through sporulation. These bodies contain few proteins in a form of crystalline and are greatly toxic to few species of insects. These toxins are called δ-endotoxins, and must be ingested by larvae to cause toxicity, generally occur as protoxins in the parasporal body which functions by proteolysis in gut. The functioning toxins demolish epithelial cells of midgut, and rapidly degrade the insects.

Bt cotton is successful because of easily grown at large scale and media availability. HD1 strain of Bt subsp. is mostly used and it produces 4 endotoxin proteins enclosed into parasporal body. An isolate named as ONR60A of *B. thuringiensis* subsp. is extremely toxic to black fly and mosquito larvae. Based on *B. thuringiensisis raelensis* various commercial products are used for controlling the nuisance and vector mosquitoes and black flies (Federici, 2005). Another isolate of Bt developed the DSM2803 and it liberates a parasporal cuboidal body toxic to several coleopterous insects and used to control beetle pests.

For example, *Bacillus popilliae* kills Japanese beetle larvae but less effective against annual white grubs. *Bacillus licheniformis*, *Pseudomonas fluorescenes*, *Bacillus* spp. and *Rhodotorula minuta* have been reported efficiently effective to control anthracnose of mango (Lacey et al., 2001).

13.6.6.3 VIRAL PATHOGENS

There are majorly six groups of insect viruses but out of them three are adequately diverse from human viruses to be considered safe and these are: the nuclear polyhedrosis virus (NPV), the granulosis virus (GV), and the cytoplasmic polyhedrosis virus (CPV).

These three are the obstruct viruses which means encompass of virus particles in a shell which is proteinaceous in nature and has a paracrystalline structure know as inclusion body. Approximately 120 types of NPV have been isolated from Orthoptera (locusts and grasshoppers), Diptera (mosquitoes and cranefly and) Hymenoptera and the Lepidoptera (moths and butterflies).

Approximately 800 species of mites and insects have been obtained; CPV and baculovirus have biological activities hence, used as biocontrol agents. Successful use of these agents in IPM depends upon optimal dose attainment, virulence of viruses, response of host cell to infection plus the determination and communication of the viruses in population of pest and its surroundings (Peters, 1996).

Baculoviruses. Few strains of CPVs and baculoviruses are commercially accessible for use as insecticides. Various entomopoxviruses, involving the

Choristoneura fumiferana have been used in small scale field trials. Members of the RNA virus and densovirus have potential to control species of Aedes.

Baculoviruses are dsDNA pathogenic viruses present generally in insects and arthropods. Applications of baculoviruses are précised in horticulture and agriculture field where minimization of threshold is performed to control pest damage. Baculoviruses cause death in the larval stage only in Lepidoptera group. Baculoviruses are consumed by larvae to begin the infection, enters via midgut, and spread throughout the body. Baculoviruses exist into two groups as granuloviruses (GVs) and nucleopolyhedroviruses (NPVs). Occlusion bodies in case of NPVs consist of various virus particles, however, in GVs contains single virus particle. Baculoviruses are occluded, i.e., embedment of virus particle into the protein layer, and these bodies allow the survival of the virus outside the host (Cory, 2000).

The majority of Baculoviruses are isolated from Diptera, Hymenoptera, Coleoptera, and Lepidoptera. Genome of virus ranges from 80 to 200 kb in size. Baculovirus generally have shorter host range and *Autographa californica* nuclepolyhedron-virus (AcMNPV) is mostly studied Baculovirus.

Baculovirus has no negative impacts on mammals, plants, fish, and birds. Baculovirus has various merits over insecticides which make them highly suitable controlling agents. They possess the capability to persist in the atmosphere therefore can be used for the development of biologically long term based control strategies.

However, Baculovirus activity against their hosts can be improved by introducing the specific insect toxins (Mazid, Kalida, and Rajkhowa, 2011).

13.6.6.4 ENTOMOPATHOGENIC NEMATODES

Nematodes are commonly known as roundworms having colorless unsegmented body and lacking appendages. Nematodes enter in insects through natural openings and liberate a bacterium which causes death of the host in 48 hours. Nematodes complete their life cycle inside the host using bacterium as a nutrient source and leave the insect to enter in new host for next generation. Various species of entomopathogenic nematodes, involving *Steinernema riobravis, Steinernema carpocapsae, Steinernema feltiae,* are commercially sold for bio-active management of fungus gnats and are safe to animals, ground insects, people, and earthworms.

Diseases of animals, plants, and humans caused by various parasitic species and the remaining species attacks pests by sterilizing their hosts.

Nematodes of *Heterorhabditidae* and *Steinernematidae* families are used to inhibit insect pest population in agriculture and shown positive effect on crop yield (Stuart et al., 2006).

For the use of nematodes as biopesticide, these are produced in a mass using *in-vitro* or *in-vivo* methods. *In-vivo* production of biopesticides requires low technology, low cost of startup, low cost efficiency whereas, *in-vivo* production increased by innovations in streamlining and mechanization. Liquid culture method in case of *in-vitro* culture is a most significant method but high maintenance and startup cost for the production of biopesticide.

Moreover, it has been noticed that the advantageous nematodes consisting *Heterorhabditis indica* and *Steinernema carpocapsae* were successful in killing and infecting pupal stages of flea beetle, soil dwelling larva. Because of better ability and high virulence activity of *Steinernema carpocapsae,* it is used as a potential biological agent of *Dacus ciliates* (Shokoofeh et al., 2013).

The types of nematodes are:

1. ***Heterorhabditis bacteriophora*:** This nematode decrease adult emergence of thrips flowers by 40%, production of progeny number that survives and ability to transfer virus diseases. *Heterorhabditis bacteriophora* have high searching activity and in deeper soil, profile attacks the western thrips pupae. Thrips control cost is high and commercially available.
2. ***Thripinema nicklewoodii*:** This nematode is contagious to adult thrips and larva. Nematodes infective stage makes bores into the body of thrips, reproduces inside its abdominal cavity, and excreted out with its frass. In adult thrips female, nematodes avoid egg production.

The features of economically valuable species are:

1. ***Steinernema carpocapsae*:** This species involves some important features such as simplicity of production and capacity to formulate in dry state which increase shelf life for many months at room temperature. *S. carpocapsae* is mostly effective against the larvae Lepidoptera, different cutworms, webworms, wood-borers, girdlers, and some weevils. It is principally effective against insects which are liquid surface adapted.
2. ***Steinernema feltiae*:** This species is mainly valuable against undeveloped dipterous insects, consisting fungus gnats, tipulids,

and mushroom flies, as well as few larvae of lepidopterous larvae. It is exclusive in preserving contagion at soil temperatures below 10°C.

3. ***Steinernema glaseri*:** This species is among the major entomopathogenic nematodes and useful mainly against scarabs of coleopterous larvae. *S. glaseri* is vastly responsive and mobile to the host volatiles of extensive-range.
4. ***Steinernema riobrave*:** This species is pathogenic and effective against the numerous orders of insects, repressing citrus root weevils. This nematode is effective in insects killing above 35°C soil temperature. Its growth is excellent under semi-arid parameters, an attribute improved by the high levels of lipid founds in infectious juveniles. Its minute size provides maximum yields either using *in-vitro* or *in-vivo* methods.

The nematodes as biological control agents are:

1. **Mermithid Nematodes:** One of the largest nematodes attacking insects and adult females are 5 to 20 cm in length. In parasitic insects like larvae of various aquatic insects, the developing stages of nematodes can be seen inside the hemocoel where they viewed as thin, long, white worms. Mermithids have been observed from various insect orders and other arthropods like crustaceans and spiders. *Romanomermis iyengari* and *R. culicivorax* species are considered valuable for used as biological controlling agents.
2. **Steinernematid and Heterorhabditid Nematodes:** These are tiny terrestrial nematodes which occur naturally and act as parasites for inhibiting the soil insects. These are unusual and develop a mutualistic connection with bacteria, which anchorages inside their alimentary canal and after invading insects, body nematodes kill the insects (Stuart et al., 2006).

13.6.6.5 PROTOZOA

Protozoa are unicellular eukaryotic organisms, can exist as free living or parasitic which makes them varying in size and shape. They show features usually related to animals, heterotrophy, and particular mobility and belongs to Protista kingdom. Protozoan takes nutrition from organic matter like organic tissues or other microbes and debris.

13.6.6.5.1 Nosema

The insect protozoan diseases are characteristically uniform and consist of necessary role in controlling insect populations. Nosema species are generally slow acting and host specific commonly causes chronic infections. The biological activities of Nosema are composite and growth of few species requires an intermediary host and in living hosts only. Microsporida species are generally observed species and chief advantages are readily recycling and persistence in host cell and debilitate effect on fitness and reproduction of target insects (Henry and Oma, 1981).

First demonstrated microsporidium, i.e., *Nosema bombycis*, is pathogenic to silkworm pebrine, which endures in North America, Europe, and Asia. Pebrine is epidemic and causes economical losses in silk manufacturing countries like China. About 1000 species of protozoan, attack various insects like helithine moths, grasshoppers, and invertebrates. Spore production by protozoans is conragious phase in certain susceptible insects.

Assimilation and development of these spores occur in host midgut and afterwards, from sporoplasm these spores are liberated and attack target cells of host. Again, beginning of sporulation course from infected tissues and stimulates apizootic disease (Becnel and Andreadis, 1999) (Table 13.2).

TABLE 13.2 Properties of Basic Recognition Strategies of Plant Pathogens

	Product	Target
Fungus		
Phlebiopsis gigantea	Rotstop	*Heterobasidium annosus*
Gliocladium virens	SoilGard	Numerous plant diseases root pathogens and Damping off
Candida oleophila	Aspire	*Penicillium* spp. and *Botrytis* spp.,
Trichoderma harzianum	Trichoderma 2000	*Sclerotium rolfsii*, *Rhizoctonia solani*,
Trichoderma harzianum	Trichodex	Fungal diseases, e.g., *Botrytis cinerea*
Cryptococcus albidus	YIELDPLUS	*Penicillium* spp., *Botrytis* spp.
Gliocladium catenulatum	Primastop	Various plant diseases
Fusarium oxysporium	Biofox C	*Fusarium moniliforme, F. oxysporium*

TABLE 13.2 *(Continued)*

	Product	Target
Beauveria bassiana	Mycotrol	Pine caterpillar, European corn borer
Culicinomyces clavisporus	Mycar	Mosquito larvae
Verticillium lecanii	Vertalec	Aphids, thrips, greenhouse whitefly
Beauveria bassiana	Boverin	Colorado potato beetle
Bacteria		
Bacillus thuringiensis var. kurstaki	Bactur, Bactospeine, Dipel, Futura, Topside, Tribactur	Caterpillars (butterflies and moths larvae)
Bacillus thuringiensis var. tenebrinos	Foil, Trident, Novardo	Larvae of elm leaf beetle adults
Bacillus thuringiensis var. israelensis (Bt)	Aquabee, Larvx, Skeetal, Mosquito attack	Larvae of *Psorophora* and *Aedes* mosquitoes, fungus gnats, black flies
Bacillus popilliae and Bacillus lentimorbus	Doom, Grub attack, Milky spore Disease	Japanese beetle larvae
Bacillus thuringiensis var. aizawai	Certan	Wax moth caterpillars
Bacillus sphaericus	Vectolex	Larvae of *Psorophora, Culex*
Viruses		
Pine sawfly NPV	Neochek-S	Larvae of pine sawfly
Gypsy moth nuclear plyhedrosis (NPV)	Gypchek virus	Gypsy moth caterpillars
Codling moth granulosis virus (GV)		Codling moth caterpillars
Tussock moth NPV	TM Biocontrol-1	Tussock moth caterpillars
Entomogenous Nematodes		
Heterorhabditis heliothidis	Nowadays accessible on a extensive basis for large scale processes	Larvae of boring insects and a broad range of soil dwelling
Steinernema feltiae	Biosafe, vector	Larvae of boring insects and a broad range of soil dwelling
Protozoa		
Nosema locustae	Grasshopper attack	Mormon crickets, grasshoppers

Source: Le and Malan (2013).

13.6.7 PESTICIDES

The humans are coexisting with over 1 million types of insects and other arthropods, a lot of which act as pest, plant pathogens. A pest can be any living organism that destroys the food crops or compete for food or spread disease. Pesticides are natural or synthetic compounds which are lethal and can destroy the pests. Pesticides have been considered as essential pest control tool since 19th century when a fungicide (sulfur-containing compound) was developed followed by the development of arsenic-containing inorganic pesticides (lead arsenate) to target the fruit and vegetable attacking pests. These inorganic pesticides were extremely poisonous.

In the early 1950s first synthetic organochlorine pesticide dichlorodiphenyltrichloroethane (DDT) was introduced. The use of pesticides became an integral part of agriculture by the mid-1950s to increase crop yield. Presently the use of over 17,000 pesticides and more than 800 active compounds has been reported in the United States. In order to increase the crop yield pesticides not only targeting the insect pest but also affecting the other living things, soil, water, and air. Pesticides can be classified as natural (botanicals or plant-based pesticides) and synthetic (organic and inorganic) pesticides on the basis of their chemical composition (Lorenz, 2009; Edwards, 1987; Ehi-Eromosele et al., 2012).

13.6.7.1 INORGANIC PESTICIDES

These are simple, crystalline, non-carbon, and environmentally stable compounds of mineral origin which are readily soluble in water. Inorganic insecticides mainly include the compounds of copper, boron, antimony, fluorine, selenium, thallium, sulfur, mercury, and zinc, and elemental sulfur and phosphorus. Antimony potassium tartrate is generally used as a toxic ant poison and to target the thrips.

1. **Arsenic:** It is a natural element possessing the chemical and physical properties of both metals and non-metals. Arsenic containing compounds are extensively used inorganic pesticides for example copper arsenate, Arsenic trioxide, Sodium arsenite, lead arsenate, zinc arsenate, calcium arsenate, and Paris green; these are known to inhibit sulfhydryl (-SH) containing enzymes by uncoupling the oxidative phosphorylation. These compounds are also used as

fungicide, rodenticide, and herbicide. The heavy spray of Paris green by airplane to control malaria has also been reported.

2. **Sulfur:** It is the oldest known very efficient insecticide. The use of sulfur by burning sulfur candles for bedbug fumigation has been registered. The elemental form of sulfur is generally used as fungicide, miticide, or acaricide on vegetable, orchard, cereals, and ornamental crops. Sulfur dust and sprays are toxic against powdery mildews. Oxidized form (gaseous sulfur oxides) of sulfur can be irritating to eyes and respiratory tract and ingested sulfur powder can cause catharsis.
3. **Aluminum Phosphide (AlP):** It is extremely toxic, widely used as a fumigant against moles and rodents but it can also be used as an insecticide for stored cereals. AlPs get converted into lethal phosphine gas by the action of acids in the digestive tract of rodents and poison them.
4. **Boric Acid:** It is considered as a competent and safe insecticide to target household cockroaches and ants and can also be used to control termites, fleas, silverfish, fire ant, and other insects by affecting their metabolism and damaging their exoskeleton. In combination with ethylene glycol, it can be used to treat dry and wet rot in timbers. Boric acid is used in agriculture to prevent or treat boron deficiencies in plants.
5. **Diatomaceous Earth:** Or diatomite is used as an insecticide because of its physico-sorptive and abrasive properties. It absorbs lipids from the waxy exoskeleton of insects causing dehydration and they die because of water pressure deficiency. This fine powder can effectively target arthropods and gastropods for protecting stored grains (Sarwar, 2016).

13.6.7.2 ORGANIC PESTICIDES

The chemicals having carbon in their molecular structure are more complex organic pesticides and typically not easily soluble in water.

1. **Organochlorines:** It is the earliest discovered and used synthetic organic pesticide best known for its broad spectrum action and long residual effects. Organochlorine insecticides like DDT, aldrin, endosulfan, dieldrin, and gamma-HCH or gamma-BHC. BHC and DDT are now banned and replaced by other pesticides because

they are highly persistent and causes long term contamination of environment, gradually get accumulated in animals and enters into food chain. When the organochlorine present in the fat reaches the maximum level, the breakdown of fats releases chemicals into the blood causing poisoning and death of the organism. Organochlorine plays important role in mosquito control.

2. **Organophosphates (OPs):** The organophosphate (OPs) insecticides are characterized by their multiple mode of action and their ability to target broad range pests. They can not only used as stomach and nervous poison but also as contact poison and fumigant. They affect nervous system by disturbing acetylcholine regulating enzyme cholinesterase. OPs provide slow pest resistance and these are biodegradable. Some widely used OPs are Temephos, diazinon fenitrothion, dichlorvos, dimethoate, malathion, pirimiphos-methyl, and glyphosate.
3. **Carbamates:** These are derived from carbamic acid and their structure and working principle are similar to OPs by inhibiting the activity of acetylcholinesterases enzyme but here the enzyme inhibition can be reversed and insect pest can recover if a too low dose is given. They are used as fumigants as well as stomach and contact poison. Carbamates are formulated with the toxic OPs being available as granules. Carbamate insecticides include carbofuran, aminocarb, carbaryl, propoxur, and bendiocarb.
4. **Synthetic-Pyrethroids:** These are more stable than their natural analog pyrethrins derived from *Pyrethrum cinearaefolium* flowers. Synthetic pyrethroids have elevated contact action, short persistence, readily biodegradable, and can effectively target lepidopterous larvae. Their introduction in the mid-1970s has proved them a powerful insecticide because of their low dosage requirements, short term toxicity, and no persistence in the environment. Their effect is frequently extended because of their ability to repel the insects. Synthetic pyrethroids include deltamethrin, bifenthrin, permethrin, cyfluthrin, cyhalothrin but cypermethrin and permethrin are the most widely used synthetic-pyrethroid pesticides (Yadav and Devi, 2017; Ehi-Eromosele et al., 2012).

13.6.7.3 BOTANICAL PESTICIDES

Herbivorous insects are majorly responsible for destroying about one-fifth of the world's overall annual crop production predominantly in tropics and sub-tropics with extremely favorable conditions for a variety of insect pests. Various kinds of volatile active compounds (secondary metabolites) produced by plants can effectively control the insects by altering their physiological characteristics or they may act as insect repellents, antifeedants, chemosterilants, attractants, toxicants or as growth retardants (Chengala and Singh, 2017). The types of botanical insecticides are:

1. **Alkaloids:** These are widely used natural insecticides, having antifeedant and larvicidal activity by making the treated plant material unpalatable or indigestible and the insect will starve to death. Quinolone and Furocoumarin isolated from the leaves of *Ruta chalepensis*, are effective against larva of *Spodoptera littoralis*; piperidine and pipernonaline against mosquito larva whereas *Arachis hypogaea* extracts are effective against chikungunya and malaria.
2. **Essential Oils:** These extract from aromatic plants are significantly replacing synthetic insecticides having various effects on insects like growth inhibitors, ovicides, oviposition inhibitors, repellent, and antifeedants. Essential oils extracted from *Zingiber officinale* rhizomes and *Piper cubebaberries* demonstrates antifeeding and insecticidal action against *Sitophilus oryzae* and *Tribolium castaneum* and essential oil from *Melaleuca alternifolia* act as toxic fumigant to *Sitophilus zeamais*. Essential oil from rosemary, eucalyptus, yarrow, mint oil, and oregano are reported as effective against various insects, as fumigant for controlling cockroaches. Oregano essential oil is a potent repellent against *Supella longipalpa*. The chief basil oil component eugenol show strong repellent effect against mosquitoes and linalol has toxic effects on *Bruchid zabrotes* sub*fasciatus* and other storage pest insects. The malaria vector *Anopheles arabiensis* is strongly affected by *Juniperus procera* essential oil. The active compounds in Eucalyptus essential oil are citronellal, p-cymene, eucamalol, citronellyl acetate, 1,8-cineole, Terpinen-4-ol, eucamalol, linalool, α-pinene, limonene, α-terpineol, aromadendrene, ϒ-terpinene and alloocimene, are active against mosquito bites, antifeedant, and can be used to lessen the tick-borne infections (Chengala and Singh, 2017; El-Wakeil, 2013).
3. **Esters and Fatty Acids:** Fatty acids methyl esters (FAME) extracted from *Solanum lycocarpum* exhibits larvicidal action against *Culex quinquefasciatus* and saturated fatty acids act as antifeedants or as repellents to target horn flies, horn flies and malaria vector *Anopheles*

funestus, whereas linoleic acid prevents larval development of *Spodoptera littoralis.*

4. **Flavonoids:** and isoflavonoids are known to protect plants from pests by altering their growth, development, and behavioral pattern. Quercetin and rutin glycosides in *Pinus banksiana* and peanuts increases the mortality of *Lymantria dispar* and *Spodoptera litura* (tobacco armyworm) respectively. Flavone glucosides in rice act as feeding deterrents in *Nilaparvata lugens* and other herbivores. The use of azadirachtin as safe, biodegradable, and potent insecticide has been registered. Some polyphenolic flavonoids (flavanone naringenin and flavonol quercetin) can be used as insecticides to control *Acyrthosiphonpisum* (pea aphid). The larvicidal and feeding deterrent action of *Arachis hypogaea* flavonoids against chikungunya and malarial vectors has been studied.
5. **Glycosides:** Cyanogenic glucosides in the plants are considered to be involved in major plant defense mechanisms against phytophagous insects. The cardenolide glycoside extracted from *Calotropis procera*; neem oil and azadirachtin from *Azadirachta indica*; digitoxin and a cardiac glycoside from *Digitalis purpurea* posses larvicidal, antifeedant, and growth inhibitory activity. Cyanohydrins are effectively used as fumigants and anthraquinones from *Cassia* species as antimalarial compounds (El-Wakeil, 2013; Wafaa et al., 2017).

13.6.7.4 INSECT GROWTH REGULATORS (IGR)

Although the use of chemical pesticides has reduced the crop damage by pests up to 5% to 30% of potential yield, but they poses a number of problems in agriculture like killing the beneficial pests, pest resurgence, secondary pest outbreak, and toxicity to all animals and humans. The biological magnification of these persistence pesticides in the food chain can affect the nontarget organisms and destroy the organisms at each trophic level. It led to the discovery of new active biomolecules such as insect growth regulators (IGR) with ecological properties, nontoxicity to other beneficial organisms, and with new mode of action against the target. IGRs hamper the metamorphosis and reproduction in insect pests; they disrupt the process of embryonic, nymph, and larva development, preventing exoskeleton formation in insects. IGRs are highly selective against arthropods and insects killing through interrupting definite biochemical pathways crucial for their growth and development to impede them reaching adulthood and declining the pets population (Dhadialla, Retnakaran, and Smagghe, 1995; Hasan and Nedim, 2004).The discovery of IGRs was dependent on growth, development, and behavioral pattern of insects and they may come from a mixture of synthetic chemicals or from natural plant sources. IGRs include the following classes:

1. **Chitin Synthesis Inhibitors (CSIs):** This inhibits cuticle sclerotization at the stages of growth and interferes with the physical development of insects. These compounds inhibit the chitin formation which is an integral structural carbohydrate polymer in the exoskeleton of insects. After treating the insect with CSI, they will grow normally up to the molting stage and dies when it molts because CSI causes abnormal deposition of uridine diphospho-N-acetylglucosamine monomers thereby interferes with chitin synthesis and polymerization. CSI toxicity also involve mortality due to decreased locomotion, swollen appendages, dislocalized mandibles, and repressed fecundity and egg viability. The first CSI was benzoyl phenyl urea (diflubenzuron), discovered in early seventies, and considered to be active pesticide against a variety of lepidopteran insect larva by inhibiting the proteolytic activation of zymogens thereby inhibiting chitin synthesis. Besides BPU, there are some other commonly used IGRs such as lufenuron, hexaflumuron, chlorfluazuron, and tefluenzuron.
2. **Juvenile Hormone (JH) and Antijuvenile Hormone Analogs:** In 1967, Carrol Williams suggested the use of insect juvenile hormones (JH) as insecticides and proposed the term "third generation pesticide" to the insect JH that are ecologically benign and no risk of pest resurgence. In the insect developmental process PTTH (prothracicotropic hormone) from the brain activate the secretion of a molting hormone (ecdysteroid), which causes either molting or metamorphosis depending upon the stage of the insect, and in particular presence of a third hormone, JH.

JH analogs control the process of development, reproduction, embryogenesis, and metamorphosis. The elevated level of JH in the larva or nymph inhibits the pupa formation, development, delay maturation and ultimately result in death. In absence or low amount of JH, epidermis is programmed for metamorphosis instead of larval molt. Generally, in adult females JH is present for production of eggs but JH remains absent in pupae and hence represses pupa formation but stimulates vitellogenesis. The effectiveness of Juvenoid IGRs depends on the timing of application. Generally, JH application does not affect adult insects but some of them become sterile in presence of high JH level. Treating the pupal stage with unusually high level of JH can result in another larval stage, larval-pupal intermediates or super larva formation but no normal adult will develop. It can cause sterility when applied to eggs or can trigger dormant stages hence it can effectively control the insect population.

Precocene is an anti-juvenile hormone (AJH) analog which gets converted into highly reactive intermediate to destroy the corpora allata gland and cause precocious molting into a nonfunctional adult within few days to weeks. A number of compounds like methoprene, triflumuron, fenoxycarb, and pyriproxyfen have been effectively used into vector management programs principally malaria, dengue, and in products used for targeting cockroaches, ants, fleas, and other household pests.

Dibenzoylhydrazine (DBH) compounds are ecdysteroid agonist, used as synthetic insecticides in the fields by causing incomplete and premature molting by binding to a nuclear ecdysteroid receptor (EcR) thereby resulting in death of insect (Coleoptera and lepidoptera) within few days to week (Dent, 2000; Dhadialla, Retnakaran, and Smagghe, 1995; Hasan and Nedim, 2004).

13.6.7.5 SEMIOCHEMICALS

Chemicals that mediate communications among organisms either intraspecific interactions through *pheromones*, or interspecific interactions using *allelochemicals*, which can be further categorized as sex pheromones, alarm pheromones and epideitic or aggregation pheromones and allelochemicals as kairomones, allomones, synomones, and apneumones. Among them, pheromones of lepidopteran insects are widely studied because of their use as attractants, repellents, antifeedants, and stimulants. Semiochemicals are not direct alternatives for chemical pesticides but they can be used as killing agents with high potency, species specificity, and low toxicity (Dent, 2000).

There are different types of behavior-modifying semiochemical involved in insect communication. Normally, the females release the pheromones for a definite time period to attract the males the process of mating. Various insects release epideitic pheromones which permit the aggregation of insects to an extent needed to proficiently exploit a habitat or host. Such aggregation pheromones are frequently found in some species of beetles such as *Gnathotricus sulcatus*, *Dendroctonus* spp., and *Trypondendron lineatum*. Alarm pheromones are produced by the insects under threat or attacked by predators or parasitoids and evoke an escape response to prevent their chances of being attacked, hence they can be used in controlling the insect pest. For instance, (E)-b-farnesene have been used to enhance the aphids mobility so that contact poison(pyrethroids) are quickly picked up, whereas simple components from a honey bee sting and mandibular gland can be used to prevent valuable for agingbees from oilseed rape throughout the period of pesticide applications (Dent, 2000; Tamoghna and Chandran, 2017).

1. **Mass Trapping:** In this, a huge number of pheromone traps are used to catch a large number of pest populations. But this method has been proven futile for a variety of insect pest management belonging to Coleoptera and Lepidoptera because sex pheromones attract only males therefore highly efficient trap models are required to catch an adequate number of male insect to interrupt their mating with females. This method is more efficient with aggregation pheromones because they attract both males and females. The use of mass trapping to target spruce bark beetle and cotton boll weevil has been reported. Aggregation pheromone component 2-methyl-3-buten-1-ol has been used in a huge mass trapping program (600,000 traps) and captured nearly 7.4 billion of beetles in 2 years and the damage to the crop was significantly reduced.
2. **Mating Disruption:** In this technique, the surrounding of the females is flooded with sex pheromone so that the male would not be able to detect the female insects in the nearby area. This mating disruption might be because of some factors like direct competition, habituation, or sensory adaptation. In direct competition the males tend to fly toward the pheromone source instead of flying toward females, thereby prevents successful mating. Because of the long term exposure to the sex, pheromones the olfactory sensory neurons become adapted and cannot sense the presence of pheromones whereas in habituation the insects stop reacting to stimulus when proper results were not according to the early responses. The presence of pheromone antagonists can also be a cause of mating disruption. To understand the mating disruption the study of behavior to be modified and sensory inputs transforming these reactions is required. This technique has been effectively employed pest management,

e.g., to control grape berry moth (*Eupaecilia ambiguella*), to control the pink bollworm (*Pectinophora gossypiella*) and artichoke plume moth (*Platyptilac arduidactyla*). This method is absolutely non-toxic to higher animals, natural enemies, and to human beings (Dent, 2000; Tewari1 et al., 2014).

13.7 CONCLUSION

The excessive reliance on chemical pesticides led to the increase of resistance among pests to pesticides, outbreaks of secondary pests and pathogens/ biotypes, and accumulation of pesticidal residues in the food chain. To overcome such circumstances and reduce damage to human and animal health, numerous organizations have started advocating the concept of IPM with better profits. This chapter aims to discuss the importance of integrated pest management strategies involving traditional practices, physical, mechanical, biological, and chemical control methods to target insect pests and diseases associated with economically important crops. The use of chemical pesticides is hazardous to the environment because of their toxicity, high persistency, non-specificity, and non-biodegradability. Chemical pesticides should not be given the utmost priority, should be estimated along with additional management strategies by considering cost, efficiency, long-term sustainability, environmental, and health impact. Pests will keep on flourishing therefore novel technologies are required to target the pest populations.

KEYWORDS

- **anti-juvenile hormone**
- **biological control**
- **cytoplasmic polyhedrosis virus**
- **host plant resistance**
- **integrated pest management**
- **pesticides**

REFERENCES

Alastair, O., (2003). Integrated pest management for resource-poor African farmers: Is theemperor naked? *World Development, 31*, 831–845.

Becnel, J. J., & Andreadis, T. G., (1999). Microsporidia in insects. In: Wittner, M., & Weiss, L. M., (eds.), *The Microsporidia and Microsporidiosis* (pp. 447–501). ASM Press, Washington, DC.

Benz, G., (1987). Integrated pest management in material protection, storage, and food industry. In: Delucchi, V., (ed.), *Integrated Pest Management, Protection Integrée: Quo Vadis?* (pp. 31–69). An International Perspective. Parasitis 86, Geneva.

Blackshaw, R. P., & Thompson, D., (1993). Comparative effects of bark and peat based composts on the occurrence of vine weevil larvae and the growth of containerized polyanthus. *Journal of Horticultural Science, 68*, 725–729.

Butt, T. M., Jackson, C., & Magan, N., (2001). Introduction-fungal biological control agents: Problems and potential. In: Butt, T. M., Jackson, C., & Magan, N., (eds.), *Fungi as Biocontrol Agents: Progress, Problems and Potential* (pp. 1–9). CAB International, Wallingford.

Chengala, L., & Singh, N., (2017). Botanical pesticides-a major alternative to chemical pesticides: A review. *Int. J. of Life Sciences, 5*, 722–729.

Coaker, T. H., (1987). Cultural methods: The crop. In: Burn, A. J., Coaker, T. H., & Jepson, P. C., (eds.), *Integrated Pest Management* (pp. 69–88). Academic Press, London.

Cory, J. S., (2000). Assessing the risks of releasing genetically modified virus insecticides: Progress to date. *Crop Prot, 19*, 779–785.

Dent, D., (2000). *Insect Pest Management* (2ndedn.), CABI Publishing, Wallingford, New York.

Dhadialla, T. S., Retnakaran, A., & Smagghe, G., (1995). *Insect Growth-and Development-Disrupting Insecticides* (pp. 55–115). Elsevier.

Dhaliwal, G. S., & Ramesh, A., (1996). *Principles of Insect Pest Management*. National Agricultural Technology Information Centre, Ludhiana.

Domsch, K. H., Gams, W., & Anderson, T. H., (1980). *Compendium of Soil Fungi* (pp. 413–415). Academic, London.

Edwards, C. A., (1987). The environmental impact of insecticides. In: Delucchi, V., (ed.), *Integrated Pest Management, Protection Integàee Quo Vadis?* (pp. 309–329). An International Perspective. Parasitis 86, Geneva, Switzerland.

Ehi-Eromosele, C. O., Nwinyi, O. C., & Ajani, O. O., (2012). *Integrated Pest Management* (pp. 105–117). Intech Open

El-Wakeil, N., (2013). *Botanical Pesticides and Their Mode of Action* (Vol. 65, pp. 125–149.). Springer.

Evans, D. E., (1987). Stored products. In: Burn, A. J., Coaker, T. H., & Jepson, P. C., (eds.), *Integrated Pest Management* (pp. 425–461). Academic Press, London.

Federici, B. A., (2005). Insecticidal bacteria: An overwhelming success for invertebrate pathology. *J. Invertebr. Pathol., 89*, 30–38.

Flinn, P. W., Hagstrum, D. W., & Muir, W. E., (1997). Effects of time of aeration, bin size, and latitude on insect populations in stored wheat: A simulation study. *Journal of Economic Entomology, 90*, 646–651.

Hagstrum, D. W., & Flinn, P. W., (1990). Simulations comparing insect species differences in response to wheat storage conditions and management practices. *Journal of Economic Entomology*, 2469–2475.

Hasan, T., & Nedim, U., (2004). Insect growth regulators for insect pest control. *Turk J. Agric. For., 28*, 377–387.

Henry, J. E., & Oma, E. A., (1981). Pest control by *Nosema locustae*, a pathogen of grasshoppers and crickets. In: Burges, H. D., (ed.), *Microbial Control of Pests and Plant Diseases* (pp. 573–586). Academic, London.

Hodges, R. J., & Surendro, (1996). Detection of controlled atmosphere changes in CO_2-flushed sealed enclosures for pest and quality management of bagged milled rice. *Journal of Stored Product Research, 32*, 97–104.

Isley, D., (1951). *Methods of Insect Control* (p. 134). Part I. Burgess Publ., Minneapolis, Minnesota, U.S.A.

Israel, P., (1969). *Integrated Pest Control for Paddy* (Vol. 6, pp. 45–53). *Oryza* (Calcutta, India).

Jones, E. R., & Mackie, D. B., (1913). The locust pest. *Philipp. Agric. Rev., 6*, 5–22.

Khader, K. H., (1996). Integrated pest management and sustainable agriculture. *Farmers and Parliament, 30*, 15–17.

Khan, M. Q., (1967). Control of paddy stem borers by cultural practices. In: *Major Insect Pest of the Rice Plant* (pp. 369–389). Johns Hopkins Press, Baltimore, USA.

Kuwayama, S., (1963). Notes on two curculionid beetles attacking the rice plant in Japan. *Plant Prot. Bull. (Japan), 5*, 156–161.

Lacey, L. A., Frutos, R., Kaya, H. K., & Vails, P., (2001). Insect pathogens as biological control agents: Do they have future? *Biol. Contr., 21*, 230–248.

Le, V., & Malan, A. P., (2013). An overview of the vine mealybug (*Planococcus ficus*) in South African vineyards and the use of entomopathgenic nematodes as potential biocontrol agent. *S. Afr. J. Enol. Vitic., 34*, 108–118.

Leather, S. R., (1993). Influence of site factor modification on the population development ofthe pine beauty moth (*Panolisflammea*) on a Scottish lodgepole pine (*Pinus contorta*) plantation. *Forest Ecology and Management, 59*, 207–223.

Lorenz, E. S., (2009). *Potential Health Effects of Pesticides* (pp. 1–8). Agricultural Communications and Marketing.

Malena, C., (1994). *Gender Issues in IPM in African Agriculture*. Socio-economic series no. 5. Chatham: Natural Resources Institute.

Maxwell-Lefroy, H., (1907). Practical remedies for insect pests. *Agric. J. India, 2*, 356–363.

Mazid, S., Kalida, J. C., & Rajkhowa, R. C., (2011). A review on the use of biopesticides in insect pest management. *Int. J. Sci. Adv. Technol., 1*,169–178.

Mnyone, L. L., Koenraadt, C. J. M., Lyimo, I. N., Mpingwa, M. W., Takken, W., & Russell, T. L., (2010). Anopheline and culicine mosquitoes are not repelled by surfaces treated with the entomopathogenic fungi *Metarhizium anisopliae* and *Beauveria bassiana. Parasite Vectors, 3*, 80.

Nafiu, B. S., Dong, H., & Cong, B., (2014). Principles of biological control in integrated pest management. *International Journal of Applied Research and Technology, 3*, 104–116.

Norton, G. A., & Mumford, J. D., (1993). *Decision Tools for Pest Management*. CAB International, Wallingford, UK.

Peters, A., (1996). The natural host range of *Steinernema* and *Heterorhabditis* spp. and their impact on insect populations. *Biocontrol. Sci. Technol., 6*, 389–402.

Ramakrishna, A. T. V., (1933). Some important insect problems connected with the cultivation of rice in South India. *Agric. Livest. India,3*, 341–351.

Ranga, R. G. V., Desai, S., Rupela, O. P., Krishnappa, K., & Suhas, P., (2009). *Wani International Crops Research Institute for the Semi-Arid Tropics (ICRISAT) Patancheru 502 324*. Andhra Pradesh, India.

Raymundo, S. A., (1986). Traditional pest control practices in West Africa. *Intl. Rice Res. Newsl., 11*, 24.
Riba, G., & Silvy, C., (1989). *Combattre Les Ravageurs Des Cultures Enjeux Et Perspective.* INRA, Paris.
Robert, L. M., & William, H. L., (1994). *Introduction to Insect Pest Management* (p. 266). New York: John Wiley and Sons, Inc.
Roberts, D. W., & Leger, R. J. S., (2004). *Metarhizium* spp., cosmopolitan insect-pathogenic fungi: Mycological aspects. *Adv. Appl. Microbiol., 54*, 1–70.
Santoso, T., Sunjaya, D. O. S., Halid, H., & Hodges, R. J., (1996). Pest management in psocids in milled rice stores in the humid tropics. *International Journal of Pest Management, 42*, 189–197.
Sarwar, M., (2016). Inorganic insecticides used in landscape settings and insect pests. *Chemistry Research Journal, 1*, 50–57.
Shelton, A. M., Zhao, J. Z., & Roush, R. T., (2002). Economic, ecological, food safety, and social consequences of the deployment of Bt transgenic plants. *Annu. Rev. Entomol., 47*, 845–881.
Shokoofeh, K., Javad, K., Mojtabi, H., Requel, C. H., & Larry, W. D., (2013). Biocontrol potential of the entomopathogenic nematodes *Heterorhabditis bacteriophora* and *Steinernema carpocapsae* on cucurbit fly, *Dacusciliatus* (Diptera: Tephritidae). *Biocontrol. Science and Technology, 23*, 1307–1323.
Smith, H. S., (1991). On some phases of insect control by the biological methodd. *Journal of Economic Entomology, 12*, 288–292.
Soderstrom, E. L., & Brandl, D. G., (1982). Anti-feeding effect of modified atmospheres on larvae of the navel orange worm and Indian meal moths (Lepidoptera: Pyralidae). *Journal ofEconomic Entomology, 75*, 704–705.
Stewart, H. R., (1934). Entomology. *Rep. Dept. Agric. Punjab, 1932, 1933*(Part 1), 35–39.
Strong, D. R., Lawton, J. H., & Southwood, R., (1984). *Insects on Plants: Community Patterns and Mechanisms*. Blackwell Scientific Publications, Oxford, UK.
Stuart, R. J., Barbercheck, M. E., Grewal, P. S., Taylor, R. A. J., & Hoy, C. W., (2006). Population biology of entomopathogenic nematodes: Concepts, issues, and models. *Biological Control, 38*, 80–102.
Tabashnik, B. E., & Mau, R. F. L., (1986). Suppression of diamondback moth (Lepidoptera: Plutellidae) oviposition by overhead irrigation. *Journal of Economic Entomology, 79*, 189–191.
Tamoghna, S., & Chandran, N., (2017). Chemical ecology and pest management: A review. *International Journal of Chemical Studies, 5*, 618–621.
Tewari1, S., Tracy, C. L., Nielsen, A. L., Piñero, J. C., & Rodriguez-Saona1, C. R., (2014). *Use of Pheromones in Insect Pest Management, with Special Attention to Weevil Pheromones* (pp. 141–168). Elsevier.
Van, D. R. G., Lyon, S., Sanderson, J. P., Bennett, K. C., Stanek, E. J., & Zhang, R. T., (2008). Reenhouse trials of Aphidiuscolemani (Hymenoptera: Braconidae) banker plants for control of aphids (Hemiptera: Aphididae) in greenhouse spring floral crops. *Florida Entomologist, 91*, 583–591.
Waage, J. K., (1993). Making IPM work: Developing country experiences and prospects. In: Srivastava, J. P., & Alderman, H., (eds.), *Agriculture and Environmental Challenges. Proceedings of the Thirteenth Agricultural Sector Symposium* (pp. 119–134). World Bank, Washington DC, USA.
Wafaa, M., Rowida, S., Baeshen, M., Hussein, A. H., & Said-Al, A., (2017). Botanical insecticide as simple extractives for pest control. *Cogent Biology, 3*, 1–16.

Wakeil, N. E., Saleh, M., Gaafar, N., & Elbehery, H., (2017). *Conservation Biological Control Practices* (pp. 41–69). Intech Open.

Wheatley, A. R. D., Wightman, J. A., Williams, J. H., & Wheatley, S. J., (1989). The influence of drought stress on the distribution of insects on four groundnut genotypes grown near Hyderabad, India. *Bulletin of Entomological Research, 79*, 567–577.

Yadav, I. C., & Devi, N. L., (2017). Pesticides classification and its impact on human and environment. *Environ. Sci. and Engg., 6*, 141–158.

Yokoyama, V. Y., & Miller, G. T., (1996). Response of walnut husk fly (Diptera: Tephritidae) to low temperature, irrigation and pest-free period for exported stone fruits. *Journal of Economic Entomology, 89*, 1186–1191.

Zimmermann, G., (1993). The entomopathogenic fungus *Metarhizium anisopliae* and its potential as a biocontrol agent. *Pest Manag. Sci., 37*, 375–379.

CHAPTER 14

Impact of Stress Factors on Plants for Enhancing Biomass Generations Towards Biofuels

DIPANKAR GHOSH, SHRESTHA DEBNATH, and KAMALENDU DE

Microbial Engineering and Algal Biotechnology Laboratory, Department of Biotechnology, JIS University Agarpara, 81 Nilgunj Road, Kolkata, West Bengal – 700109, India, Phone: +91-7872882337, E-mails: dghosh.jisuniversity@gmail.com, d.ghosh@jisuniversity.ac.in (D. Ghosh)

ABSTRACT

Global warming and overpopulation are two thresholds of the last few decades that bring a challenge to all living beings on this earth. Global warming welcomes devastation in this earth initially by creating environmental stress conditions both biotic and abiotic levels. These environmental stressors affect on plant's growth and yield by single, multiple individually or by combined. All these can bring modification at the epigenetic level. On the contrary, overpopulation demands more food and bio-fuels to maintain civilization. So it is a joint venture of all expert namely plant physiologist, plant pathologist, molecular biologist, synthetic biologist, metabolic engineers, and genetic engineers can mitigate the burning issues doing work together at the level of plant-microbe interactions, regulation on plant cell wall biosynthesis, plant phytohormone levels, plant metabolism, etc. The latest advancements in genetic engineering, plant cellular biology, and molecular biology enable us to detect complex signaling networks of plant growth and development. By using this knowledge we can exploit it in the field of biomass production of plant namely by plant-microbe interaction, by regulating cell wall biosynthesis or by manipulating phytohormones level under abiotic and biotic stress conditions. To this end, current book chapter we will try to high light an update on recent studies focusing on how to improve biomass production in plants under abiotic stress condition as well as positive interaction of multiple stresses condition. Genetic engineering can invent a transcriptional circuit to prepare transgenic plants that have been genetically engineered having new characteristics to endure any kind of adverse climatic condition. These kinds of transgenic plants are designed with a view to tolerate stress and enhance biomass production. At the same time, we will focus on the kind of transgenic plant in which the antioxidant defense mechanism is involved for scavenging of ROS (reactive oxygen species) and H_2O_2 from the cell compartment which are the effect of environmental stressors by SOD, CAT, APX, and GPX. Finally, the current chapter will briefly summarize the transformation of these plant biomass towards biofuels generations.

14.1 INTRODUCTION

Plant biomass contains lignocellulosic constituent which has been considered as a major precursor for biofuels generations. Plant lignocellulosic biomass consists of cellulose, hemicellulose, and lignin including a small amount of pectin, nitrogen compounds, and mineral residues. The proportion of these constituents alters from one plant species to another (Ezeilo et al., 2017; De Gonzalo et al., 2016). Plant-based biomass utilization enhances worldwide to support the increasing bioenergy crisis in the growing global population. At present, biomass-based energy plants are the primary source of large scale electricity and heat productions towards biofuels regenerations (Kocar and Civas, 2013). However, plant-derived biomass supply towards biofuels generations entirely depends on advanced agricultural practices, plant biomass biosynthesis, and its corresponding signaling networks in response to environmental impacts. Amelioration of plant-based biomass deals with the plant growth considering an increase in irreversible plant overall size, plant cell division, and an increase in cell sizes. Global warming and overpopulation are two major driving forces that bring challenges to all living entities on this earth. Global warming encounters devastation of plant-derived biomass biosynthesis through generating environmental selective pressures on both biotic and abiotic levels. These environmental stress agents influence plant growth and biomass yield through individual or combined efforts at the epigenetic level. Environmental stresses on plants have been classified into two major segments likely abiotic and biotic stress. Abiotic environmental stresses involve salinity, drought, extreme temperature, heavy metals, radiations, etc., whereas biotic stresses include pathogenic attack likely fungi, bacteria, oomycetes, nematodes, and herbivores, etc., (Verma et al., 2013). Drought situation affects the plant biomass generation in diverse biophysicochemical aspects. Drought involves in the closing of stomata, reducing transpiration rate, decrease osmosis, up-regulation of respiration efficacy, inhibition of photosynthesis, and growth rate in plant system. Abscisic acid (ABA)-dependent or ABA-independent signaling module has been induced through drought stress in plants (Sarwat et al., 2013). Cold stressor alters plant cell membrane lipid composition. Plant cellular DNA replication, spliceosome, protease biosynthesis, and mismatch repair pathways are greatly influenced via cold stress. However, cold stress influences plant nuclear and mitochondrial gene expression from time to time (Rong et al., 2011; Naydenov et al., 2010). Salinity stress reduces the water uptake capacity in plants which reduces plant metabolic rate and growth rate simultaneously. Salt stress also results in photosynthetic efficiency in the leaf area of the plant system. Moreover, salinity stress upregulates signal transducing molecules, i.e., jasmonic acid (JA), etc., (Walia et al., 2006). In a similar fashion, heat stress changes plant physiology and development through influencing plant cellular component biosynthesis and plant cellular metabolism (Qin et al., 2008; Mittal et al., 2009; Bita et al., 2011). Plant biomass yield has been diminished by flooding which causes plant hypoxia. Flooding influences carbon-amino acid central metabolism, transcriptional-translational regulations, signal transduction, protein homeostasis, photosynthesis, etc. On the contrary flooding down-regulates gene products related to cell wall biosynthesis, secondary metabolites synthesis, secondary metabolites transport, cellular arrangements, and chromosomal structural biogenesis (Zou et al., 2010; Nanjo et al., 2011). Plant biomass synthesis needs heavy metal likely iron, copper, cobalt, molybdenum, manganese, and zinc as potential trace elements which act as cofactors of plant metabolic networks. However, higher content of heavy metals are detrimental to plant biomass biogenesis. Higher content of heavy metals induce reactive oxygen species (ROS) and inactivates of plant metabolic enzyme activities. Moreover, heavy metal stress involves in pathways related to plant photosynthesis, chaperones biosynthesis, and transcriptional-translational regulations (Zhao et al., 2011). On the other hand, plant biomass biosynthesis has also been influenced through biotic stress factors which cause severe infection by pathogens (including bacteria, fungi, viruses, and nematodes) and attack by herbivore pests (Atkinson and Urwin, 2012).

Practically, recent advanced agricultural approaches cannot ignore these stressors. Hence, one of the primary challenges in today's agricultural system is to enhance plant biomass yields and productivities under the aforementioned adverse environmental stressors. In the molecular level, plant biomass growth, yield, and productivities can be promoted through regulations on following important aspects likely plant microbe's interaction, an antioxidant defense mechanism in plants during stress condition, the role of phytohormones in stress condition, cell wall biosynthesis, plant cell cycle regulations, regulation on plant hormones, regulation of transcription, regulation on photosynthesis, regulation over plant central metabolism, and regulation over micro RNA in plants. To this end, the current book chapter will focus on different genetic engineering strategies on diverse ranges of stressors for regulating stress responses to ameliorate plant biomass production and its retransformation into potential biofuels.

14.2 GENETIC ENGINEERING STRATEGIES ON STRESS RESPONSES TOWARDS IMPROVING PLANT BIOMASS GENERATION

Adaptation of plants under different stressors and its corresponding stress responses are very complicated networks. This complex network indulge with cascade of genes and their corresponding regulatory cross talks under diverse environmental factors during the entire plant developmental cycle (Kumari et al., 2009; Sharan et al., 2017; Lakra et al., 2017). Therefore, complete understanding on environmental stressors, genetic regulatory network, and phenotypic response of plant are foremost vital platforms to improve the plant biomass, biomass yield and biomass productivities (Figure 14.1).

14.2.1 PLANT MICROBE'S INTERACTION

Plant microbial interaction provides the stressors tolerance of plants to mitigate biomass productions under biotic or abiotic stress environmental situations. Plant growth promoting bacteria (PGPB) are the major candidates which accelerate essential interaction with plant systems in the rhizosphere zone. Usually, PGPB grows and accumulates around the root and enters into the root system to establish endophytic population in the rhizosphere. PGPB can be categorized into two major classes. Class I PGPB prevents the mode of action of phytopathogens and indirectly enhances plant growth. In contrary, Class II PGPB stimulates direct plant growth through improving nitrogen fixation, phytohormone biogenesis, and mineral availabilities. In a study, it has been shown that PGPB likely *Achromobacter piechaudii*, *Burkholderia caryophylli*,

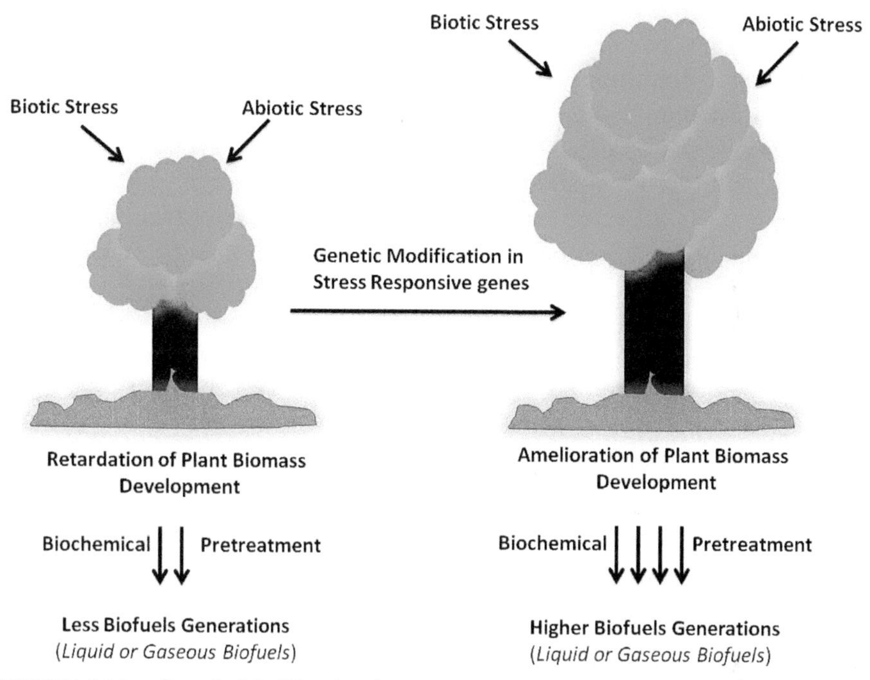

FIGURE 14.1 Genetic Modification in stress responsive genes under biotic and stress conditions towards improving plant biomass yields and productivities for biofuels generations.

and *Pseudomonas* sp, naturally reduces endogenous phytohormone (ethylene) levels in plants through the catalytic activities of 1-aminocyclopropane-1-carboxylic acid deaminases (ACCD) to accelerate plant root growth, salinity, and drought stresses (Wu et al., 2009). In rhizosphere, *Trichoderma atroviride* and/or *Pseudomonas putida* provides growth promotion and stress tolerance in tomato plant (*Solanum lycopersicum*) through higher synthesis of indole acetate (IAA) and ACCD enzyme catalytic activities (Gravel et al., 2007). In similar fashion, down regulation of ACCD gene expression in a fungal species *Trichoderma asperellum* causes plant growth acceleration and salt tolerance in *Brassica napus* (Rapeseed plant) (Viterbo et al., 2010).

14.2.2 ROLE OF PHYTOHORMONES IN STRESS CONDITION

Intermolecular cross-talk amongst various phytohormones signaling networks ameliorate plant growth and biomass yield under diverse range of environmental stress conditions. A broad range of phytohormones, i.e., auxins, cytokinins (CKs), gibberellins (GAs), brassinosteroids (BRs), and ABAs regulate plant growth and biomass generations under biotic and abiotic stress conditions (Gupta et al., 2017; Gonzalez et al., 2009). In general, CKs show impact on shoot apical meristem development, panicle branching, leaf area expansion, and delayed senescence, etc. GAs accelerate hypocotyls elongation, axillary bud development, and increased branching. BRs affect on shoot apical meristem development and ABAs enhance plant biomass yield through shoot developments. In rice plant (*Oryza sativa*), knocking down of cytokinin oxidase (CKX2) gene expression results in higher biomass yield under salinity stress through proper maintenance of photosynthetic rates, higher panicle branching (Joshi et al., 2018). In peanut or groundnut (*Arachis hypogaea*) and rice plant, controlled expression isopentenyl transferase (IPT) reflects higher plant biomass generations through retarded leaf senescence, higher photosynthetic activity at drought stress environment (Peleg et al., 2011; Qin et al., 2011). In Alfalfa plant (*Madia sativa*) and *Arabidopsis thaliana*, down-regulation of squamosa promoter binding-like protein-8/9 transcription factor (SPL8/9) biomass yield under salt and drought stress conditions. In the molecular level, SPL8/9 controls over GAs signaling networks in the alfalfa plant (Gou et al., 2018; Yamaguchi, 2008). In *Arabidopsis thaliana,* overexpression of histone deacetylase (HDC1) enhances the plant biomass productivities through the minimization of salt sensitivity and ABA signaling network dependence (Haak et al., 2017). In another study, the tomato plant attains cold, drought, and salt stress tolerances through the overexpression on the ABA-responsive complex (ABRC1) from the barley plant which ultimately ameliorates tomato plant biomass yield (Lee et al., 2003). Moreover, 9-cis-epoxy-carotenoid dioxygenase (NCED) overexpression enhances ABA accumulation in leaves of peanut and tomato plants which accelerates drought and salt tolerance towards improving biomass yields (Wan et al., 2006; Tung et al., 2008). In *M. sativa* and *A. thaliana* plants, knock out of serine/threonine protein kinases (GSK1) gene results in enhanced biomass production and reduction in abiotic stress sensitivity (Koh et al., 2007).

14.2.3 CELL WALL BIOSYNTHESIS

The plant cell wall is the first line of defense against abiotic and biotic stresses. The primary constituents of plant outermost cell wall contain cellulose, hemicellulose, and lignin. However, the biochemical compositions of those cell wall constituents are highly unique from one to another cell types in the plant system. Moreover, their proportion and composition are highly variable during developmental stages in response to several biotic and abiotic stresses. It has been evident that plant enhances the rate of pathway enzymes in lignin biosynthesis during abiotic stress conditions (Pauly et al., 2008). In contrary knockdown of 4-coumarate-coenzyme, a ligase results in the reduction of lignin content in plant secondary cell walls (Wagner et al., 2009). Overexpression of cellulose synthase (CesA1) in *Arabidopsis thaliana* enhances cellulose accumulation in the plant cell wall and improves stress tolerance (Rao et al., 2017; Xie et al., 2011). In *A. thaliana* and *O. sativa*, integration of the BR receptor corresponding gene segment enhances drought tolerance and maximizes higher root-shoot biomass accumulations (Sánchez-Rodríguez et al., 2017; Placido et al., 2013). In another study, while sucrose synthase (Susy) and UDP glucose pyrophosphorylase (UDPase) overexpresses in *Gossypium* sp. it improves cellulose accumulation towards enhancing plant biomass yield along with drought tolerance (Le Gall et al., 2015). Overexpression of xyloglucan endotransglucosylase/hydrolases (XTH3) genes, MYB41, MPS transcription factors (TFs) and TOR1/2 (Ser/Thr kinase of the phosphatidylinositol 3-kinase-related kinase family proteins) improves xyloglucan biogenesis, cell wall biosynthesis along with increasing water stress tolerance, salts stress tolerance and osmotic stress tolerance respectively (Cho et al., 2006; Schmidt et al., 2013; Deprost et al., 2007). In *Glycine max* EPXB2 (beta-Expansin encoding gene) overexpression improves the root system under insufficient water stress conditions (Guo et al., 2011). Moreover, silencing or knocking down of NAC2 and EXPA4 genes in the rose plant (*Rosa* sp.) causes a reduction in drought resistance (Dai et al., 2012). Finally, knockdown of CCR gene in *Zea mays* and overexpression of transcription factor MYB46 in *Medicago sativa* inhibits cell wall extensibility at water stress and improves lignin deposition in plant cell wall under abiotic stress situations, respectively (Fan et al., 2006; Haak et al., 2017).

14.2.4 ANTIOXIDANT DEFENSE MECHANISM DURING STRESS CONDITION TO IMPROVE PLANT BIOMASS

Environmental stresses including intense light, heat, drought, UV, cold, and herbicides increase ROS production. ROS attack biomolecules causing

DNA mutation, protein denaturation, and membrane lipid peroxidation. This oxidation phenomenon disturbs normal cellular metabolism and causes molecular cellular damage. Molecular O_2 is relatively non-reactive in its normal state but during the period of environmental stress, O_2 can generate reactive excited states such as free radicals and its derivatives. These reactive derivatives are singlet oxygen ($^1O_2$), superoxide radical (O_2^-), hydrogen peroxide (H_2O_2), and hydroxyl radical (OH^-) (Suzuki et al., 2012; Miller et al., 2009). In normal conditions, the plant attains a high degree of ROS toxicity and utilizes ROS as signal molecules for initiating defense mechanisms to sustain ROS homeostasis. ROS is naturally neutralized by antioxidant molecules, i.e., superoxide dismutase (SOD), catalase (CAT), ascorbate peroxidase (APX), glutathione peroxidase (GPX) and non-enzymatic defense includes several compounds (like water-soluble ascorbate, glutathione, phenolic compounds, flavonoids, and lipid-soluble carotenoids and Vitamin E), etc., (Scandalios, 2005; Willekens et al., 1995; Noctor et al., 1998; Asada, 1992). All these antioxidants protect the plant cell by scavenging the ROS or RNS (reactive nitrogen species) and convert them into less reactive species by donating electrons. Several SOD genes have been cloned in a variety of plant species such as *A. thalina*, maize, tobacco, tomato, and rice. Recently few SODs have been characterized for their physiological role in protecting plant cells from toxic ROS and also against environmental stresses like drought, salinity, cold, etc. ROS ameliorates an abundance of mRNA transcript in response to biotic and abiotic stress while oxidative stresses which play a vital role in stress tolerance and enhance biomass production. Transgenic plant overexpressing various SOD isoforms exhibit enhanced tolerance to oxidative stress and to various environmental stress factors. It has already been trailed mainly in rice, potato, alfalfa, Poplar, *Arabidopsis*, tobacco, etc. Moreover, APXs catalyze the reduction of H_2O_2 to water by using ascorbate as an electron donor. APX isoenzymes are distributed in four different plant cellular compartments namely cytosol (cAPX), mitochondria (mitAPX), chloroplast stroma (sAPX), peroxisomes, glyoxisomes (mAPX) and chloroplast thylakoids (tAPX) (Racchi, 2013). APX plays a co-operative role in protection of is organelle. With the advent of genetic engineering knowledge these stress responsive genes can be introduced into plants through compatible helper microbial vector which activity can enhance stress tolerance and increase biomass production in transformed plant. In this way CATs, GPX antioxidant can be used to reduce stress in plants and enhance plant biomass generations (Racchi, 2013).

14.2.5 REGULATION BASED STRATEGIES ON IMPROVING PLANT BIOMASS UNDER STRESS CONDITIONS

Plant biomass growth refers to an irreversible increase in the size of plant cells, plant cell division. There are several regulatory strategies involve towards improving plant biomass under diverse stress conditions. Most promising genetic candidates for the regulatory checkpoint biomolecules include cell cycle genes, hormones, transcription regulatory genes, photosynthesis-related factors, metabolism associated factors, miRNAs, etc., (Marcelo et al., 2017). Plant cell cycle regulatory components involve an increase in cell number which ultimately expands the ultimate plant tissue or organelle size. Therefore, understanding of plant growth associated cell-cycle network is a very important aspect on plant biomass and yield enhancements (Marcelo et al., 2017). Plant hormones are another typical factor towards ameliorating plants over all biomass. Plant hormones generally function as regulators of seed germination, leaf expansion, stem elongation, flower, and fruit developments. Hence, the optimal level of plant hormonal biosynthesis and secretion secures the plant biomass developments (Hedden and Sponsel, 2015). Plant transcription is the first and foremost important segment of different gene expressions related to plant biomass enhancements. Plant transcription machineries and regulatory factors control over plant biomass accumulation. Hence, it is the utmost necessity to understand the cross-talk amongst different plant TFs related to gene expression for plant biomass and yield ameliorations (Joshi et al., 2016). Photosynthesis and its machineries involve in plant biomass accumulation. Light uptake (photosynthetic pigments), electron transport (around photosystems), and carbon dioxide sequestration (through Calvin-Benson Cycle) are the major regulatory compartment which directly or indirectly participate for increasing plant growth and biomass yield (Marcelo et al., 2017). All the above regulatory networks require energy supply for biomass development. Plant retains a very organized and complicated metabolic reaction networks which favor and fulfill these demands. However, plant metabolism accelerates plant biomass accumulation considering proper nutrient uptake and its critical balance at the molecular level, i.e., carbon source (C), nitrogen source (C), C/N molar ration, and trace elements (Zheng, 2009). A class of naturally existing small noncoding RNA biomolecule (miRNAs) carries out a very pivotal activity towards improving plant biomass, yields, and productivities in response to environmental biotic and abiotic stressors (Borges and Martienssen, 2015). To this end, miRNAs (i.e., MiR156 in angiosperms of plants) expression alterations (knock out or knockdown) could generate a novel avenue to accelerate plant biomass accumulation (Morea et al., 2016). Aforementioned all regulation based strategies are depicted in Figure 14.2.

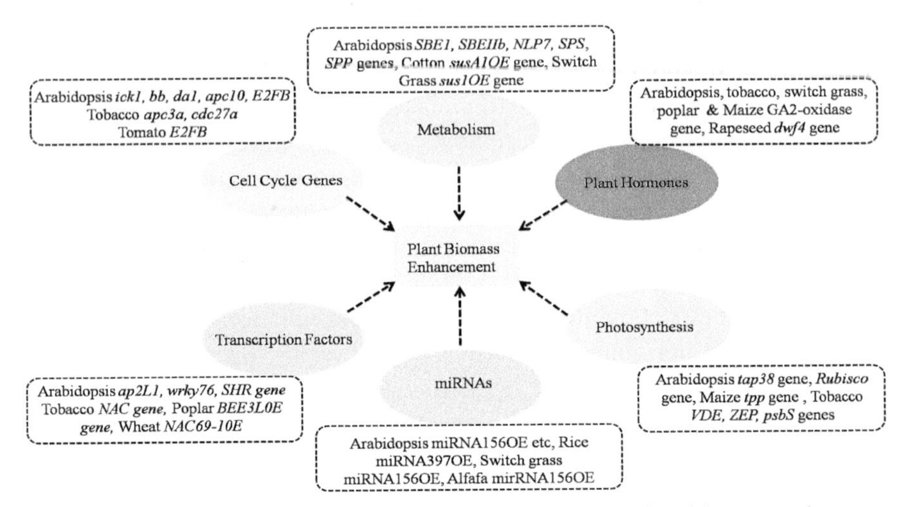

FIGURE 14.2 Genetic engineering strategies for improving plant biomass under stress conditions (Joshi et al., 2016).

14.3 CONCLUSION

In this book chapter, the current scenario has clearly been depicted that there are several strategies applies to improve the plant biomass productions. Recently advancements in molecular biology, biotechnology, and genetic engineering have more potential to generate higher plant-derived biomass to value-added products likely biofuels. Many of the plant growth-enhancing genes, transcripts, and enzymes play a pivotal role to ameliorate plant biomass accumulation under diverse biotic and abiotic stress conditions. Usages of transgenic and genetic engineering avenues allow direct or indirect alteration of plant genotype considering specific stress-responsive genes. However, the present experimental platform is not quite well established enough to implement these techniques on the practical field towards improving plant biomass. Though genetic and metabolic regulatory mechanisms are less complicated and partially established on model plant systems but are not effective enough on most of the field crops. It happens mainly due to allelic diversity, difficulty in genetic transformation and modification, genetic silencing factors in many field crops. To this end, more advanced genetic tools, i.e., RNAi, transposon insertion, and CRISPR/Cas9 system could be applied to generate stable transgenic plants towards biofuels generations (Bortesi and Fischer, 2015; McCarty et al., 2005). These approaches allow plant biotechnologists, metabolic engineers, synthetic biologists, and plant biotechnologists to generate transgenic stable allele varieties in plant systems for enhancing biomass yield.

On the other hand, global climate change, growing populations, and limited cultivable land raise further issues to improve the plant-derived biomass yield. To this end, the current book chapter also emphasizes a detailed description on the effects of biotic and abiotic stress factors upon plant biomass accumulation. Moreover, several options have been explained to explore plant-microbe interactions for enhancing plant biomass in marginal arable land spaces. Though most of these interactions are not pretty much clear as far molecular mechanisms. Therefore, metabolically, or genetically engineered plant and compatible symbionts molecular cross-talks will open up new avenues to modify plants towards higher biomass synthesis in the presence of different environmental stressors. In a similar fashion, plant cellular differentiation, photosynthesis rate, antioxidative defense mechanisms, respiration rate, cell wall development, phytohormone regulations and biosynthesis, etc., need to be explored in more details to find out the genetic and metabolic knowledge gap in the presence of external stressors for the amelioration of plant biomass. Hence it will also be highly essential to elucidate the plant responses upon combinatorial or multiple biotic and abiotic stress factors which mimic similar kind of field conditions.

While a higher amount of plant biomass has been achieved, it needs to be converted into some value-added biomolecules likely biofuels (gaseous biofuels and/or liquid biofuels). As a far technical aspect, plant biomass has been converted into biofuels through different kinds of biological and chemical treatments. Where these potential pretreatments breakdown the complex plant biomass into simple accessible sugar moieties towards biofuels generations using naive or metabolically engineered biofuels producers (Ghosh and Hallenbeck, 2012; Hood, 2016; Kerckhoffs and Renquist, 2013; Sticklen, 2006). As a take-home message, current genetic tools which have already been applied to plant could be useful for algal biomass maximization considering homologs of plant stress-responsive genes and its regulatory networks in algae (blue-green or green) towards higher algal biomass generations (Voloshin et al., 2016) as a sustainable feedstock for potential biofuels generations in near future.

KEYWORDS

- **abiotic stress**
- **biofuels**
- **biotic stress**
- **genetic engineering**
- **plant biomass**
- **stress responses**

REFERENCES

Asada, K., (1992). Production and scavenging of active oxygen in chloroplasts. In: Scandalios, J. G., (ed.), *Current Communications in Cell and Molecular Biology 5; Molecular Biology of Free Radical Research Scavenging Systems* (pp. 173–192). Cold Spring Harbor Laboratory Press: New York, NY, USA.

Atkinson, N. J., Lilley, C. J., & Urwin, P. E., (2013). Identification of genes involved in the response of *Arabidopsis* to simultaneous biotic and abiotic stresses. *Plant Physiology, 162*, 2028–2041.

Bita, C. E., Zenoni, S., Vriezen, W. H., Mariani, C., Pezzotti, M., & Gerats, T., (2011). Temperature stress differentially modulates transcription in meiotic anthers of heat-tolerant and heat-sensitive tomato plants. *BMC Genomics, 12*, 384.

Borges, F., & Martienssen, R. A., (2015). The expanding world of small RNAs in plants. *Nature Reviews Molecular Cell Biology, 16*(12), 727–741.

Bortesi, L., & Fischer, R., (2015). The CRISPR/Cas9 system for plant genome editing and beyond. *Biotechnology Advances, 33*(1), 41–52.

Cho, S. K., Kim, J. E., Park, J. A., Eom, T. J., & Kim, W. T., (2006). Constitutive expression of abiotic stress-inducible hot pepper CaXTH3, which encodes a xyloglucan endotransglucosylase/hydrolase homolog, improves drought and salt tolerance in transgenic *Arabidopsis* plants. *FEBS Lett., 580*, 3136–3144.

Dai, F., Zhang, C., Jiang, X., Kang, M., Yin, X., Lü, P., Zhang, X., Zheng, Y., & Gao, J., (2012). RhNAC2 and RhEXPA4 are involved in the regulation of dehydration tolerance during the expansion of rose petals. *Plant Physiol., 160*, 2064–2082.

De Gonzalo, G., Colpa, D. I., Habib, M. H. M., & Fraaije, M. W., (2016). Bacterial enzymes involved in lignin degradation. *Journal of Biotechnology, 236*, 110–119.

Deprost, D., Yao, L., Sormani, R., Moreau, M., Leterreux, G., Nicolaï, M., Bedu, M., Robaglia, C., & Meyer, C., (2007). The *Arabidopsis* TOR kinase links plant growth, yield, stress resistance and mRNA translation. *EMBO Rep., 8*, 864–870.

Ezeilo, U. R., Zakaria, I. I., Huyop, F., & Wahab, R. A., (2017). Enzymatic breakdown of lignocellulosic biomass: The role of glycosyl hydrolases and lytic polysaccharide monooxygenases. *Biotechnology and Biotechnological Equipment, 31*, 4, 647–662.

Fan, L., Linker, R., Gepstein, S., Tanimoto, E., Yamamoto, R., & Neumann, P. M., (2006). Progressive inhibition by water deficit of cell wall extensibility and growth along the elongation zone of maize roots is related to increased lignin metabolism and progressive stellar accumulation of wall phenolics. *Plant Physiol., 140*, 603–612.

Ghosh, D., & Hallenbeck, P. C., (2012). Advanced bioethanol production. In: Hallenbeck, P. C., (ed.), *Microbial Technologies in Advanced Biofuels Production* (pp. 165–181).

Gonzalez, N., Beemster, G. T., & Inzé, D., (2009). David and Goliath: What can the tiny weed *Arabidopsis* teach us to improve biomass production in crops? *Curr. Opin. Plant Biol., 12*, 157–164.

Gou, J., Debnath, S., Sun, L., Flanagan, A., Tang, Y., Jiang, Q., Wen, J., & Wang, Z. Y., (2018). From model to crop: Functional characterization of SPL8 in M. truncatulaled to genetic improvement of biomass yield and abiotic stress tolerance in alfalfa. *Plant Biotechnol. J., 16*(4), 951–962.

Gravel, V., Antoun, H., & Tweddell, R. J., (2007). Growth stimulation and fruit yield improvement of greenhouse tomato plants by inoculation with *Pseudomonas* putida or *Trichoderma* atroviride: Possible role of indole acetic acid (IAA). *Soil Biol. Biochem., 39*, 1968–1977.

Guo, W., Zhao, J., Li, X., Qin, L., Yan, X., & Liao, H., (2011). A soybean beta-expansin gene GmEXPB2 intrinsically involved in root system architecture responses to abiotic stresses. *Plant, J., 66*, 541–552.

Gupta, B., Joshi, R., Pareek, A., & Singla-Pareek, S. L., (2017). In: Pandey, G. K., (ed.), *Mechanism of Plant Hormone Signaling Under Stress* (pp. 533–567). John Wiley & Sons, Inc.; Hoboken, NJ.

Haak, D. C., Fukao, T., Grene, R., Hua, Z., Ivanov, R., Perrella, G., & Li, S., (2017). Multilevel regulation of abiotic stress responses in plants. *Front. Plant Sci., 8*, 1564.

Hedden, P., & Sponsel, V., (2015). A century of gibberellin research. *Journal of Plant Growth Regulation, 34*(4), 740–760.

Hood, E. E., (2016). *Plant-Based Biofuels* (Vol. 5, No. 15, pp. 1–9). F1000Research.

Joshi, R., Sahoo, K. K., Tripathi, A. K., Kumar, R., Gupta, B. K., Pareek, A., & Singla-Pareek, S. L., (2018). Knockdown of an inflorescence meristemspecific cytokinin oxidase–OsCKX2 in rice reduces yield penalty under salinity stress condition. *Plant Cell Environ., 41*(5), 936–946.

Joshi, R., Wani, S. H., Singh, B., Bohra, A., Dar, Z. A., Lone, A. A., et al., (2016). Transcription factors and plants response to drought stress: Current understanding and future directions. *Frontiers in Plant Science, 7*(1029).

Kerckhoffs, H., & Renquist, R., (2013). Biofuel from plant biomass. *Agronomy for Sustainable Development, 33*, 1–19.

Kocar, G., & Civas, N., (2013). An overview of biofuels from energy crops: Current status and future prospects. *Renewable and Sustainable Energy Reviews, 28*, 900–916.

Koh, S., Lee, S. C., Kim, M. K., Koh, J. H., Lee, S., An, G., Choe, S., & Kim, S. R., (2007). T-DNA tagged knockout mutation of riceOsGSK1, an orthologue of *Arabidopsis* BIN2, with enhanced tolerance to various abiotic stresses. *Plant Mol. Biol., 65*, 453–466.

Kumari, S., Nee, S. V. P., Kushwaha, H. R., Sopory, S. K., Singla-Pareek, S. L., & Pareek, A., (2009). Transcriptome map for seedling stage specific salinity stress response indicates a specific set of genes as candidate for saline tolerance in *Oryza sativa* L. *Functional Integration Genomics, 9*, 109–123.

Lakra, N., Kaur, C., Anwar, K., Singla-Pareek, S. L., & Pareek, A., (2017). Proteomics of contrasting rice genotypes: Identification of potential targets for raising crops for saline environment. *Plant Cell Environment*, 10.1111/pce.12946.

Le Gall, H., Philippe, F., Domon, J. M., Gillet, F., Pelloux, J., & Rayon, C., (2015). Cell wall metabolism in response to abiotic stress. *Plants, 4*, 112–166.

Lee, J. T., Prasad, V., Yang, P. T., Wu, J. F., David, H. T. H., Charng, Y. Y., & Chan, M. T., (2003). Expression of *Arabidopsis* CBF1 regulated by an ABA/stress inducible promoter in transgenic tomato confers stress tolerance without affecting yield. *Plant Cell Environ., 26*, 1181–1190.

Marcelo, D. F. L. M. F., Eloy, N. B., Batista, D. S. J. A., Inzé, D., Hemerly, A. S., et al., (2017). Molecular mechanisms of biomass increase in plants. *Biotechnology Research and Innovation, 1*, 14–25.

McCarty, D. R., Settles, A. M., Suzuki, M., Tan, B. C., Latshaw, S., Porch, T., et al., (2005). Steady-state transposon mutagenesis in inbred maize. *Plant Journal, 44*(1), 52–61.

Miller, G., Suzuki, N., Ciftci-Yilmaz, S., & Mittler, R., (2009). Reactive oxygen species homeostasis and signaling during drought and salinity stresses. *Plant Cell Environ., 33*, 453–467.

Mittal, D., Chakrabarti, S., Sarkar, A., Singh, A., & Grover, A., (2009). Heat shock factor gene family in rice: Genomic organization and transcript expression profiling in response to high temperature, low temperature, and oxidative stresses. *Plant Physiol. Biochem., 47*(9), 785–795.

Morea, E. G. O., Da Silva, E. M., Silva, G. F. F, E., Valente, G. T., Rojas, C. H. B., Vincentz, M., et al., (2016). Functional and evolutionary analyses of the miR156 and miR529. families in land plants. *BMC Plant Biology, 16*(1), 40.

Nanjo, Y., Maruyama, K., Yasue, H., Yamaguchi-Shinozaki, K., Shinozaki, K., & Komatsu, S., (2011). Transcriptional responses to flooding stress in roots including hypocotyl of soybean seedlings. *Plant Molecular Biology, 77*, 129–144.

Naydenov, N. G., Khanam, S., Siniauskaya, M., & Nakamura, C., (2010). Profiling of mitochondrial transcriptome in germinating wheat embryos and seedlings subjected to cold, salinity, and osmotic stresses. *Genes Genet. Syst., 85*(1), 31–42.

Noctor, G., & Foyer, C. H., (1998). Ascorbate and glutathione: Keeping active oxygen under control. *Annu. Rev. Plant Physiol. Plant Mol. Biol., 49*, 249–279.

Pauly, M., & Keegstra, K., (2008). Cell wall carbohydrates and their modifications a resource for biofuels. *Plant J., 54*, 559–568.

Peleg, Z., Reguera, M., Tumimbang, E., Walia, H., & Blumwald, E., (2011). Cytokinin-mediated source/sink modifications improve drought tolerance and increase grain yield in rice under water-stress. *Plant Biotechnol. J., 9*, 747–758.

Placido, D. F., Campbell, M. T., Folsom, J. J., Cui, X., Kruger, G. R., Baenziger, P. S., & Walia, H., (2013). Introgression of novel traits from a wild wheat relative improves drought adaptation in wheat. *Plant Physiol., 161*, 1806–1819.

Qin, A. D., Wu, H., Peng, H., Yao, Y., Ni, Z., Li, Z., Zhou, C., & Sun, Q., (2008). Heat stress-responsive transcriptome analysis in heat susceptible and tolerant wheat (*Triticum aestivum* L.) by using wheat genome array. *BMC Genomics, 9*, 432.

Qin, H., Gu, Q., Zhang, J., Sun, L., Kuppu, S., Zhang, Y., Burow, M., et al., (2011). Regulated expression of an isopentenyl transferase gene (IPT) in peanut significantly improves drought tolerance and increases yield under field conditions. *Plant Cell Physiol., 52*, 1904–1914.

Racchi, M. L., (2013). Antioxidant defenses in plants with attention to *Prunusand citrus* spp. *Antioxidants, 2*, 340–369.

Rao, X., & Dixon, R. A., (2017). Brassinosteroid mediated cell wall remodeling in grasses under abiotic stress. *Front. Plant Sci., 8*, 806.

Rong, E., Zhao, Z., Ma, J., Zang, W., Wang, L., Xie, D., & Yang, W., (2011). Wheat cold and light stress analysis based on the *Arabidopsis* homology protein-protein interaction (PPI) network. *J. Med. Plant Res., 5*(23), 5493–5498.

Sánchez-Rodríguez, C., Ketelaar, K., Schneider, R., Villalobos, J. A., Somerville, C. R., Persson, S., & Wallace, I. S., (2017). Brassinosteroid insensitive 2 negatively regulates cellulose synthesis in *Arabidopsis* by phosphorylating cellulose synthase 1. *Proc. Natl. Acad. Sci. U.S.A., 114*, 3533–3538.

Scandalios, J. G., (2005). Oxidative stress: Molecular perception and transduction of signals triggering antioxidant gene defenses. *Br. J. Med. Biol. Res., 38*, 995–1014.

Schmidt, R., Schippers, J. H., Mieulet, D., Obata, T., Fernie, A. R., Guiderdoni, E., & Mueller-Roeber, B., (2013). Multipass, a rice R2R3-type MYB transcription factor, regulates adaptive growth by integrating multiple hormonal pathways. *Plant J., 76*, 258–273.

Sharan, A., Soni, P., Nongpiur, R. C., Singla-Pareek, S. L., & Pareek, A., (2017). Mapping the 'two-component system' network in rice. *Science Rep., 7*, 9287.

Sticklen, M., (2006). Plant genetic engineering to improve biomass characteristics for biofuels. *Current Opinion in Biotechnology, 17*, 315–319.

Suzuki, N., Koussevitzky, S., Mittler, R., & Miller, G., (2012). ROS and redox signaling in the response of plants to abiotic stress. *Plant Cell Environ., 35*, 259–270.

Tung, S. A., Smeeton, R., White, C. A., Black, C. R., Taylor, I. B., Hilton, H. W., & Thompson, A. J., (2008). Over-expression of LeNCED1 in tomato (*Solanum lycopersicum* L.) with the rbcS3C promoter allows recovery of lines that accumulate very high levels of abscisic acid and exhibit severe phenotypes. *Plant Cell Environ., 31*, 968–981.

Verma, S., Nizam, S., Verma, P. K., Sarwat, M., et al., (2013). *Stress Signaling in Plants: Genomics and Proteomics Perspective* (Vol. 1, pp. 25–48). Doi: 10.1007/978–1-4614–6372–6_2, Springer Science Business Media New York.

Viterbo, A., Landau, U., Kim, S., Chernin, L., & Chet, I., (2010). Characterization of ACC deaminase from the biocontrol and plant growth-promoting agent *Trichoderma* asperellum T203. *FEMS Microbiol. Lett., 305*, 42–48.

Voloshin, R. A., Margarita, V., Rodionova, M. V., Zharmukhamedov, S. K., et al., (2016). Review: Biofuel production from plant and algal biomass. *International Journal of Hydrogen Energy, 41*(39), 17257–17273.

Wagner, A., Donaldson, L., Kim, H., Phillips, L., Flint, H., Steward, D., Torr, K., Koch, G., Schmitt, U., & Ralph, J., (2009). Suppression of 4-coumarate-CoA ligase in the coniferous gymnosperm *Pinus radiata*. *Plant Physiol., 149*, 370–383.

Walia, H., Wilson, C., Wahid, A., Condamine, P., Cui, X., & Close, T. J., (2006). Expression analysis of barley (*Hordeum vulgare L*) during salinity stress. *Funct. Integr. Genomics, 6*(2), 143–156.

Wan, X. R., & Li, L., (2006). Regulation of ABA level and water-stress tolerance of *Arabidopsis* by ectopic expression of a peanut 9-cis-epoxycarotenoid dioxygenase gene. *Biochem. Biophys. Res. Commun., 347*, 1030–1038.

Willekens, H., Inzé, D., Van, M. M., & Van, C. W., (1995). Catalases in plants. *Mol. Breed., 1*, 207–228.

Wu, C. H., Bernard, S. M., Andersen, G. L., & Chen, W., (2009). Developingmicrobe-plant interactions for applications in plant-growth promotion and disease control, production of useful compounds, remediation and carbon sequestration. *Microb. Biotechnol., 2*, 428–440.

Xie, L., Yang, C., & Wang, X., (2011). Brassinosteroids can regulate cellulose biosynthesis by controlling the expression of CESA genes in *Arabidopsis*. *J. Exp. Bot., 62*, 4495–4506.

Yamaguchi, S., (2008). Gibberellin metabolism and its regulation. *Annu. Rev. Plant Biol., 59*, 225–251.

Zhao, L., Sun, Y. L., Cui, S. X., Chen, M., Yang, H. M., Liu, H. M., Chai, T. Y., & Huang, F., (2011). Cd-induced changes in leaf proteome of the hyper accumulator plant *Phytolacca* Americana. *Chemosphere, 85*(1), 56–66.

Zheng, Z. L., (2009). Carbon and nitrogen nutrient balance signaling in plants. *Plant Signaling and Behavior, 4*(7), 584–591.

Zou, X., Jiang, Y., Liu, L., Zhang, Z., & Zheng, Y., (2010). Identification of transcriptome induced in roots of maize seedlings at the late stage of waterlogging. *BMC Plant Biology, 10*, 189.

Index

B

C

D

E

N

O

P

Q

R

T